SPACE VEHICLE ELECTRONICS

Lunar Excursion Module landing on the moon. *Courtesy*, LFE Electronics.

PRINCIPLES OF
GUIDED MISSILE DESIGN

Editor of the Series

GRAYSON MERRILL, CAPTAIN, U.S.N. (Ret.)

President, PRD Electronics, Inc.

GUIDANCE—*by* Arthur S. Locke and collaborators

AERODYNAMICS, PROPULSION, STRUCTURES AND DESIGN PRACTICE—*by* E. A. Bonney, M. J. Zucrow, and C. W. Besserer

OPERATIONS RESEARCH, ARMAMENT, LAUNCHING—*by* G. Merrill, H. Goldberg, and R. H. Helmholz

MISSILE ENGINEERING HANDBOOK—*by* C. W. Besserer

DICTIONARY OF GUIDED MISSILES AND SPACE FLIGHT—*Edited by* Grayson Merrill

SPACE FLIGHT—*by* K. A. Ehricke
Vol. I: ENVIRONMENT AND CELESTIAL MECHANICS
Vol. II: DYNAMICS Vol. III: OPERATIONS

SYSTEMS PRELIMINARY DESIGN—*by* J. J. Jerger

AIRBORNE RADAR—*by* D. J. Povejsil, R. S. Raven, and P. Waterman

INERTIAL NAVIGATION—*by* Richard H. Parvin

INSTRUMENTATION FOR AEROSPACE RANGES—*by* J. J. Scavullo

SPACE VEHICLE ELECTRONICS—*by* David B. Newman

PRINCIPLES OF
GUIDED MISSILE DESIGN

Edited by

GRAYSON MERRILL, CAPTAIN, U.S.N. (Ret.)

DAVID B. NEWMAN
Institute of Naval Studies
Center of Naval Analyses of the Franklin Institute

SPACE

VEHICLE

ELECTRONICS

D. VAN NOSTRAND COMPANY, INC.

PRINCETON, NEW JERSEY · TORONTO · NEW YORK · LONDON

D. VAN NOSTRAND COMPANY, INC.
120 Alexander St., Princeton, New Jersey (*Principal office*)
24 West 40 Street, New York 18, New York

D. Van Nostrand Company, Ltd.
358, Kensington High Street, London, W.14, England

D. Van Nostrand Company (Canada), Ltd.
25 Hollinger Road, Toronto 16, Canada

Published simultaneously in Canada by
D. Van Nostrand Company (Canada), Ltd.

Library of Congress Catalogue Card No. 64—19828

PRINTED IN THE UNITED STATES OF AMERICA

FOREWORD

Space Vehicle Electronics is another volume in the series entitled PRINCIPLES OF GUIDED MISSILE DESIGN. The purpose of this series as a whole is to give a basis for instruction to graduate students, engineers, and technical officers so they can become better grounded in the technology of guided missiles and space flight. Other titles published in the series are *Guidance; Aerodynamics, Propulsion and Structures; Operations Research, Armament, and Launching; Space Flight*—Volume I, *Environment and Celestial Mechanics*—Volume II, *Dynamics; Missile Engineering Handbook; Dictionary of Guided Missiles and Space Flight; Systems Preliminary Design; Inertial Navigation;* and *Airborne Radar*. Titles to be published are *Space Flight*—Volume III, *Operations; Handbook of Satellite and Space Travel;* and *Instrumentation for Aerospace Ranges*.

The present book shows the reader how, in light of experience and the current state-of-the-art, he can select an electronic system most likely to insure the success of a space vehicle mission, given the specific environment and objectives of that mission.

Criticisms and constructive suggestions are invited. By this means and by keeping abreast of the state-of-the-art, we hope to make timely revisions in the text.

Grateful acknowledgment is made to the many authors and publishers who kindly granted permission for the use of their material in this series and to the Department of Defense whose cooperation made possible meaningful texts without violation of security.

The opinions or assertions contained herein are the private ones of the author and editor and are not to be construed as official or reflecting the views of any government agency or department.

GRAYSON MERRILL
EDITOR

Brooklyn, New York
April, 1964

ACKNOWLEDGMENTS

In the preparation of this book, material from the following sources was used substantially in its original form. The number in this book of the figure or table is given below following the citation. The author is deeply indebted to all parties for permission to use such matter.

S. Chapman, "Sun Storms and the Earth: The Aurora Polaris and the Space Around the Earth," *American Scientist*, September, 1961: Figure 2-20.

R. Luders, "Satellite Coverage for Continuous Zonal Coverage," *ARS Journal*, February, 1961: Figures 8-10, 8-11, 8-12, and 8-13.

M. Gilleo and S. Kash, "Propulsion by Composite Beams of Negative and Positive Ions," *ARS Journal*, May, 1961: Figures 7-10 and 7-11.

K. Akasofu and S. Chapman, "New Theory of the Aurora Polaris," *ARS Journal*, June, 1961: Table 2-6.

R. Saunders, "Geomagnetic and Geoelectric Activity as a Function of Magnetic Latitude," *ARS Journal*, October, 1961: Figures 2-1 and 2-2.

A. Mikhailov, "The other side of the moon," *Astronautics*, March, 1961: Figure 2-33.

R. Parks, "The U. S. planetary exploration program," *Astronautics*, May, 1961: Figure 5-11.

N. Renzetti, "DSIF in the Ranger project," *Astronautics*, September, 1961: Figures 5-29, 5-30, 5-31, and 5-32.

M. Yarymovych, F. Weiss, and R. John, "Feasibility of arc-jet propelled spacecraft," *Astronautics*, June, 1962: Figure 7-9.

W. Stone, "A List of Generalized Laplace Transforms," *Iowa State Journal of Science*, Volume 22, Ames: Iowa State University Press, Table 6-3.

J. McNaul, "Electronic Components for Space Instrumentation," *IRE Transactions on Military Electronics*, April-July, 1960: Figures 4-24, 4-25, and 4-26.

S. Perlman, W. Russell, and F. Dickson, "Concerning Optimum Frequencies for Space Vehicle Communications," *IRE Transactions on Military Electronics*, April-July, 1960: Figures 2-14, 2-15, 2-16, 2-17, and 2-18.

H. Fox and R. Roberts, "Fuel Cells," *IRE Transactions on Military Electronics*, January, 1962: Table 7-1.

G. Lozier, "New Materials for Primary Batteries," *IRE Transactions on Military Electronics*, January, 1962: Figures 7-2, 7-3, 7-4, and 7-5.

N. Snyder, "Solar-Cell Power Systems for Space Vehicles," *IRE Transactions on Military Electronics*, January, 1962: Figure 7-1.

J. Judge, "NASA Finds Solar Probe Is Feasible," *Missiles and Rockets*, May 8, 1961: Figure 5-18.

M. Aucreamanne, "The Ionosphere Beacon Satellite, S-45," *NASA Technical Note D-695*, January, 1961: Figure 2-19.

R. Gallet, "The Very Low-Frequency Emissions Generated in the Earth's Exosphere," *Proceedings of the IRE*, February, 1959: Figure 2-25 and Tables 2-7 and 2-8.

R. Helliwell and M. Morgan, "Atmospheric Whistlers," *Proceedings of the IRE*, February, 1959: Figures 2-21, 2-22, 2-23, and 2-24.

A. Smith, "Extraterrestrial Noise as a Factor in Space Communications," *Proceedings of the IRE*, April, 1960: Figures 4-6 and 4-7.

B. Kovit, "Active radar augmenters," *Space/Aeronautics*, September, 1960: Figure 5-19.

"Aerospace in Perspective: Investigation," *Space/Aeronautics*, January, 1964: Figure 2-3.

Space Log, September 1963, Redondo Beach, California: Space Technology Laboratories, Inc., a subsidiary of Thompson Ramo Woolridge, Inc.: Table 1-2.

J. Ragazzini and L. Zadeh, "The Analysis of Sampled-data Systems," *Transactions AIEE*, Volume 71, Part 2, 1952: Table 6-4.

C. Allen, *Astrophysical Quantities*, 2nd Edition, London: The Athlone Press, The University of London; New York: Oxford University Press, Inc., 1963: Tables 2-28, 2-29, 2-30, and 2-32.

H. Black, *Modulation Theory*, Princeton, N. J.: D. Van Nostrand Co., Inc., 1953: Figures 4-28, 4-29, 4-30, 4-31, and 4-32.

G. Kuiper, *The Sun*, Chicago: The University of Chicago Press, 1953: Table 2-19.

S. Singer, "Some Considerations of Expected Radiation Belts of Planets Mars and Venus," in *Advances in the Astronautical Sciences*, Volume 6, New York: American Astronautical Society, 1960: Tables 2-23 and 2-24.

Coast and Geodetic Survey, United States Department of Commerce: Figure 1-7.

Laboratory for Electronics, Inc.: Figures 9-5, 9-8, 9-9, 9-10, 9-11, and Frontispiece.

SOVFOTO: Figure 1-13.

National Aeronautics and Space Administration, Figure 5-28.

National Aeronautics and Space Administration and Mr. V. J. Oliver, United States Weather Bureau: Figure 1-3.

PREFACE

Space Vehicle Electronics is a technical field whose breadth and depth are limited only by the horizons of the professionals engaged. In keeping with the systems approach of the Principles of Guided Missile Design series, this volume has stressed technical areas of basic necessity to preliminary design. The text is aimed at the graduate student and professional engineer. Liberal use has been made of periodical references so that further source material may be obtained as required. The systems approach taken in this volume is that of first ascertaining requirements, environment, the nature of free-flight motion in space and the means for sensing such motion accurately and with timeliness. Communication, radar, guidance, control and power techniques are then covered from the viewpoint of systems performance in this context. Payload considerations such as operating laws, limitations, required capabilities and inter-relations with other system elements are treated. Space, weight, power and reliability budgets are considered in conjunction with installations requirements. Finally, the current state-of-the-art leading to a preliminary design for the APOLLO lunar landing radar is appraised.

In such a fast-moving field as Space Vehicle Electronics new subjects and different ways to treat them continually occur to the author. It is sincerely hoped that readers will comment on and criticize this text such that a supplementary volume may someday prove worthwhile.

I am greatly indebted to many individuals and organizations for use of material; a list of most of these precedes this preface. However, I could not let the opportunity pass to acknowledge here the immeasureable contribution made to *Space Vehicle Electronics* by other books in the series of which it is a part: *Missile Engineering Handbook* by C. W. Besserer, *Airborne Radar* by Donald Povejsil, Robert S. Raven, and Peter Waterman, *Inertial Navigation* by Richard Parvin, *Guidance* by Arthur S. Locke and his collaborators, and, most notably, Kraft Ehricke's *Space Flight*, Volume 1. Not only have many figures and tables been chosen from the aforesaid volumes (particularly the last) but the text matter itself often reflects the magnitude of their contribution. While the responsibility for this book is necessarily my own, I am greatly indebted to each of these other authors because without such help my own task would have been substantially more difficult.

Much credit is due to my long-suffering wife, my secretaries, my colleagues, and the generous cooperation of the Laboratory for Electronics and the Fairchild Stratos Corporation for their support of this project.

DAVID B. NEWMAN

Wayland, Massachusetts

CONTENTS

CHAPTER 1

SPACE VEHICLE MISSIONS

1.1 INTRODUCTION

Space vehicle missions dictate, in the final analysis, the nature of the vehicle system and electronic sub-systems, subject to the restraints of physical laws and environments, and budgetary expectations. In this chapter these missions have been subdivided into broad primary and secondary categories, to illustrate the importance of knowing one's objective or problem before attempting to develop a system or solution. The differences among missions and their ramifications will make it further evident that each mission will have its own peculiar optimum solution at any point in time, and, though similarities will exist among some systems, these should result from similarities in the particular mission or requirement definitions rather than hardware state-of-the-art. That is, system design, while cognizant of hardware potentials, must be geared initially to the mission or requirements and their fulfillment, and only thereafter to the state-of-the-art. To design a space vehicle system from a "hardware available" point of view without an understood mission is as non-sensical as assembling words from letters without a thought to be expressed.

Space Vehicle Missions are usually divided into three categories, Military, Civil, and International. Military missions comprise a large and urgent portion of the mission spectrum, are generally needed by a specified date and thus are much influenced design-wise by applied or directed research as well as by operating mission. Civil missions are either of a basic research nature or involve operations on a commercial or subsidized basis for the general welfare. International missions pertain to geo-political and cooperative ventures.

The reader will appreciate that a discussion of space missions necessarily embodies a great deal of speculation. The author has indulged in such speculation, not with a view toward prophecy but rather to stimulate the imagination of those who, in years to come, may be called upon to understand such missions and provide electronic devices for their accomplishment.

1.2 MILITARY SPACE MISSIONS

Military missions in space have been defined by the various governments, and, characteristically, will be re-defined periodically in the future. The purpose of this chapter is to present logically for the student the mission requirements spectrum for military space vehicles.

Military action is based on definition of an enemy, and a national policy to oppose this enemy strategically and/or tactically. The success of such opposition will be based on the technical, logistic, positional, psychological, numerical, and decisional strengths of the competing military organizations. Space vehicles offer peculiar technical advantages in the conduct of terrestrial war. Therefore, control of space and vehicles operating therein will evolve as another facet of warfare with generations of space vehicles and combat techniques to be conceived. Fundamental to engaging an enemy is knowledge of his nature; not only who he is, but his qualitative and quantitative capabilities. Secondly, one must ascertain and predict the character of the terrain, climate and environment to assess the desirability of (or probability of success of) proposed actions. The use of this information is restricted or enhanced by the communication system available. The action or reaction of the military commander is then embodied in the striking power which is used on his decision. The immediately following sections of this chapter deal with space vehicle requirements for six functions:

1.2.1 Reconnaissance—the inspection of the enemy.

1.2.2 Surveillance—the measurement of the operating environment.

1.2.3 Communication—the transfer of information.

1.2.4 Aids to Navigation—the deployment of weapons, forces, and materials.

1.2.5 Data Transport and Recovery—the delayed accumulation of strategic data.

1.2.6 Strategic Weapons—long-range striking power.

Beyond strategic weapons which may carry explosive, incendiary, radiation, bacteria, and chemical payloads, one progresses to the more exotic warfare that may reach fruition in the space age. Such things are considered in the second group of sections:

1.2.7 Weather control

1.2.8 Tactical weapons

1.2.9 Space control

1.2.10 Lunar control

1.2.11 Planetary control

The foregoing sections do not deal with the requirements for military control, which are taken as axiomatic, but consider various guidance, control, and communication problems which arise in accomplishing these missions.

Section 1.2.12 deals with the ever necessary supporting military research which appears to be always with us, not because civilians cannot administer such research, but because purely research-oriented activities are directed wherever the research workers interest leads, whereas military research must investigate militarily significant physical laws and environments in a timely fashion. That is, military development activities must be provided with the necessary background to perform their tasks on a useful schedule to support military operations and maintain the national military stature on a competitive basis with foreign powers.

1.2.1 Reconnaissance

Of inestimable value to the military is reconnaissance of the enemy—the ability to know his activities, and through intelligence, interpret their significance. Space vehicles are expected to play a major role in this field, particularly, because they may be orbited over enemy territory with a minimum risk of interdiction. The difficulties associated with such interdiction stems from their high energy requirements.

Reconnaissance vehicles carry a payload which generally (although not always) will be passive in nature. It may generally be typified as either an

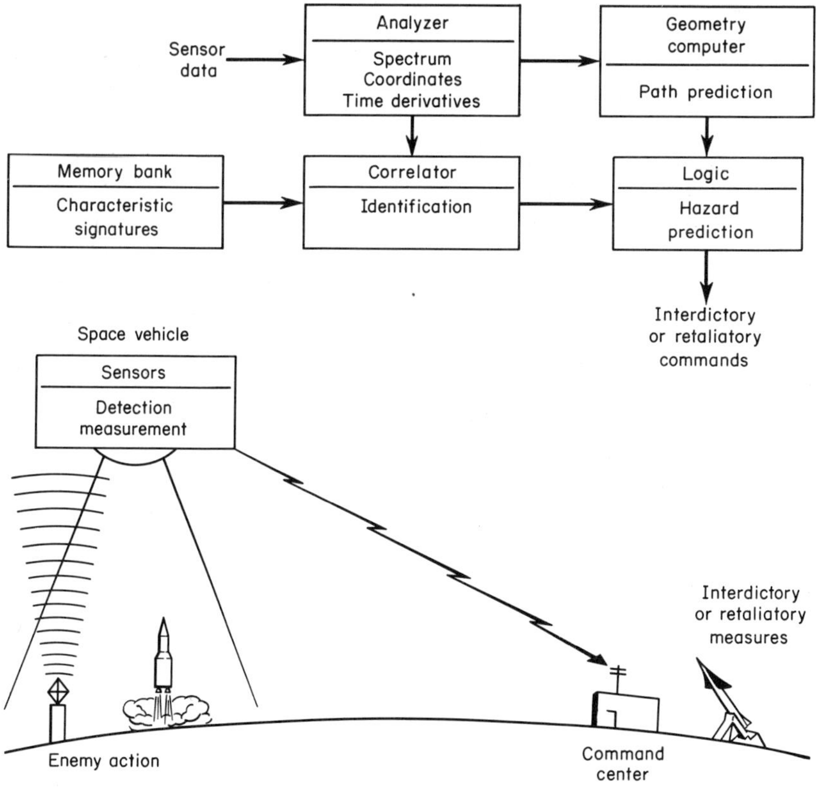

FIG. 1-1. Reconnaissance System Functional Schematic.

energy-collecting device (visual optics, infrared, microwave, ultraviolet) or a field measuring device (radio). The portion(s) of the electromagnetic spectrum, the scan methods, and the satellite path(s) employed in reconnaissance will be dependent on (1) the electromagnetic "windows" or propagation spectrums available, (2) the type of activity to be detected and its radiation characteristics (or "signature"), (3) the amount of specific information necessary to establish that a significant level of interesting activity is occurring, (4) the interpretability of the data taken, and (5) the data recovery or transfer means employed.

A generalized Reconnaissance System Functional Schematic is depicted as Fig. 1-1. This schematic shows the inter-relations between enemy actions, sensors to detect and measure this action, analysis, memory, correlation, logic, communication and counter-weapon measures. The schematic possibilities depicted in Fig. 1-1 actually may be arbitrarily or physically limited for various reasons at some point in time. Automatic retaliatory action is psychologically unacceptable to most people (at least, in the U.S.A.). Automatic interdictory action may be uneconomical if the enemy has the capability to saturate or decoy the defense complex. The communication link may be temporarily (or permanently) disrupted by design and horizon limits or by enemy action. Communication link(s) may be inserted between various other functional boxes to transfer various functions to ground (or other) sites. However, the fundamental functional relations portrayed in Fig. 1-1 should hold true, even if human operators are employed in some functions (e.g. as correlators or decision elements).

Reconnaissance vehicles may be typified as probes, satelloids, and satellites. These are shown pictorially in Fig. 1-2. The reconnaissance probe may be employed by tactical units for a "quick look" at surrounding areas, and may radio back data or record information for recovery. The satelloid is a probe that circles the Earth once and is recovered, thus permitting observation of any area on Earth with complete data recovery

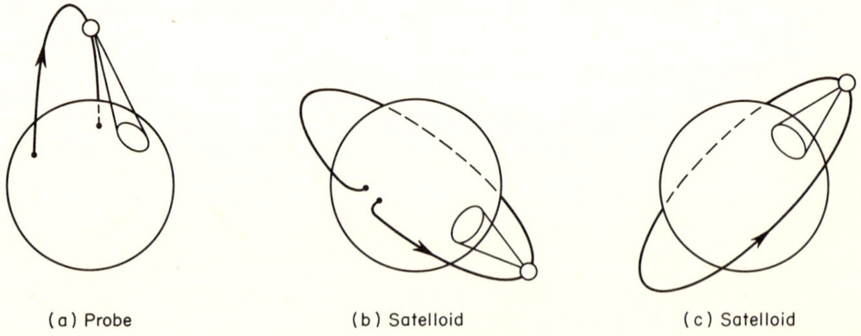

FIG. 1-2. Reconnaissance Orbits.

and minimum vulnerability to interdiction. The satellite will likewise orbit the globe, but will have to transmit its information by radio or some sort of recoverable data capsule. Without going further into the military merits of these three vehicles, we can say, from an electronics point of view, that the following problems exist and must be solved as related to any single vehicle concept:

1. Reconnaissance payload orientation, stabilization, synchronization (as in camera frame triggering), motion compensation (as with scanning systems), and exposure control.
2. Data recording and/or transmission (including film or other recording media), processing, and scanning.
3. Vehicle navigation, guidance, and control.
4. Electrical power generation, distribution, and regulation.
5. Recovery sequence operation.
6. Orbit or trajectory determination including external surface-mounted tracking devices.
7. Data recovery by radio receiving equipment and/or capsule or vehicle locating systems.

Electronic radiation analysis is a further item of reconnaissance beyond the pictorially oriented camera or the electro-magnetic scanning device. Radiation analysis can be employed to ascertain the communication and electromagnetic radiation activity. In addition to reception and analysis of transmitted radiations, passive detection may be made of natural infrared and radiometric radiation (Ref. (*1*)). The implication of this is that radiation characteristic changes or local "hot spots" in enemy territory may indicate significant enemy maneuvers or activity such as large-scale bomb tests or large rocket firings. The significance for space vehicles is that reconnaissance satellites, following orbital paths such as depicted in Fig. 1–2 should be able to detect such radiations. A command center is required with sufficiently sophisticated analysis equipment to perform a signature correlation between the detected signal and a memory bank of characteristic signatures as indicated in Fig. 1–1. A further requirement is the tracking of signals with velocity characteristics and the prediction of flight trajectory with a warning of impact area and the call out of interdictory or retaliatory armament in accordance with some pre-determined logic pattern. This includes both space and ground-based systems such as the current MIDAS (Missile Defense Alarm System), BMEWS (Ballistic Missile Early Warning System), and SPASUR (Space Surveillance System). Specific characteristic "signature" recognition systems will generally be highly classified in nature. However, it should be evident that fundamentally the signature concept is that of correlating with sufficient precision an observed signal with some stored signal in a manner analogous to identifying fingerprints; the signature being defined in terms of frequencies (wavelengths), phases, modulation types, etc.

1.2.2 Surveillance

The measurement of the operating environment is of critical importance to the military. Target coordinates must be established with sufficient accuracy to enable an ICBM to place its payload within lethal distance of the true target. For this reason it is necessary to obtain accurate map data throughout the world and to correlate the various maps with each other. Gravitational anomalies and magnetic field distortions must be established and measured to permit a prediction of, and compensation for, their effect on inertial and magnetic guidance systems. Meteorological conditions must be surveyed to determine and predict climatic conditions in operating areas. Knowledge of world target and battlefield weather conditions and trends can play a decisive role in warfare provided it is accurate, timely, and presented concisely for use. The changes in weather conditions within a single satellite orbit period may be of sufficient significance as to require constellations of weather satellites.

Aerial photo-mapping is an established art which revolves around taking a continuous chain of overlapping pictures with a camera noted for its high resolution film, minimum distortion lens system, and wide angular or area coverage. Mapping cameras are often vertically stabilized by a suitable mount and reference to minimize image blur. Particular flight paths are flown with respect to visually or electronically identified artifacts or land marks. Film overlap is set for greater than 50 per cent between adjacent frames to permit use of stereo techniques to identify and extrapolate artifact coordinates in three dimensions down the chain of films. Recent advances in radar echotime measurement circuitry and barometric altimeter methods have resulted in the development of ground profile recorders which can trace a cross-cut of the Earth's surface defining surface altitude with respect to sea level as well as radar echo altitudes. This technique provides object scaling information and a cross-check and accuracy enhancement of the stereo-artifact technique employed with mapping cameras. Electronic ranging on beacons is used to tie together known bench marks of different surveying datum line systems, as between continents and islands. These techniques may and will be employed in mapping satellites to provide the military with accurate mapping and coordinate data. The coverage areas depicted in Fig. 1–2 apply to mapping cameras as well as to other reconnaissance devices. Coverage for electronic ranging systems will be horizon limited so that the coverage available from a single satellite is that depicted in Fig. 1–2. This figure indicates the broad surface coverage available from high altitude satellites where system requirements are horizon-limited, primarily as in the case of electronic triangulation or trilateration systems that might use a large balloon reflecting satellite or an active transmitter or transponder.

Gravitational and magnetic anomalies must be established to determine their effect on guidance systems which sense and utilize these fields. Perturbations in circular orbits can be employed to establish the existence of and character of gravity anomalies if such perturbations are measured carefully during the affected orbit period. An anomalous field may deviate a satellite orbit from an intended circle to an ellipse. Such deviations in uncontrolled vehicles (or, at least vehicles with no subsequent path correction means) will cause undesired orbits and trajectories.

Meteorological measurements from satellites offer a unique opportunity to study cloud formations (as with cameras). In addition, radiation, infrared, ozone, and heat exchange measurements offer quite a large body of information on which to predict weather and its patterns. Coupled with radiosonde measurements within the atmosphere, such data will doubtless facilitate superior predictions.

The measurement of cloud patterns is somewhat analogous to reconnaissance and mapping. The orbits desired will be of the types depicted in Fig. 1–2. Sensors other than the cloud-mapping camera will quite likely be carried in the same satellite. The problem will be to accumulate

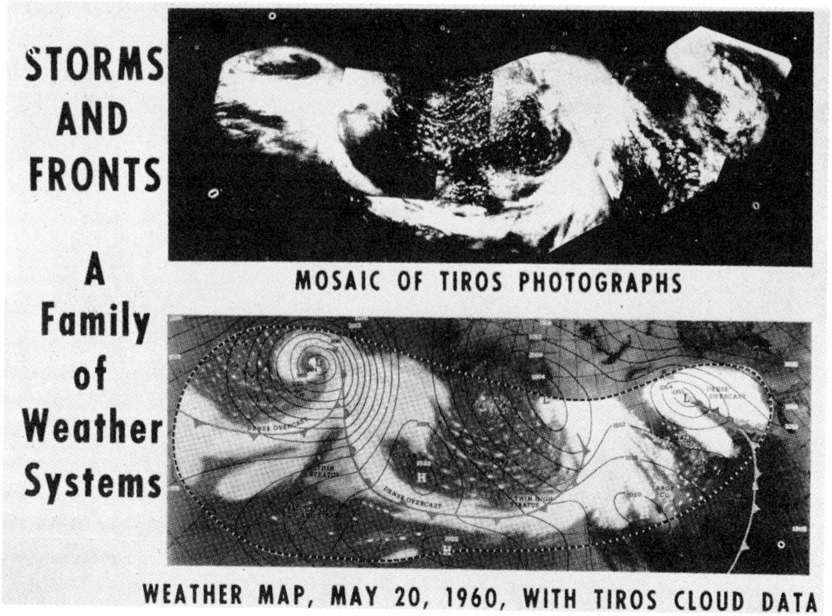

Fig. 1-3. TIROS I Meteorological Photographs and Meteorological Front Analysis.

cloud maps and other data (heat exchange rate pattern, for example) and to relay these data rapidly to Earth in an easily identifiable and integratable form. Quick use must be made of meteorological information or the data will soon become stale. This requirement has led to the employment of television and scanning infrared sensors with tape data storage and radio-link playback on command in the NASA TIROS program (Ref. (*2*)). Figure 1–3 illustrates a series of cloud formation photographs obtained from the TIROS I satellite, and a portrayal of these cloud formations on the corresponding meteorological frontal analysis. From this figure it is evident that cloud conditions and visibility data are immediately available from such a satellite system. The vertical (or layer) distribution of the cloud system is not immediately evident from the TV photograph, nor is the pressure situation. However, this information ultimately may be derived from comparisons of successive pictures of the same area or from comparisons with simultaneous data from other radiometric sensors operating at preset or scanned wavelengths.

The so-called "fixed" or "24-hour" satellite is of particular interest for the observation of meteorological conditions in the equatorial zone. This satellite derives its concept from the intuitive idea that it would be extremely useful to maintain observations of weather conditions from a station (satellite) with fixed geographic coordinates. Such a satellite obviously must be in a circular equatorial orbit and have a period identical to that of the Earth.

It is expected that the "stationary" satellite must be kept in position by means of some reference sensing device operating a closed-loop feed-back vernier propulsion system in order to minimize both the excursions from the reference point and the expenditure of control energy (perhaps by an altitude correction). It is clear that a constellation of equatorial satellites could provide continuous weather information up to some sensor-resolution limited latitude (a horizon limit will allow more than 80 per cent of the Earth's weather to be observed continuously [Fig. 1–4]). Polar weather

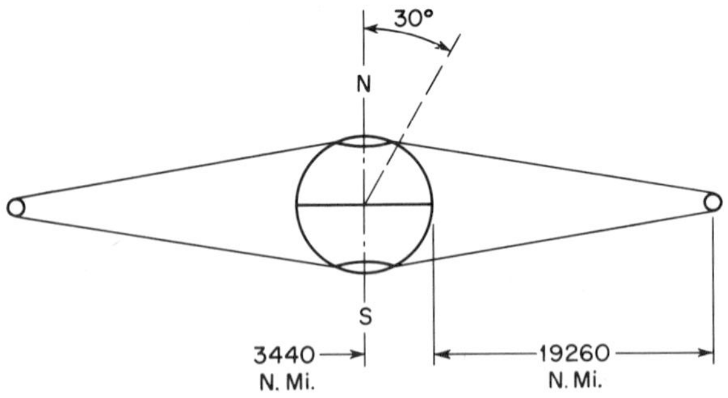

FIG. 1-4. Equatorial Satellite Surveillance Coverage.

coverage may be obtained either by supplementing the equatorial constellation with a polar orbiting constellation, or by providing sufficient inclination to the orbit planes to obtain polar coverage. The latter would no longer be a true equatorial satellite, but with a 24-hour period would describe a figure-eight orbit centered on one longitude.

In general surveillance vehicle system orbits may be categorized as similar to the reconnaissance systems depicted in Fig. 1–2. Also, the electronic system requirements will be similar to those previously cited for reconnaissance systems. The primary difference between surveillance and reconnaissance is in the nature of information being obtained. Reconnaissance is more exploratory in nature and is aimed at determining specific new information (where none previously existed) on enemy emplacement, deployment and actions. Surveillance is more comprehensive in nature and indicates the maintenance of continuous or periodically sampled quantitative information on the physical characteristics of interest; unpredictable deviations in these characteristics provide a form of reconnaissance.

1.2.3 Communication

Military requirements are historic in the need for communications in both directions via the chain-of-command. Satellite communication vehicles offer this possibility in many ways as reflectors or repeaters. Another communication requirement for space-vehicles is that of data transfer or recovery at large information rates. This latter requirement indicates recovery of data capsules.

In command communication systems the usual requirement is for rapid communication capability in both directions; however, special conditions may require or permit communications in one direction only. Examples of this might be a submarine which wishes to transmit without revealing its position, and does so by sending up a rocket to broadcast a taped message. Another case might be a national strategic warfare direction or war initiation system where the importance of sending out simple (in the sense of information bandwidth content) messages (such as a code word to execute a plan from several pre-determined alternatives), without listening for replies because of the uncertainty of answering service in the presence of enemy interruptive or jamming action, and the need for maintaining secret (hence, quiescent) locations for mobile field units.

Large passive satellites offer a potential for universal communication between any two parties in view of the satellite. A large inflated structure is conceivable such as the NASA Echo satellite. Such a satellite may be thought of as a sort of artificial ionosphere or radio wave reflective device. However, its usefulness is limited by the power, range, and bandwidth requirements for the communication problem of interest. Also, such a satellite is so universal that it may be used by the enemy for his own communications and to intercept yours. Therefore it is more likely to be useful as a non-military device.

Military communication satellites can be expected to require high gain directive ground antennas which track the satellite automatically while transmitting and receiving. It is conceivable that such ground stations might acquire, track, and communicate with or relay via satellites during each or chosen passings. Upon coded command, communication satellite antennas may be oriented towards particular locales or directed to acquire and track particular ground beacons so as to minimize the field equipment power requirements and the likelihood that an enemy can listen to the

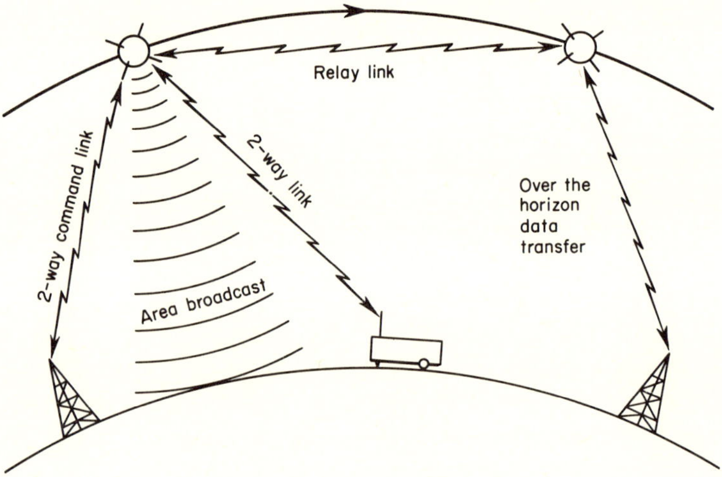

FIG. 1-5. Military Satellite Functions.

transmission or take control. Fig. 1–5 portrays schematically the various functions that can be performed in a military satellite communication system. These functions may be tabulated as:

(1) General command broadcast to an area
(2) 2-way or 1-way link between discrete stations
(3) Over-the-horizon relay via multiple (2 or more) satellite system
(4) Stored data transfer on program or command

Probes and satelloids (a globe girdling probe or once-around satellite) may be used where only short-term communication service is necessary or desired (for security) or possible (because of interdiction). Communications will become a military necessity for any extra-terrestrial military operation. Some examples are portrayed in Fig. 1–6.

Figure 1–6a shows both direct and relayed communications between the Earth and the Moon or a planet. Relay stations have been depicted as

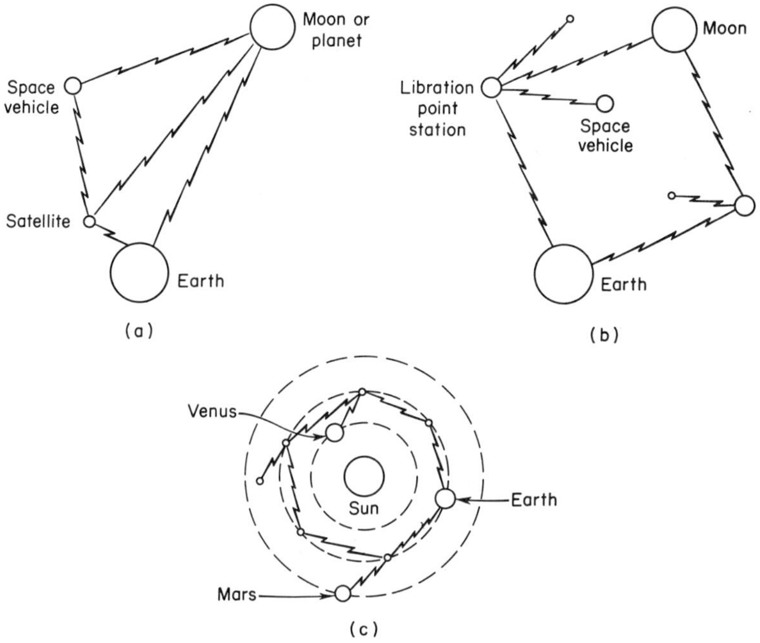

FIG. 1-6. Extra-Terrestrial Communication System Examples.

satellites or libration point* stations to minimize their energy requirements and to provide predictable locations (from orbit or Earth-Moon ephemeris knowledge). Propagation problems must be considered; particularly atmospheric "windows" (wavelengths at which minimum attenuation is experienced), ambient noise conditions, transmitter power, receiver sensitivity, antenna gain and/or collecting powers, and antenna orientation schemes.

Figure 1–6b portrays a particular scheme for establishing libration point relay stations in the Earth-Moon system. These stations can be in communication with both the Earth and Moon (and incidentally with greater surface area than is available by direct contact). The libration point stations should provide a good predictable relay link for cislunar† space vehicles because the relay station-to-space vehicle link is much shorter than direct vehicle-to-Earth links and the relay station can relieve the space vehicle of the high-power, "window" wavelength link equipment

*A libration point is an equilateral coplanar point at which any two-body system may have third bodies added, and at which equilibrium may be maintained without the expenditure of energy. A primary example is the Trojan asteroids which are clustered at the equilateral coplanar points in the Sun-Jupiter system.

†Cislunar space is that space in the Earth-Moon system in which the gravitational fields of both bodies are significant.

requirement, thus permitting assignment of broadband channels for vehicle-to-relay station service. Also, the libration point station can serve as an information filter and storage center, transmitting priority messages to Earth, and retaining most of the relayed information for transmittal via its logistic support vehicle system.

Figure 1–6c portrays an interplanetary communications relay scheme in which a ring of sun-orbiting relay stations might be established to permit communication with planets on the far side of the sun, and with sun-orbiting space vehicles operating near the ecliptic plane. One can envision a group of such solar-orbiting relay station rings operating at various radii from the sun to provide communication service throughout the solar system. Also, one can envision supplemental communication relay rings operating in various planes off the ecliptic.

1.2.4 Navigational Aids

Satellites, libration point stations, and artificial planetoids offer unique potential as aids to Navigation. Precise knowledge of a satellite orbit coupled with a measurement of one's position with respect to that orbit can provide a navigational "fix." Such a system is currently under development as the TRANSIT program. It utilizes master ground stations to track and compute the satellite's orbit, transfers the orbit parameter information by radio to the satellite for area re-broadcast and permits

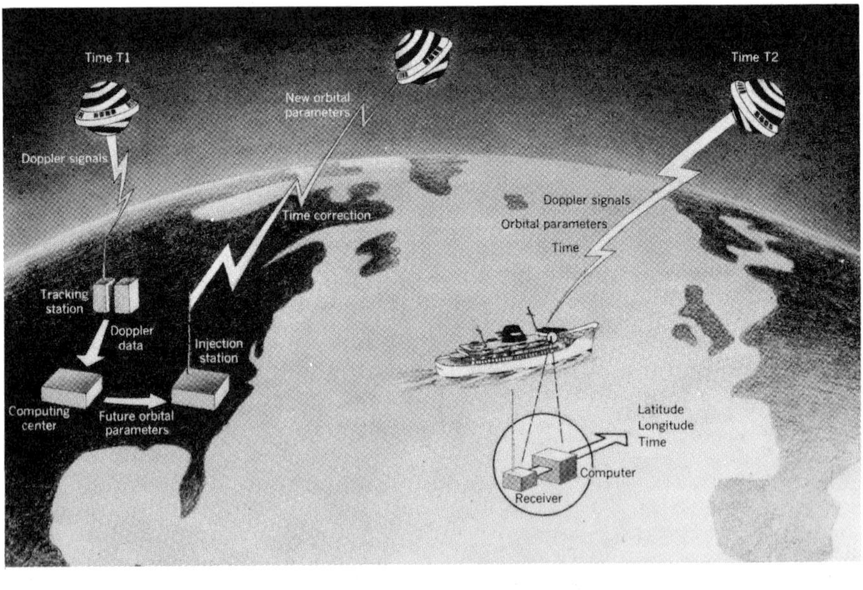

FIG. 1-7. Transit System.

any surface observer to receive the orbit parameters and, by a doppler measurement of the satellite's passing, ascertain his geocentric coordinates. Figure 1–7 illustrates this system. In addition to single-pass doppler techniques, satellite reference systems may be developed which employ precision clocks and distance measurements to make a position fix by simultaneous distance measurements on a constellation of satellites. A variation of this technique is to make range difference measurements—a satellite-borne Loran system.

In cislunar and interplanetary space man should be able to recognize the stellar constellations and navigate by star sightings in a manner analogous to traditional ship navigation but without atmospheric refraction and weather encumbrances.

For navigation on or near a lunar or planetary surface some local body reference system should be employed. Navigation in cislunar space will quite likely be accomplished using an Earth-Moon base-line reference, and, eventually, libration point station locations as supplementary references. Operations in interplanetary space must logically use a solar-centered reference system with stellar constellation sightings for "fixes." The communication relay systems portrayed in Fig. 1–6 each have potential as navigational reference systems for space navigation provided only that the satellite, libration point station, or artificial asteroid orbits are known accurately; these sites can be contacted through some part of the electromagnetic spectrum, and either a suitable constellation grouping or a precise common time reference is available. Of course, the planets themselves can be used similarly; also the traditional astronomers clock, the eclipses of the moons of Jupiter. The navigation problem resolves eventually into one of accurate measurements. With energy surpluses available from nuclear sources military space vehicles will become interested in transit time minimization rather than minimum energy trajectories. This indicates that mid-course navigation precision requirements may be relaxed somewhat and that destination tracking and lead-orbit courses should become a major military terminal navigation requirement.

1.2.5 Data Transport and Recovery

The use of wide-bandwidth information channels in space communication and for space-to-Earth communications will necessarily be restricted by physical phenomena (i.e., noise, channel availability, information form, ability to translate fully as in scanning a film) and security requirements (quiescent operation to conceal terminal locations and information content). These requirements therefore indicate clearly that space data transport and recovery systems will be required in military operations. Such requirements may obviously be met by the same means as human transportation systems. However, the recovery of large amounts of data from unmanned satellites dictates that automatic recovery systems be developed, as is currently being done in the SAMOS program. Tactical probes, satelloids and satellites may be designed to recover stored data as well as to radio high priority information as indicated in Figure 1–8. The requirements of these systems will generally dictate a precise navigation or guidance and control system to bring the data capsule to a pre-determined recovery area or a radio-commanded destination. If sufficient accuracy is not available for this to be accomplished on internal program in any or all phases of the recovery sequence, a radio beacon may be required as an aid for both the ground tracking, receiving and command stations, and detection and acquisition by the capsule recovery team.

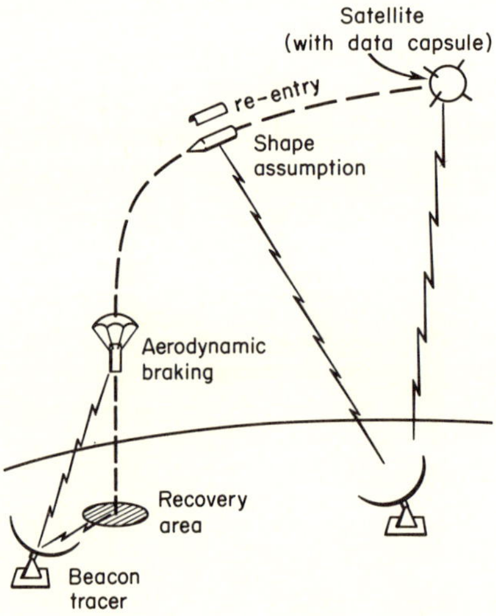

FIG. 1-8. Data Recovery System.

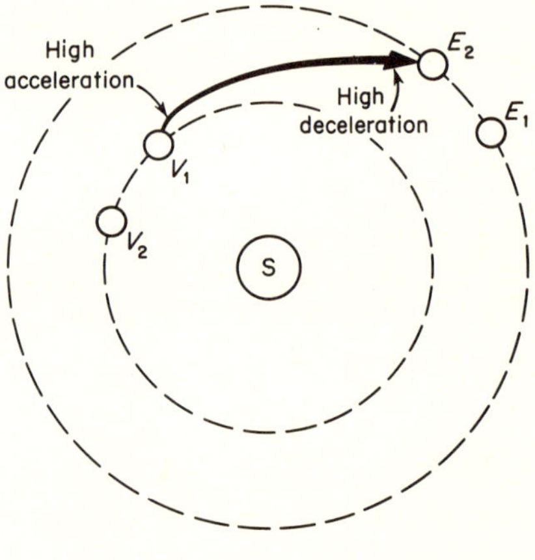

FIG. 1-9. Data Capsule Rocket System.

In cislunar and interplanetary communications, when the information transfer rate required is too high for radio relay, and the priority or timeliness requirement cannot be met by normal logistics operations, data capsules may be rocketed through space as depicted in Fig. 1-9. The advantages of a data capsule rocket system are (1) that the space and weight necessary to sustain and transport a human are not necessary thus increasing the terminal velocity available from a particular rocket, and (2) a data capsule rocket can be accelerated (and decelerated) at much higher values than a human can tolerate. This enables the system designer to achieve a transfer in a much shorter time by using high acceleration values with no increase in energy (hence, fuel) expenditure.

1.2.6 Strategic Weapons

Strategic weapons of the ICBM class have provided an essential tool for initial space exploration and operations through the development of large rocket motors capable of providing sufficient impulse velocity* to place significanc payloads in orbit. Strategic weapons have their place in the military scheme as the major offensive or retaliatory force a country possesses. Figure 1–10 depicts various forms of ballistic missiles. These missiles may be launched from fixed or mobile sites including surface pads, underground silos, railway cars, ships, submarines and aircraft. The distinction between intermediate range and intercontinental is merely one of range. Ballistic missiles may be placed in orbit, and called down on command (or lack of staying command) to a designated target area.

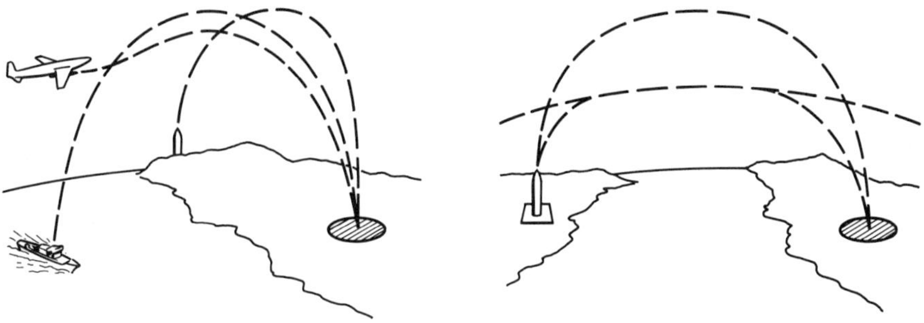

(a) Intermediate range ballistic missiles (surface, air, ship and submarine launched)

(b) Intercontinental and orbiting ballistic missiles

FIG. 1-10. Ballistic Missiles.

Ballistic missiles are generally (at the present time) multi-staged vehicles, and are usually ballistic only after the final stage burn-out. Control devices are generally employed during the motors' burning to fit the subsequent ballistic trajectory to the desired target. Trajectory establishment may be by means of radio (radar) tracking from ground stations or by inertial devices or some combination of these. Control is effected by thrust direction or orientation control and stage burn-out time control.

The orbiting ballistic missile is one which is placed in orbit with sufficient fuel and a space startable retro-rocket so that on command (or lack of staying command) the rocket engine may be started and the missile brought out of orbit, and down to some target. Target coordinates may be preset in a storage system, and up-dated by radio command.

Enemy take-over must be considered and suitable precautions taken if such a missile is to respond to radio commands. This might be accomplished

*The primary characteristic of a rocket is its impulse velocity—that is, the incremental velocity change the rocket can impart to the payload.

by coding schemes (including spectral signature recognition), or by energizing command reception channels only over friendly areas.

Bombing from satellites is a variation of the orbiting ballistic missile in that the basic device (manned or unmanned) is placed in orbit, and the bombs are in reality ballistic missiles retro-fired from the satellite.

The accuracy of a strategic weapon is usually expressed as so many meters CEP (Circular Error Probable). This term is used to define the hitting accuracy as the radius of a circle drawn about the mean point of impact which includes 50 per cent of all firing trajectory terminations and excludes 50 per cent. Thus the accuracy is expressed in terms of the 50 per cent probability that a single shot will terminate within the stated distance from the target (assuming all random as distinct from system or bias errors). In test work, one must distinguish between the CEP from the mean (or average) hitting point and the bias of the mean from the desired (or target) hitting point.

Strategic weapons CEP must be related to the payload or warhead characteristics in order to appraise lethality. Various possible warheads include high-explosive, incendiary, nuclear, chemical and bacterial. The high-explosive and incendiary warheads are economical only if extremely tight CEP's can be obtained. The remaining three are useful, depending on the target terrain and meteorological conditions, with more nominal CEP's.

Electronic problems associated with ballistic missiles will generally revolve around the following:

1. Target location
2. Ascent guidance, control and communications
3. Orbital communications, orbit measurement and correction, call-down trajectory prediction, and call-down commands
4. Terminal guidance (if any)
5. Warhead action initiation (fuzing).

1.2.7 Weather Control

As knowledge of meteorological causes and effects increases, it is to be expected that weather control may become a tactical or strategic weapon. Although weather is an atmospheric characteristic, it is dependent to some extent on heat exchange relations with outer space (Ref. (*3*)), and some localizing of heat exchange or distribution to a particular area or time era may become feasible in the near future.

1.2.8 Tactical Weapons

Space operation may be expected to develop tactical as well as strategic weapons. Specific weapon possibilities include nuclear detonations for many purposes, destruction of vehicles, pre-detonation of enemy warheads, ionosphere and, hence, communication disruption, etc. Exotic weapons include synthetic ball lighting and high-velocity neutron beams. As extra-terrestrial warfare develops it can be expected that still more exotic systems will be proposed and exploited. These developments may or may not be electronic in nature (e.g., laser beam weapons); however, it may generally be expected that the fuzing and detonation mechanisms of these devices, and some defenses against them or their fuzing action will be electronic. Electronic identification and recognition devices in space warfare may be expected to develop along lines similar to aerial warfare techniques or else space vehicles may suffer interdiction for failure to "file" suitable navigation plans with all the appropriate control points.

1.2.9 Space Control

As man develops avenues into space, the commercial and military values of space and extra-terrestrial bases will dictate the need for military control of space activities just as naval control of the high seas became a necessity during past generations. Tremendous booster rockets may be expected to extend man's "reach" into outer space, his "lifting" capability in cislunar space, and the development of truly maneuverable space vehicles that can change missions and trajectories significantly without refueling stops. Space control systems may logically be organized as depicted in Fig. 1–11. Space control systems revolve around a Control Center which accepts flight plans of friendly trajectories and surveillance data on unknown trajectories. At the Space Control Center(s) the nature and imminence of a military threat must be evaluated and the command decision made pertaining to the need for, and nature of, interdictory action. Space is so vast a domain it is logical to expect it eventually to be divided into military control zones much as the United States and the

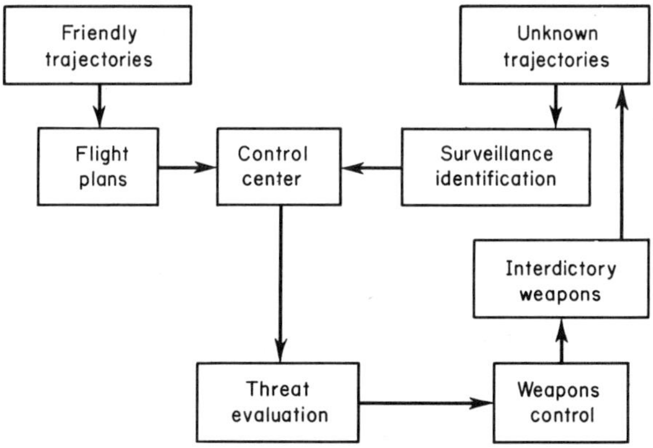

FIG. 1-11. Space Control System Functions.

World are presently divided into Army, Navy, and Air Force zones. Also, because of the unique nature of space flight, the zones may be expected logically to be body-oriented with the intervening space controlled in large sectors either from the nearest bodies or from free-flight military stations such as might be established at libration points, in satellites, or in artificial planetoids.

Friendly activities will of necessity file flight plans with the controlling military bodies, quite likely with the military control at the point-of-origin for take-off permission and communication to enroute and destination control points.

Unknown bodies or trajectories may be anticipated in a military control system. These bodies may be natural objects (e.g., meteorites), hostile objects or "tramps" which have somehow progressed without filing appropriate flight plans. Depending on the prevailing state of war psychology, the military may either (1) engage all unknowns in combat because the potential danger of the unknown will be so great that it would be suicidal not to destroy it, or (2) attempt to survey the unknowns further to identify and classify them as to nature (meteorite, weapon, transport vehicle, logistic vehicle, decoy, etc.), and the potential risk in allowing them passage through the control sector. Risks may be assigned on the basis of points of trajectory origin and destination trajectory proximity to vulnerable military targets (e.g., communication relays), velocity and maneuverability characteristics (hence ease of interdiction and hazard in delaying interdiction), radiation characteristics (e.g., jamming, temperature, reconnaissance attributes which may indicate mission), and control exhibited (whether tumbling without sensible control or attitude stabilization or whether in-flight control impulses are evident from comparing actual with free-flight trajectories). These characteristics will, in large measure, dictate the need for and nature of interdictory action.

Ballistic missile interdiction is the first type of space vehicle interdictory action. Some schemes for this action are depicted in Fig. 1–12. These include, (a) detection and tracking of the ballistic missile by satellite and/or ground installations, the assignment of defensive weapons, and the destruction or disablement (deflection or disarming) of the missile on descent. A second variation (b) depends upon knowledge of the general location of the launch site and the ability to place interdictory weapons in the general vicinity of that site. With good ocean surface and sub-surface traffic control such a system might prove very effective in neutralizing submarine-launched ballistic missile threats. Such a system could detect and track the ballistic missile in the ascent phase, and commit a fast acting weapon to destroy or disable it during ascent, possibly before motor burnout to capitalize on motor burning radiations and achieve a destruction by exploding the missile engine or propellants. A third concept (c) is a satellite base which can detect ballistic missiles on the ascent and commit weapons to interdiction action directly. A fourth concept (d) is an orbiting anti-missile which could be called into play by a command link operating from a control center. The choice and implementation of these (and possibly other) systems will depend on state-of-the-art in hardware development (e.g., detection sensors, rocket engines, control reference accuracies, etc.) and the permissible decision time available with each system and, above all, the economics compared to kill probability. Call-back capability is desirable, but a quick-acting system, which permits maximum utilization of all intelligence, would seem to be the most logical choice generally. While Fig. 1–12 depicts anti-missiles as the interdictory agents, other methods are being considered (Ref. (*4*)). Warheads may be nuclear or high-explosive, ablative meteoritic types, gases, plasmas, high energy beams of fundamental particles, coherent light or radio-frequency radiation, or various types of artificially generated fields. Thus the method of warhead delivery may be by a conveyance such as a missile, by a focusing device (lens or mirror equivalent) for high energy beams, or by the warhead itself as with artificial field generation. Similarly, the action of the warhead

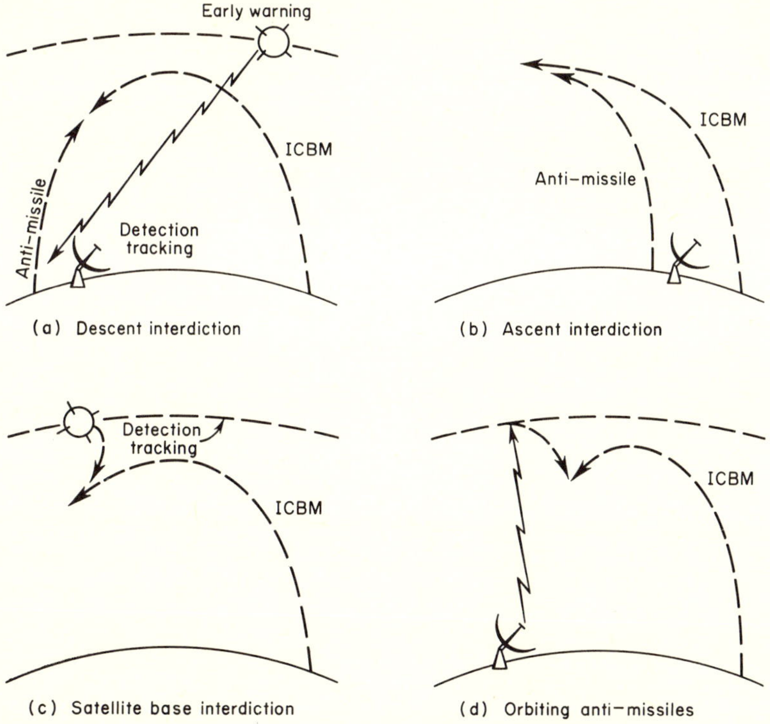

FIG. 1-12. Ballistic Missile Interdiction.

may be destructive by direct or sympathetic action, neutralizing (as in disarming a bomb), or deflecting (as with field effects on field reference devices or artificial decoys to generate false constellation references for celestial reference systems).

Satellite interdiction is somewhat similar to ballistic missile interdiction in terms of detection, conveyances and warhead operation near the apogee of a trajectory. However, satellites may not be "fair game" in all instances and satellites are not directly a menace until called-down or activated in some way. It may be necessary progressively to inspect the unknown satellite for pertinent characteristics which will determine whether it is a sufficiently hazardous threat to warrant further interdiction. Thus satellites may be interrogated and observed by space and ground stations to ascertain their orbit parameters, the constancy of these parameters, volume, size, shape (the design for re-entry is particularly important), and orientation (tumbling may indicate out-of-control). Thermal radiations may indicate whether a satellite is manned or not. Satellite inspection systems may be able to classify vehicles as to type, mission, and performance of mission by decoding radiated signals and examining them for intelligence (e.g., reconnaissance data). If command links and coordinate reference systems can be ascertained, command take-over or reference "spoofing" may be attempted when such action is indicated. For those satellites whose potential cannot be ascertained adequately, an orbiting inspection system may be employed to make a closer inspection (and, if necessary to accomplish disablement). Such an orbiting system might be made co-orbital with the unknown satellite, and by optical (or TV), radiometric, doppler radar and other techniques ascertain (by virtue of the resolution and lack of interference resultant from physical proximity) more precisely the nature, purpose, and threat of the unknown satellite, and further by direct contact engage the other satellite (if not tumbling out of control) to explore its nature, the value and safety of retrieving it for further detailed inspection, and the need for destruction, disablement, or deflection.

Space combat between various types of space vehicles will almost certainly become a reality just as aerial combat developed from the World War I airplane reconnaissance era. Such combat will require the essential ingredients of detection, tracking, identification, threat evaluation, vulnerability assessment, weapon choice, and assignment. The reaction time factor will depend on the relative length of one's detection and weapon reach with primary attention to an estimate of the enemy's potential in the same area. Maneuverability in space will not play the same role as in aerial combat (where an air mass is used as a soft reaction surface to effect course changes) until a major energy surplus is available with the means for expending it at high rates under suitable control (rate or duration) or some new method of effecting reactive control (e.g., some break-through in maneuvering by means of gravitational or magnetic field strengths or gradients) is developed. However, the conduct of space combat by small military vehicles using high performance missiles or focused beams as weapons may be expected with the interesting perturbations resulting from operating in an environment in which free-flight trajectories are conic sections and the action distances may become so great that relativistic time distortion may become important in determining aiming vectors. Missiles may be thought of as having some combination of fire control, mid-course, and terminal guidance. Focused beams will not enjoy the advantages inherent in terminal seeker guidance for missiles, but will have to rely on target tracking and prediction carried on from great distances, and will thus require considerable accuracy in pointing.

1.2.10 Lunar Control

Control of the Moon as a military operation is a distinct possibility and certainly has some (salient) advantages. The "Panama" theory holds that a nation must secure and hold sufficient key territorial rights to enable that nation to operate independently in the future, exploiting the natural advantages available through these territorial rights. Thus, Spain in holding the Isthmus of Panama was able to operate and secure a Pacific Coast empire extending from Cape Horn to above San Francisco bay. In space flight, the Moon may turn out to be a similar key installation. That is, if significant raw materials can be found and exploited on the Moon, it may easily become the prime "refueling" stop in all interplanetary space travel, and if one nation does secure and hold the significant lunar areas militarily, that nation will hold the key to an interplanetary empire. Further arguments for lunar control or lunar military bases revolve around the ability to detect military actions on or near the Earth, and to take effective military action (e.g., a ballistic missile) from such a lunar base. The accuracies, energy requirements and availabilities, Earth-action visibilities, etc. required for these concepts will be extremely demanding in a system design. However, it would seem that by virtue of the "Panama" theory alone, lunar military control will become a military objective within the next generation of space vehicles.

To achieve lunar control will require a definition first of what is to be controlled. That is, what area of the Moon, what constitutes control, and the amount of control necessary. Thus for lunar control we may desire control of an area because:

1. It has a good observation vantage point—always facing the Earth.
2. It has a good supply of raw materials from which we may fabricate fuels, oxidizers, empennages, structures, food, shielding, medicines, water, alcohol, etc.

3. It has a good position for communication with Earth, Earth satellites, space bases (i.e., at libration points, in lunar orbit, in solar orbit).
4. It may have good terrain for vehicle landing and take-off.
5. It is particularly well concealed for stealth and surprise operations (as the back side of the Moon).
6. It is particularly well suited as a military and/or civilian research experimental station or astronomical observatory which may hold the key to interstellar flight or communications.

Whether or not competitive nations can or will operate side-by-side on the Moon for any or all purposes indicated above remains to be seen. The military requirements for controlling these areas stem from the nature and form of man's competitiveness. Control may be necessary for one or more of the following:

1. Traffic in and out
2. Mineral exploitation
3. Protection for exploratory parties
4. Communications
5. Noise sources which may interfere with communications or experiments
6. Potential enemy reconnaissance and surveillance
7. Life-lines or transportation and communication routes between lunar bases and with Earth.

Lunar Control may be effected by satellite reconnaissance and surveillance systems, ballistic missiles, satellite communication relays or reflectors, etc. as discussed previously in connection with Earth warfare. The lack of significant atmosphere will eliminate conventional type aircraft, however maneuverable rocket craft may be developed. The lack of atmosphere will have a major impact on warhead effectiveness; nuclear, high explosive, chemical and bacterial warheads may take new forms. The reduced gravitational effect will have its impact on these warheads as well, and will also affect the design of all lunar vehicles, e.g. ballistic missiles. The nature of the lunar surface may permit surface wave communication in addition to satellite communication links. Ground exploration teams will require mobile surface (and under-surface) vehicles for operation on the lunar plains and through the lunar dust. Such vehicles may be brought into play militarily and may require weapons and fire control devices as well as communication, navigation, recording and experimenting equipment. Although most lunar bases will likely be underground (e.g., caves), certain installations will necessarily be above ground and subject to military reconnaissance, surveillance and other actions. Such items as antennas (although dust immersion might be a partial, non-radiometric, camouflage), telescopes, launch and landing pads, satellites, etc. will be subject to observation and attack.

1.2.11 Planetary Control

Control of the small planets and of the moons of the large planets may become a military requirement. It may be accomplished in much the same way as for the Moon allowing for the significant temperature differences that may require extreme thermal insulation and control or design for thermal expansion and contraction.

The two sister planets, Mars and Venus, are particularly promising and interesting because of their atmospheres which may be employed for aerodynamic flight and maneuverability in surveying these two planets, and which will have some significant effect on warhead design and usage and communications. In fact, it is expected that Venus and Mars may have military communications and observation satellites eventually with call-down ballistic missiles or focused-beam weapons. Surveillance of Venus must be accomplished with radar or other electromagnetic radiations outside the visible spectrum because of the dense cloud cover. Whether or not these other planets will become principle "refueling" stops or colonies in themselves is pure conjecture at this time. Nevertheless, the moons of Mars and of the large planets particularly will offer good vantage points for observation and control of activities relative to these planets, just as in the Earth-Moon system.

Thus, planetary control of traffic, exploration, raw resources, terrain resources, weather and communications may be effected throughout the solar system eventually by means of satellites, missiles, focused-beams, artificial planetoids, etc. as suitable energy release and control devices become available.

1.2.12 Military Research

Military research may be defined as embracing two broad categories: (1) research pertaining to the intelligence, control, deployment, and effects of weapons, and (2) research necessary to compete on a timely basis with other nations. The first category would include weapon fire control from an aircraft while the second would entail the development of military aircraft to offset enemy capabilities in lieu of merely adapting weapons to civilian aircraft. Logically, the same philosophy may be extrapolated to military space activities when such activities are dictated by the course of world events. Thus, military space craft and extra-terrestrial bases must be anticipated and the supporting research for these activities becomes a military necessity (and usually of greater urgency than civil research).

Space environmental measurements and the effects of these environments on various materials, organic life, fuels, food, medicines, humans, etc. must be studied. The effects of warhead actions and protection against these actions must be studied. Physical survival of military personnel dictates the research and development of protective devices and life-boat systems. The nature of such research requires gathering much data on each trial which dictates that probes and experimental space vehicles have not only radio communication (or telemetering) links with supporting and using activities, but capsules for data transfer as well. Research may be depicted as a physical activity in many ways, but the electronic activity centers around the measurement, recording of data, sorting, filtering, and scanning of the data and transfer of data of interest to another location. We may classify military research systems as Vehicles and Missions as follows:

Vehicles

Terrestrial: bases, observatories, laboratories
Probes: experiments
Satellites: experiments, observatories, laboratories
Extra-terrestrial: bases, observatories, laboratories

Missions

Environment measurement
Environment effect measurement
Warhead and weapon characteristics
Reconnaissance and surveillance techniques
Intelligence sorting and recognition techniques
Communications techniques

Military research missions embrace broad fields which can draw upon and contribute significantly to civil research efforts.

1.3 CIVIL SPACE MISSIONS

Civil missions in space are manifold. Much research can be accomplished as to the nature of space, the actual environment, its effect on Earth, the origin of the universe, and the nature of fundamental physical laws. Exploration of surrounding bodies in a quantitative scientific manner portends great discoveries. The development of non-military transport, communication and navigational aid systems is another considerable field. Meteorological studies, forecasting and weather control systems are needed. Traffic control will be required. Automatic transmission and conveyance of data, specimens and instruments will be needed for civil operation in space.

In general, one can say that civil missions in space must be justified on an economic or subsidized basis with some net overall improvement in the general welfare. This indicates that efficiency rather than time will be of the essence in achieving a result. That is, the military requirement for timeliness in an operating system will not predominate in the civil systems.

Such systems will require timeliness in their development and operation, but not with the same impelling urgency as do military and geopolitically oriented systems. Thus, civil transport development, for example, will trade-off transit-time with energy expenditure rates in a different proportion to military transport development for force deployment.

1.3.1 Physical Measurements

Physical measurements of the aerospace environment have historically forced man to greater altitudes. The vehicles have generally been balloons, gliders, aircraft and rockets. Instruments have been of many types ranging from astronomical telescopes to energetic particle counters. Thanks to the rocket, we now have what is termed the space probe which enables us to conduct experiments and measure the physical properties of the upper atmosphere on out into solar space, with the eventual prospect of extra-solar or galactic space measurements. Probes, as opposed to orbiting bodies, are, relatively speaking, short lived. That is, an ionospheric probe returns to Earth or burns up in the atmosphere. Probes which do not intercept bodies will of course follow some Earth, moon, planet or sun centered orbit. Thus the principal distinction between a probe and a satellite or planetoid in some cases may be merely semantic, for example, related somewhat to the eccentricity of the orbit, the probing or exploratory nature of the instrumentation, the short duration of communication from the probe, or merely the name associated with the original mission. Satellites, planetoids and maneuverable orbiters will generally be used to conduct measurements over long periods of time, whereas probes will generally be employed to obtain a vertical profile of environmental characteristics and where atmospheric restrictions such as drag and heating prevent the employment of orbiters. Examples of experiments which may be accomplished by terrestrial probes and satellites are tabulated in Table 1–1.

Terrestrial satellites offer particularly good bases of operation for certain missions. An orbiting astronomical observatory can explore the heavens without the encumbrance of atmospheric attenuation, scattering, cloud blanking, and refraction. An orbiting geophysical observatory can conduct earth and environmental measurements on a global scale providing significant data on the Earth's characteristics. Measurements of the Earth's shape may be made, accurate intercontinental maps obtained, and global weather forecasts made. Through satellite and space probe experiments it is hoped to further test and develop man's knowledge of the fundamental laws of physics, and the origin of the Earth, Moon, solar system, and universe.

Similarly, satellite observatories can be expected to be placed in orbit around the moon, other planets and their satellites, and the sun to map these bodies, and chart their "weather," magnetic fields, etc. Probes will be sent throughout cislunar, inter-planetary and solar space and beyond to chart and measure the environment and its effects on men and materials, test the fundamental laws and theories of physics (e.g., sun spot and solar flare relation to local radiation pressures and high-energy charged particle

TABLE 1–1. TERRESTRIAL PROBE AND SATELLITE EXPERIMENTS

Environment Measurements:
- Pressure Profile
- Temperature Profile
- Atmospheric Density
- Chemical Composition
- Bacterial count
- Moisture content
- Ionization
- Radiation count
- High energy particle count
- Meteoric particle count
- Cosmic radio noise
- Electromagnetic radiation
- Electric and magnetic fields

Environment effects on materials:
- Structural
- Shielding
- Operating systems characteristics
- Biological

Communications tests:
- Continuity
- Noise sources
- Ionization detuning of antennas
- Propagation through plasmas and radiation belts

System and technique development:
- Recovery
- Navigation aids
- Geophysical and Astronomical observation tests

counts, the validity of the special and general theories of relativity throughout the universe and on macroscopic and microscopic scales). Impact probes will be sent to the Moon, to the planets and their satellites, and to the sun to explore the chemistry and construction of their "atmospheres" and "crusts."

Lunar exploration will be conducted initially by telescope and radar observations. Telescopic pictures of the near side of the Moon are well known. Photographs of the Moon's other side, Fig. 1-13, have been obtained by sending a vehicle to circumnavigate the moon (Ref. (*5*)). Measurements of temperatures, magnetic fields and trapped charged particle belts will be conducted from lunar satellites. Probes will test the surface structure to ascertain such mechanical properties as brittleness and dust layer thickness, surface chemical composition, "atmosphere" composition, solar radiation densities in exposed and shadowed areas. Based upon knowledge of Earth, Sun, meteors, planetary atmospheres, etc. it is expected that all of the chemical elements essential to life sustenance may be found on the Moon (Ref. (*6*)), although not in the same proportions as on Earth. Also, the extreme temperature differences within local areas because of the lack of a temperature equalizing atmosphere should make the development of thermal engines of high efficiency quite feasible to the point of providing oxygen, nitrogen, water and artificial foodstuffs from local minerals so that a "cave society" might be developed in support of lunar exploration. Indeed, with some nuclear fusion reaction breakthrough, an artificial lunar atmosphere of CO_2 to support plant life and act as a radiation shield might be contemplated as indicated by F. Zwicky in Ref. (*6*). Thus Lunar exploration may be carried out to the point of manned sub-surface bases operating in an artificially created "shirt-sleeve" environment, living off

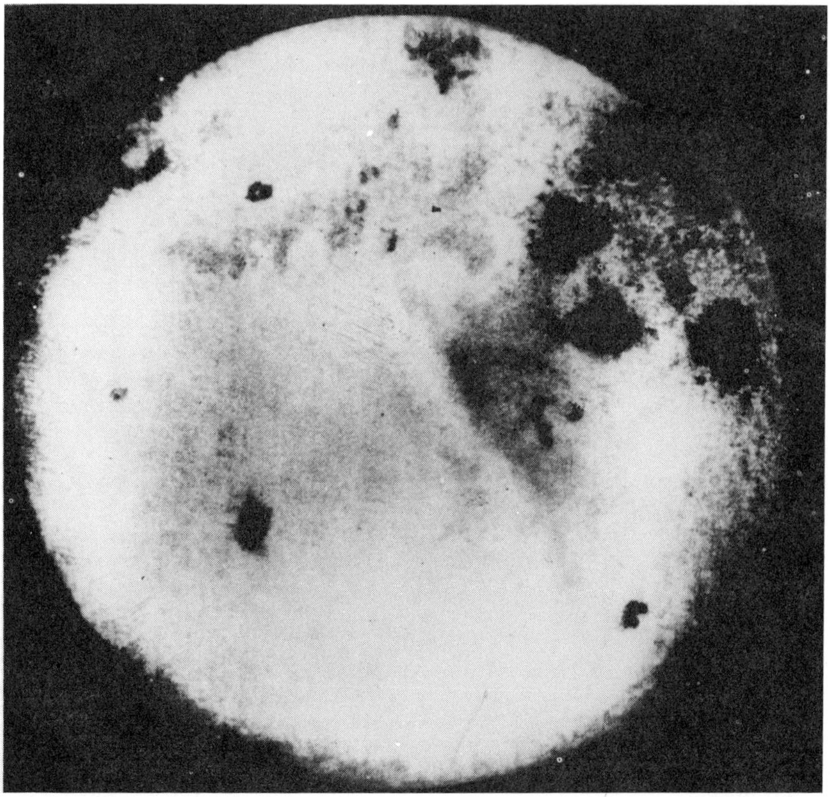

FIG. 1-13. Photograph of Far Side of the Moon. (*Courtesy*, SOVFOTO)

the land, and supporting robot and manned surface vehicles which may roam the moon in search of more hospitable environments, valuable mineral deposits (e.g. water bearing minerals), sites for astronomical observatories, lunar research stations, communication bases, rocket and propulsion (e.g., ion or photon) test bases, planetary mission refueling stations, etc.

The exploration of the near planets, Mars and Venus, will follow patterns similar to that of the Moon. That is, probes, satellites, soft landing vehicles, surface base development and ground exploration by robot and manned vehicles will be used. Again, it is to be expected that the essential chemicals will be found, and that the problem will be to locate them in useful form, to develop suitable engines operating from local energy sources, and to convert the chemicals to provide artificial (enclosed) atmospheres, foodstuffs, and fuels. With Venus and Mars there are atmospheres to be explored and to contend with in this development. The magnetic fields of these planets may provide trapped charged particle zones above the atmospheres similar to those encountered above the Earth's atmosphere. Of particular interest are the two Martian moons which might be employed as staging bases for exploratory trips to the Martian surface. One of these moons, Phobos, has been predicted as providing a "sweeping" of charged particles in the local region (Ref. (*7*)). This moon may provide a unique observation and staging base site, and the opportunity to measure precisely the energy, nature and rate of supply of the charged particles trapped in radiation belts. From the rate and direction of charged particle supply, their origin may be deduced with some precision.

The inner planet Mercury may be explored and developed in a manner quite similar to the Moon with probes, satellites and landing vehicles. Satellites of Mercury will encounter some extreme thermal and radiation problems because of the proximity to the Sun. Assuming the essential elements are present in the rocks, it is conceivable that a self-supporting base could be developed underground on the dark side of Mercury near the sun-eclipsing boundary, taking advantage of the thermal differences between the light and the dark hemispheres to provide a munificent energy source. Such a base could provide, in addition to Mercurial exploration facilities, a close-in solar observatory able to examine closely operations on the Sun and within the solar "atmosphere."

The great planets and outer planets present exploration problems peculiar in their own right. The gravity of Jupiter may restrict its exploration to unmanned vehicles. The Jovian moons will be interesting bodies to explore in themselves as will the asteroids and moons of Saturn, Uranus and Neptune. Even more natural satellites may be discovered. The high escape velocity requirements for the great and outer planets indicate the desirability of developing local energy sources which in turn may lead to development of permanent bases. Local atmospheres and the increasing distances from the sun will reduce the availability of sufficient solar-derived energy to convert local minerals into desired forms; therefore prime energy sources must be sought, e.g., fissionable materials or meteorological systems.

Extra-solar system exploration will be limited for some time to astronomical instruments, essentially telescopes operating in the various bands of the electromagnetic spectrum, e.g., radio, microwave, infrared, visual, ultra-violet, X-rays, etc. Such telescopes and observations tell us much about the nature of the galaxy and the universe and, as measurements become more precise, the information and theories deduced therefrom will become more precise. It is interesting to speculate on the additional astronomical data which will become accessible with sizable electromagnetic collectors mounted on satellites or the Moon once outside the Earth's atmosphere.

It is also interesting to speculate on the use of observational base lines not limited by the Earth's orbit around the sun. Figure 1–14 illustrates the fundamental base line available currently (even with satellite and lunar observatories). This base line is nominally two A.U.'s (Astronomical Units) and requires six months between observations to use its full length. A longer base-line, unencumbered by atmospheric effects could be used to establish more precisely the ranges to various stars and galaxies presently obtainable by measuring the doppler red-shift of their spectrum. That is, parallel measurements may establish directly, ranges to be associated with certain red-shifts, and thus corroborate expanding universe theories. A solar system base-line may be developed on the Plutonian orbit except that the Plutonian year is 247 Earth years, and measurements would require 12 generations to accomplish.

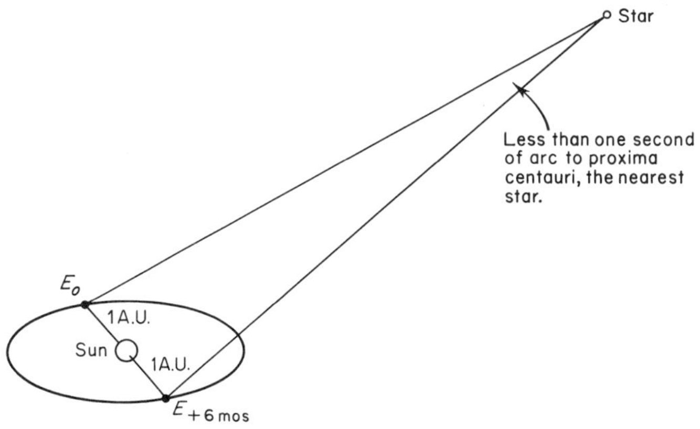

FIG. 1-14. Earth-Limited Astronomical Base-Line.

Figure 1–15 illustrates a second approach to the problem which employs observatories on two or more planets (or artificial planetoids) simultaneously. This system requires:

1. Communication between observatories
2. Synchronization of measurements
3. Precise common timing
4. Precise knowledge of base-line distances
5. Precise angular measurements.

These requirements are requisite to any single body system (e.g., Earth, Moon, Pluto), and may be solved by a variety of means. The character of astronomical observations is such that the greatest precision possible is continuously sought so that the computations based on measurements will be the more accurate.

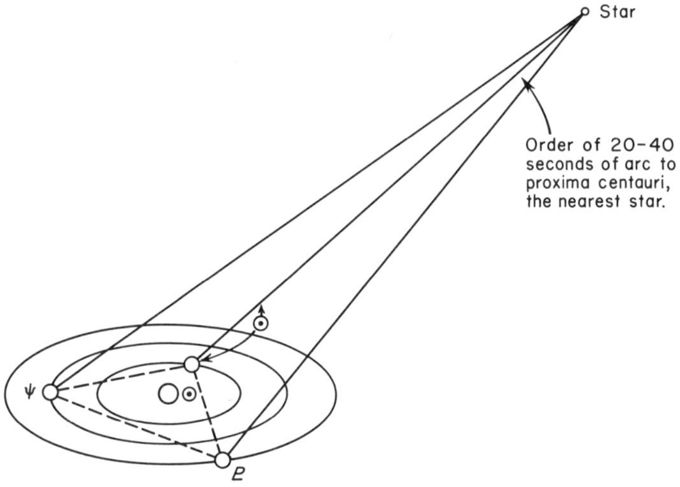

FIG. 1-15. Simultaneous Astronomical Observatory System.

The fond dream of science-fiction authors for manned extra-solar exploration will no doubt be attempted someday. However, the distances (hence, transit times) to the nearest stellar bodies are so great that return or receipt of data from these bodies will be measured in terms of many years. Unmanned probes may conceivably be sent, although the energy requirements and communication limitations are formidable by present day standards. Radio contact with other sentient beings is being attempted in a passive manner, and may be tried actively in the future when man has decided to gamble on whether such beings are within one life-span (theirs) travel distance, hostile or not, sufficiently superior to regard us as cattle, etc.

1.3.2 Transport Development

Space transport development for other than military uses must be justified on some economic basis just as all other transport is. That is, the laws of supply and demand will govern its use, at least in a free society. The use of space for transport between terrestrial terminuses will therefore depend on the urgency of transporting people and/or goods. Military urgency demands are only rarely encountered in the non-military society, so that only very high-priority transport missions might be considered until the cost of inter-terrestrial terminus space flight is reduced to something near the cost of other travel modes (particularly air travel). Intercontinental rockets may conceivably be employed for mail use, although the faster reliable communication channels available through radio and cable lines would seem to restrict rocket mail to a very small select consumer market (e.g., diplomatic mail) until the cost, reliability, and hazard situations are vastly improved over present capabilities.

A further item regarding the transport of people or goods is that speedy rocket travel will suffer from the same terminal transport limitations currently plaguing air travel. That is, if the air travel time represents one-half the total travel time between actual origin and destination, as is typical for much transcontinental travel, then reducing the air travel

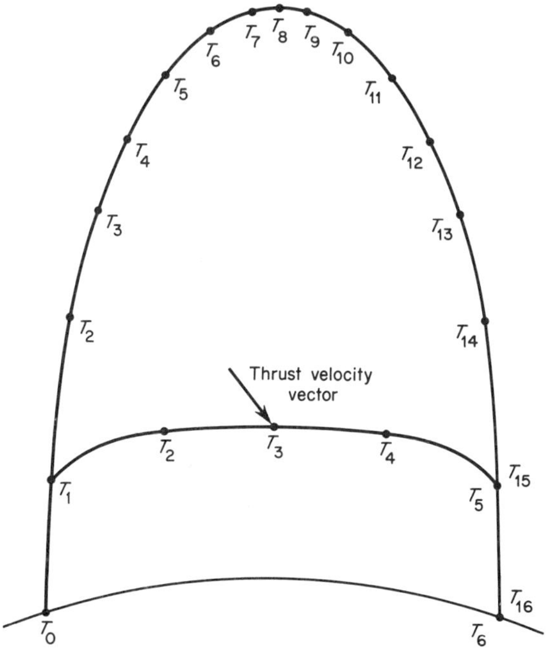

FIG. 1-16. Comparison of Ballistic with Powered All-the-Way Transit Times.

time by half with rocket travel will only reduce the total travel time by one-quarter. Such a time reduction would generally not be meaningful unless it is of the order of one hour or more and the schedule is dependable. Thus, the impetus for terrestrial rocket travel will be predominantly for transcontinental and intercontinental usage (where current schedules provide intervals over 5 hours which may be reduced by Mach 2 or 3 aircraft within a decade). The time limits on terrestrial (and all) rocket travel will be imposed by human (and goods) tolerance to acceleration and deceleration. Ballistic trajectories impose a natural time restriction except that one could conceivably conduct a powered all the way, maneuverable, rocket flight to shorten the transit time. Figure 1–16 illustrates this situation. As soon as the powered rocket leaves the atmosphere, rocket power is applied to counteract its ascent velocity and to speed the vehicle toward the destination. Mid-way, enroute, rocket power is applied again to decelerate on a similarly controlled path, the basic idea being that escape velocity may be exceeded without departing from the Earth's gravitational field. The centripetal effect is nullified by applying suitable accelerating force toward the Earth's center while travelling at supra-escape velocity. The need for this in civilian, as well as military transportation may develop as energy sources become cheaper in rocket travel.

The next steps in transport development are satelloid and satellite transports. These transports will be needed for the support, logistics and maintenance of manned and un-manned satellite bases. Peculiar to these transports will be the rendevous requirements which will ultimately dictate control over orbit plane, velocity, eccentricity, and their phasing with another orbiting body. The final approach between two co-orbital satellites will be a further requirement which will dictate the use of very precise control methods to coordinate orientation alignments and soften docking contacts.

The establishment of observation, supply and exploration bases at libration points and on the Moon, other planets, their satellites, and other bodies will require transport development for logistic and maintenance reasons. Further, the development of self-sustaining lunar and planetary colonies will require civilian transport development for social and political reasons as well as commercial and industrial. The civilian transport development in these areas will quite logically follow military transport development. That is, minimum energy trajectories will be employed for the bulk of freight transport; time conservative trajectories (when feasible) are more likely for human and high priority cargo (e.g., medicine, munitions, critical supplies).

The development of transport vehicles for transfer in space may become a critical item, particularly where the orbit parameters of a space station (satellite or planetoid) are critical and must not be disturbed or cannot be controlled. This could be the case with navigational references, astronomical observatories, or communication relays. Two space vehicles cannot engage and disengage from each other without affecting the trajectories or orbits of both; therefore transfer vehicles of very low mass characteristics may be needed for particular usages so as to create a minimum of perturbation on some vehicles' orbit characteristics. Should space travel develop to the point where vehicles may pass on skew trajectories, special high performance (in the maneuverability-acceleration sense) vehicles may be needed to effect transfer between these vehicles. (While a transfer between a Venus-Mars liner and a Moon-Ganeymede supply ship seems far-fetched now, the development of self-sustaining bases may ultimately dictate just this situation.)

Emergency transports will be needed which will be required to vacate a space-ship in trouble and broadcast its trajectory or orbit parameters in the event of some disaster. Rescue vessels will be needed to retrieve such "space life-boats", and will quite likely be of high performance and use direction finders for navigational reference.

1.3.3 Communication

Civilian communication developments currently underway indicate that large passive reflector satellites and active repeater satellites have large potential in increasing the information handling capabilities between terrestrial terminuses. Freedom from dependence on ionospheric reflection, tropospheric scatter and line-of-sight horizon effects will enable the assignment of frequency channels over a much broader portion of the electromagnetic spectrum. Relay systems may be developed parallel to those already postulated for the military depending on the traffic demand and requirements. Civilian communication networks will not be concerned with security to the same extent as the military and one may expect less coding of information (although postal and diplomatic facsimile may require some). Also, civilian communication stations will not be concerned with enemy jamming of channels or take-over of tracking and guidance systems to the same extent so that one may expect civilian communication networks to develop on a more efficient, less expensive basis than parallel military networks. It may be, however, that military and civilian communication relay stations will share common vehicle facilities for economic reasons; thus some attention to mutual interference will be in order.

1.3.4 Aids to Navigation

Current developments include the civilian use of a military deployed navigational reference satellite (TRANSIT). These transmissions are available to anyone within radio sight of the satellite, and include the satellite orbit parameters as well as sufficient doppler information to extract one's Earth surface coordinates with respect to the satellite orbit characteristics. It is expected that superior navigation reference systems will be developed employing precise ranging techniques so that one may ascertain position (including altitude) in an essentially instantaneous manner and thus reduce the navigational errors introduced by high velocities (as with aircraft) and the requirement of waiting for particular transits of doppler reference satellites. Communication relay satellites in suitable constellations may be employed as Navigational references. Range difference measurements of the Loran type may be employed, establishing one's position in space by the intersection of two hyperboloids with an altitude sphere.

Libration point bases, Earth-Moon system satellites, and solar planetoids offer Navigational reference aid possibilities for lunar and interplanetary travel. The Sun, planets, natural satellites, and asteroids offer references and station possibilities as well. One can conceive, for example, of a Loran equivalent network being set up for cislunar space, employing the Earth, (or an Earth-satellite network), the Moon and space stations located at the equilateral libration points. Such a system would be inherently useful in the Earth-Moon plane which will quite likely include all cislunar flights except those intentionally sent out of the plane for exploratory or military reasons.

Further networks for interplanetary navigational references may be conceived using the planetary orbit equilateral "Trojan" points as bases, or an independent artificial planetoid ring spaced equilaterally with the sun as depicted in Fig. 1–7. Obviously, such space bases may be expected to be employed as communication relay stations as well as navigational reference platforms.

1.3.5 Weather Control

Civilian weather control would be desired to effect climate equalization, crop improvement, forest or grass fire control etc., as opposed to military weather control which would be intended for enemy harassment, insurance of own troop deployability, etc. As far as space missions go, terrestrial weather control would be limited to observations from satellite observatories with the possibility that cloud breaking or routing devices might be deployed from orbiting stations (although the latter seems more costly than direct deployment from the Earth's surface). Space weather control as a subject becomes more interesting when one considers the task of generating more hospitable atmospheres on Mars or Venus; or even on the Moon as has been postulated by Zwicky (Ref. (*6*)). Observatories in orbit seem likely. Vehicles capable of rerouting or moving clouds and air-currents are an exotic control possibility. Whether it is more economical to deploy such vehicles from orbital or surface locations would seem to depend on their origin (i. e. place of manufacture and source of raw materials—whether native to Earth or local body). Should self-sustaining communities be developed as postulated by Zwicky, then weather control would seem to follow. In general, weather observations and control appear to be of a different order of magnitude in accuracy of formation location and definition than surface mapping missions; thus guidance, control and, communication (data transfer) requirements will be somewhat relaxed, and accomplished more easily, more quickly, or at less expense.

1.3.6 Traffic Control

As civilian space transport develops, traffic control can be expected to become a problem. Orbit and trajectory plans will be required to assign landing patterns and facilities at first (and perhaps orbit planes and altitudes if traffic density becomes too great). Extra-terrestrial traffic control is very remote as a requirement except for piracy possibilities which would call for police or military action. A solution to the problem of a lunar ship arriving without pre-planned reception facilities would seem to require parking orbits for such ships while communications are established and traffic control directs the ship to an appropriate satellite base or to surface facilities.

Traffic control will necessarily require extensive communication networks linking the control center with all surface launch and landing or recovery facilities, and all satellite staging or transfer point facilities so that control of all facilities for Earth-to-satellite, satellite-to-Earth, and Earth-to-Earth transport may be desirable for both hazard avoidance and reasons of efficiency.

1.3.7 Data, Specimen, and Instrument Recovery

Recovery of data, specimens, and instruments from exploratory probes, expeditions, and observatories will be a peculiar problem. Instrumented exploratory probes must either be returned, or their data returned via radio link or data capsule. Initially, data capsules will be recovered by firing retro-rockets while in orbit so that these capsules may be recovered by aircraft while descending by parachute or by ships upon deployment of marking beacons. With the development of satellite transport, these techniques will be refined only if they prove to be more economical than loading up the transport itself for its return trip to the surface. In general, it would seem that data and specimen capsules might be transported through space more quickly and economically than people because they can withstand higher accelerations and thus achieve higher velocities with less energy expenditure. Also, life support facilities should not be needed. However, it appears that even the transport of data capsules may not be necessary in all cases. If, for example, a man can process and interpret the raw data at a lunar base sufficiently well to screen out only the significant data he might reduce its content to an information bandwidth level which could be processed by a radio link in a useful time period (any period less than capsule transit period should be "useful" as a first approximation).

Figure 1–17 depicts a fundamental data transfer system which might include humans or be completely automatized. The key item is the significance interpreter which, according to some pre-set plan, or even by an adaptive (learning) technique will decide which data obtained are of significance and whether the residue is of sufficiently low bandwidth content to go by radio transmission, rather than by a special capsule. The significance interpreter might form his judgement based upon the difference between the data content and some presupposed content and a recollection of the data previously transmitted and subsequently transmitting new data with some code expression correlating the new data with the old. Many such systems are conceivable and feasible according to the task to be performed.

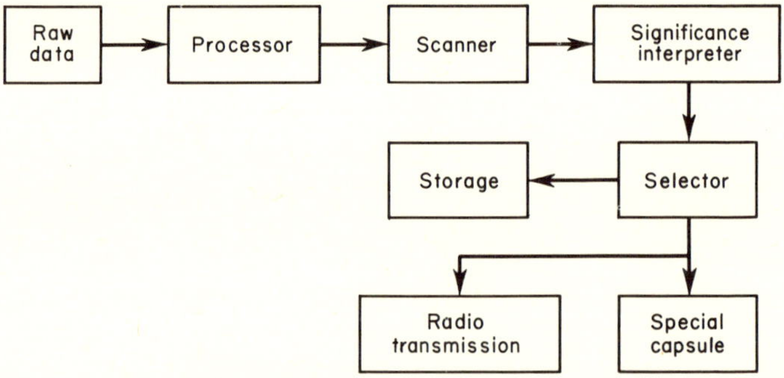

Fig. 1-17. Basic Data Transfer System.

1.4 INTERNATIONAL SPACE MISSIONS

The discussion of various military and civilian space missions has assumed either some war, near-war or peaceful era, and explored how one nation might develop its space vehicle missions in such an environment. In addition to these, there seem to be some international space missions worthy of separate mention. Specifically, these are treated as, geo-political missions, frequency control and police actions. In addition, an historical recount of space achievements to date is presented to acquaint the student with the present stage of space vehicle mission capability.

1.4.1 Geopolitical Considerations

The development of space flight under the stimulus of geopolitics is shaped by two primary objectives: (1) space achievements ahead of competitors which tend to show a competence and leadership in certain fields and which will impress "neutral" observers, and (2) space achievements which are tangible, spectacular and observable events which even the uninformed can interpret very easily. Thus, the first visually observable satellite placed in orbit indicated a certain prowess in rocket power and guidance which previously had only been a goal. The first lunar probe yielding pictures of the far side of the moon; planetary probes, solar orbiters, manned vehicles and specially instrumented data on space environment have all been used to exemplify one or another nation's superiority in the space arena. It would almost seem that space achievements have replaced territorial holdings, standards of living, and many other achievements as signs of national prowess. However, this may not truly be accepted by all concerned particularly by those in possession of some but not all of the other prowess indicators. Hence one must read more into the geopolitical implications than record seeking alone.

Some feats have indicated an ability to orbit enough mass that, had the rocket been re-directed and the payload been a nuclear weapon, it could have wiped out a large city. The repeatability of orbit parameters, orbit injection altitudes, apogee, etc. has indicated a preciseness of control that can be extrapolated into hitting accuracy. Observation platforms passing over a nation's territory and providing pictures of its terrain can obviously provide detailed coordinate information for target determination and weapon direction.

1.4.2 Frequency Control

In the international use of space, frequency assignment and control on an international basis appears to be in the best interests of all concerned during the common use of broadcasting paths. This is not to say that jamming of competitor's transmissions may not be attempted. Rather, communication links are needed and useful to all nations on an international basis. Traffic control may eventually fall into this category should traffic between competing nations develop.

1.4.3 Police Actions

While all the foregoing military space missions may be thought of as preparatory to and incident to a two-nation space war development, they may also be thought of as attempts at mutual policing between two rival camps. It may eventually develop that self-sufficient lunar or planetary colonies will become independent nations or that certain individuals or groups may establish control over enough space capability to run rampant as the pirates once did in the Caribbean. Should the latter happen, the need for military or police action on an international basis is obvious, and it is considered that such an international police force may be developed.

TABLE 1–2 INTERNATIONAL SPACE FLIGHT ACHIEVEMENTS

NAME	INT'L DESIG.	PROJECT DIRECTION	LAUNCH DATA			WEIGHT
			Date[1]	Site	Vehicle	
Sputnik 1	1957 A 1	USSR	Oct 4, 1957			184
Sputnik 2	1957 B 1	USSR	Nov 3, 1957			1120
Vanguard TV 3[2]	None	USN	Dec 6, 1957	AMR	Vanguard	3
Explorer 1	1958 A 1	USA	Jan 31, 1958	AMR	Jupiter C	31
Vanguard TV 3[2] backup	None	USN	Feb 5, 1958	AMR	Vanguard	3
Explorer 2	None	USA	Mar 5, 1958	AMR	Jupiter C	32
Vanguard 1	1958 B 2	USN	Mar 17, 1958	AMR	Vanguard	3
Explorer 3	1958 Γ 1	USA	Mar 26, 1958	AMR	Jupiter C	31
Vanguard TV 5[2]	None	USN	Apr 28, 1958	AMR	Vanguard	22
Sputnik 3	1958 Δ 2	USSR	May 15, 1958			2925
Vanguard SLV 1[3]	None	USN	May 27, 1958	AMR	Vanguard	22
Vanguard SLV 2[3]	None	USN	June 26, 1958	AMR	Vanguard	22
Explorer 4	1958 E 1	ARPA	July 26, 1958	AMR	Jupiter C	38
Able 1 (Pioneer)	None	USAF	Aug 17, 1958	AMR	Thor-Able	84
Explorer 5	None	ARPA	Aug 24, 1958	AMR	Jupiter C	38
Vanguard SLV 3[3]	None	USN	Sept 26, 1958	AMR	Vanguard	22
Pioneer 1	None	NASA	Oct 11, 1958	AMR	Thor-Able	84
Beacon 1	None	NASA	Oct 23, 1958	AMR	Jupiter C	9
Pioneer 2	None	NASA	Nov 8, 1958	AMR	Thor-Able	87
Pioneer 3	None	NASA	Dec 6, 1958	AMR	Juno II	13
Score	1958 Z 1	ARPA	Dec 18, 1958	AMR	Atlas B	8750
Lunik 1	1959 M 1	USSR	Jan 2, 1959			3245
Vanguard 2	1959 A 1	NASA	Feb 17, 1959	AMR	Vanguard	22
Discoverer 1	1959 B 1	ARPA	Feb 28, 1959	VAFB	Thor-Agena A	1300
Pioneer 4	1959 N 1	NASA	Mar 3, 1959	AMR	Juno II	13
Discoverer 2	1959 Γ 1	ARPA	Apr 13, 1959	VAFB	Thor-Agena A	1600
Vanguard SLV 5[3]	None	NASA	Apr 13, 1959	AMR	Vanguard	23

AMR: Atlantic Missile Range
PA: Point Arguello
WI: Wallops Island
VAFB: Vandenberg AFB

[1]As of August 15, 1963, if applicable
[2]TV: Test Vehicle
[3]SLV: Satellite Launching Vehicle

TABLE 1–2 (Cont.)

INITIAL CURRENT[1] ORBITAL DATA				RESULTS
Period	Perigee	Apogee	Incl.	
96.2	141	588	65.1	Decayed 1-4-58: first artificial satellite, transmitted 21 days
103.7	140	1038	65.3	Decayed 4-14-58: carried dog Laika, transmitted 7 days
—	—	—	—	Failed to orbit: lost thrust after 2 seconds
114.7 104.9	224 213	1584 1024	33.3 33.2	In orbit: transmitted until 5-23-58, discovered radiation belt
—	—	—	—	Failed to orbit: control system malfunction
—	—	—	—	Failed to orbit: unsuccessful fourth stage ignition
134.3 133.8	405 409	2462 2442	34.3 34.2	In orbit: still transmitting in sunlight on 108.023 mc, proved earth pear-shaped
114.7	117	1739	33.5	Decayed 6-28-58: radiation and micrometeorite data until 6-16-58
—	—	—	—	Failed to orbit: third stage ignition malfunction
105.8	140	1168	65.2	Decayed 4-6-60: variety of data returned up to re-entry
—	—	—	—	Failed to orbit: improper third stage trajectory
—	—	—	—	Failed to orbit: premature second stage cutoff
110.1	163	1372	50.1	Decayed 10-23-59: mapped Project Argus radiation shells until 10-6-58
—	—	—	—	Lunar probe failed: first stage lox pump stopped
—	—	—	—	Failed to orbit: upper stages fired in wrong direction
—	—	—	—	Failed to orbit: insufficient second stage thrust
—	—	—	—	Lunar probe failed: third stage thrust insufficient; 70,717-mile altitude, returned radiation, magnetic field, micrometeorite data
—	—	—	—	Failed to orbit: upper stages separated prior to burnout
—	—	—	—	Lunar probe failed: third stage ignition unsuccessful
—	—	—	—	Lunar probe failed: early first stage cutoff; reached altitude of 63,580 miles, discovered outer portion of Van Allen radiation belt
101.5	115	914	32.3	Decayed 1-21-59: first comsat, relayed messages for 13 days
450 days	.9766AU[4]	1.315AU	0.01[5]	In solar orbit: lunar probe, passed within 4660 miles of moon
125.9 125.2	347 342	2064 2046	32.9 32.9	In orbit: transmitted for 18 days, returned first cloud cover photos
94.1	184	407	90.0	Decayed 3-5-59: first polar orbit, did not carry re-entry capsule
398 days	.9871AU[4]	1.142AU	1.30[5]	In solar orbit: lunar probe, passed within 37,300 miles of moon
90.6	152	225	90.0	Decayed 4-26-59: capsule ejected on orbit 17, lost in Arctic Ocean
—	—	—	—	Failed to orbit: second stage damaged at separation

[4]Astronomical Unit
[5]Inclination from the ecliptic
[6]From local time at launch site

Period: Minutes
Perigee and Apogee: statute miles
Inclination: degrees from the equator
Statute miles to kilometers: 1.609347
Statute to nautical miles: 0.86839

TABLE 1–2 (Cont.)

NAME	INT'L DESIG.	PROJECT DIRECTION	LAUNCH DATA			WEIGHT
			Date	Site	Vehicle	
Discoverer 3	None	ARPA	June 3, 1959	VAFB	Thor-Agena A	1600
Vanguard SLV 6[3]	None	NASA	June 22, 1959	AMR	Vanguard	23
Discoverer 4	None	ARPA	June 25, 1959	VAFB	Thor-Agena A	1600
Explorer S-1	None	NASA	July 16, 1959	AMR	Juno II	92
Explorer 6	1959 Δ 1	NASA	Aug 7, 1959	AMR	Thor-Able	143
Discoverer 5	1959 E 1	ARPA	Aug 13, 1959	VAFB	Thor-Agena A	1700
Beacon 2	None	NASA	Aug 14, 1959	AMR	Juno II	10
Discoverer 6	1959 Z 1	ARPA	Aug 19, 1959	VAFB	Thor-Agena A	1700
Lunik 2	1959 Ξ 1	USSR	Sept 12, 1959			858
Transit 1A	None	ARPA	Sept 17, 1959	AMR	Thor-Able	265
Vanguard 3	1959 E 1	NASA	Sept 18, 1959	AMR	Vanguard	100
Lunik 3	1959 Θ 1	USSR	Oct 4, 1959			614
Explorer 7	1959 I 1	NASA	Oct 13, 1959	AMR	Juno II	92
Discoverer 7	1959 K 1	ARPA	Nov 7, 1959	VAFB	Thor-Agena A	1700
Discoverer 8	1959 Λ 1	USAF	Nov 20, 1959	VAFB	Thor-Agena A	1700
Atlas-Able 4 (Pioneer)	None	NASA	Nov 26, 1959	AMR	Atlas-Able A	372
Discoverer 9	None	USAF	Feb 4, 1960	VAFB	Thor-Agena A	1700
Discoverer 10	None	USAF	Feb 19, 1960	VAFB	Thor-Agena A	1700
Midas 1	None	USAF	Feb 26, 1960	AMR	Atlas-Agena A	4500
Pioneer 5	1960 A 1	NASA	Mar 11, 1960	AMR	Thor-Able	95
Explorer S-46	None	NASA	Mar 23, 1960	AMR	Juno II	35
Tiros 1	1960 B 2	NASA	Apr 1, 1960	AMR	Thor-Able	270
Transit 1B	1960 Γ 2	ARPA	Apr 13, 1960	AMR	Thor-Able Star	265
Discoverer 11	1960 Δ 1	USAF	Apr 15, 1960	VAFB	Thor-Agena A	1700
Echo A-10	None	NASA	May 13, 1960	AMR	Delta	132
Sputnik 4	1960 E 1	USSR	May 15, 1960			10,008

[1]As of August 15, 1963, if applicable [3]SLV: Satellite Launching Vehicle

TABLE 1–2 (Cont.)

INITIAL CURRENT[1] ORBITAL DATA				RESULTS
Period	Perigee	Apogee	Incl.	
—	—	—	—	Failed to orbit: second stage fired but no satellite signals received
—	—	—	—	Failed to orbit: second stage propulsion malfunction
—	—	—	—	Failed to orbit: insufficient second stage velocity
—	—	—	—	Failed to orbit: destroyed by range safety officer
768	157	26,366	47.0	Decayed 7-61: first use of solar paddles, transmitted until 10-6-59, discovered earth's electrical current system, returned first TV photo of earth
94.1	135	456	80.0	Decayed 9-28-59: capsule ejected into higher orbit on 16th pass
—	—	—	—	Failed to orbit: first stage fuel depletion, upper stage attitude control system malfunction
95.2	131	528	84.0	Decayed 10-20-59: capsule ejected on 17th orbit, recovery failed
Flight time: 33.5 hours				Impacted on moon: detected lunar ionosphere, no magnetic field or radiation
—	—	—	—	Failed to orbit: third stage ignition malfunction
130.2 129.7	317 320	2329 2311	33.3 33.4	In orbit: mapped earth's magnetic field, returned radiation and micrometeorite data until 12-11-59
16.2 days	25,257	291,439	76.8	Decayed 4-20-60: passed within 4373 miles of moon, photographed far side for 40 minutes from distance of 37,000-43,000 miles
101.2 101.1	346 342	676 668	50.3 50.3	In orbit: transmitted data on magnetic fields and storms, solar flares, radiation belt, micrometeorites until late 1961
94.6	99	519	81.6	Decayed 11-26-59: poor stabilization prevented capsule separation
103.7	120	1032	80.6	Decayed 3-8-60: capsule ejected 15th orbit, overshot recovery area
—	—	—	—	Lunar probe failed: payload shroud broke away after 45 seconds
—	—	—	—	Failed to orbit: premature first stage cutoff
—	—	—	—	Failed to orbit: destroyed by range safety officer
—	—	—	—	Failed to orbit: second stage failed to separate
312 days	.8061AU[4]	.995AU	3.35[5]	In solar orbit: interplanetary probe, measured size of solar system, returned data on magnetic field, solar wind out to 22.5M miles
—	—	—	—	Failed to orbit: apparent upper stage ignition malfunction
99.2 99.1	430 426	468 469	48.3 48.4	In orbit: first weather satellite, sent 22,952 photos up to 6-17-60
95.8 94.2	232 226	463 371	51.3 51.3	In orbit: first navigation satellite, ejected dummy pickaback payload, returned infrared data, transmitted until 7-12-60
92.3	103	375	80.1	Decayed 4-26-60: capsule ejected on 17th orbit but not recovered
—	—	—	—	Failed to orbit: second stage attitude control malfunction
91.3	194	229	65.0	Decayed 9-5-62: fragment recovered in Wisconsin; Vostok prototype, cabin ejection attempted 5-19-60, separated into 8 pieces

[4]Astronomical Unit [5]Inclination from the ecliptic

TABLE 1–2 (Cont.)

NAME	INT'L DESIG.	PROJECT DIRECTION	LAUNCH DATA			WEIGHT
			Date	Site	Vehicle	
Midas 2	1960 Ζ 1	USAF	May 24, 1960	AMR	Atlas-Agena A	5000
Transit 2A	1960 Η 1	USN	June 22, 1960	AMR	Thor-Able Star	223
Greb 1	1960 Η 2					42
Discoverer 12	None	USAF	June 29, 1960	VAFB	Thor-Agena A	1700
Discoverer 13	1960 Θ 1	USAF	Aug 10, 1960	VAFB	Thor-Agena A	1700
Echo 1	1960 Ι 1	NASA	Aug 12, 1960	AMR	Delta	166
Discoverer 14	1960 Κ 1	USAF	Aug 18, 1960	VAFB	Thor-Agena A	1700
Courier 1A	None	ARPA	Aug 18, 1960	AMR	Thor-Able Star	500
Sputnik 5	1960 Λ 1	USSR	Aug 19, 1960			10,120
Discoverer 15	1960 Μ 1	USAF	Sept 13, 1960	VAFB	Thor-Agena A	1700
Atlas-Able 5A (Pioneer)	None	NASA	Sept 25, 1960	AMR	Atlas-Able	387
Courier 1B	1960 Ν 1	USA	Oct 4, 1960	AMR	Thor-Able Star	500
None	None	USSR	Oct 10, 1960			
Samos 1	None	USAF	Oct 11, 1960	PA	Atlas-Agena A	4100
None	None	USSR	Oct 14, 1960			
Discoverer 16	None	USAF	Oct 26, 1960	VAFB	Thor-Agena B	2100
Explorer 8	1960 Ξ 1	NASA	Nov 3, 1960	AMR	Juno II	90
Discoverer 17	1960 Ο 1	USAF	Nov 12, 1960	VAFB	Thor-Agena B	2100
Tiros 2	1960 Π 1	NASA	Nov 23, 1960	AMR	Delta	280
Transit 3A/Greb 2	None	USN	Nov 30, 1960	AMR	Thor-Able Star	203/40
Sputnik 6	1960 Ρ 1	USSR	Dec 1, 1960			10,060
Explorer S-56	None	NASA	Dec 4, 1960	WI	Scout	14
Discoverer 18	1960 Σ 1	USAF	Dec 7, 1960	VAFB	Thor-Agena B	2100
Atlas-Able 5B (Pioneer)	None	NASA	Dec 15, 1960	AMR	Atlas-Able	388

TABLE 1–2 (Cont.)

INITIAL / CURRENT[1] ORBITAL DATA				RESULTS
Period	Perigee	Apogee	Incl.	
94.4 94.2	299 288	321 317	33.0 33.1	In orbit: data link quit 2nd day, 67 micrometeorite impacts reported
101.7 101.6 101.6 101.6	389 385 382 383	665 653 657 654	66.7 66.7 66.8 66.7	In orbit: transmitted geodetic data until 8-62 In orbit: first pickaback, returned solar X-ray and UV data until 4-61
—	—	—	—	Failed to orbit: second stage attitude instability
94.1	157	431	82.8	Decayed 11-14-60: capsule recovered from ocean after ejection on orbit 17, first recovery of man-made object from space
118.2 114.9	941 634	1052 1175	47.2 47.3	In orbit: first passive comsat, relayed voice and crude TV signals
94.5	113	502	79.6	Decayed 9-16-60: capsule ejected 17th orbit, recovered in mid-air
—	—	—	—	Failed to orbit: launch vehicle exploded 2½ minutes after lift off
90.7	190	211	64.9	Re-entered 8-20-60: cabin and capsule with dogs Belka and Strelka recovered on 18th orbit
94.2	125	469	80.9	Decayed 10-18-60: capsule ejected on 17th pass, sighted but not recovered
—	—	—	—	Lunar probe failed: second stage oxidizer system malfunction
106.9 106.9	586 606	767 748	28.3 28.4	In orbit: delayed and realtime communication tests made for 17 days
—	—	—	—	Mars probe failed
—	—	—	—	Failed to orbit: second stage ignited but did not go into orbit
—	—	—	—	Mars probe failed
—	—	—	—	Failed to orbit: second stage, booster failed to separate
112.7 112.3	285 262	1422 1400	50.0 49.9	In orbit: transmitted data on ionosphere until 12-27-60
96.4	113	614	81.9	Decayed 12-29-60: capsule separated 31st orbit, mid-air recovery
98.3 98.2	387 378	452 460	48.5 48.5	In orbit: returned 36,156 cloud cover photos until 12-4-61
—	—	—	—	Failed to orbit: destroyed by range safety officer
88.6	116	165	65.0	Decayed 12-2-60: re-entry cabin with dogs Pchelka and Mushka burned up as recovery attempt failed
—	—	—	—	Failed to orbit: second stage ignition malfunction
93.8	143	426	80.8	Decayed 4-2-61: capsule recovered in mid-air on 48th orbit
—	—	—	—	Lunar probe failed: exploded 70 seconds after liftoff

[1]As of August 15, 1963, if applicable

TABLE 1–2 (Cont.)

NAME	INT'L DESIG.	PROJECT DIRECTION	LAUNCH DATA			WEIGHT
			Date	Site	Vehicle	
Discoverer 19	1960 Τ 1	USAF	Dec 20, 1960	VAFB	Thor-Agena B	2100
Samos 2	1961 Α 1	USAF	Jan 31, 1961	PA	Atlas-Agena A	4100
Sputnik 7	1961 Β 1	USSR	Feb 4, 1961			14,292
Venus Probe Sputnik 8	1961 Γ 1 1961 Γ 3	USSR	Feb 12, 1961			1419 14,292
Explorer 9	1961 Δ 1	NASA	Feb 16, 1961	WI	Scout	15
Discoverer 20	1961 Ε 1	USAF	Feb 17, 1961	VAFB	Thor-Agena B	2450
Discoverer 21	1961 Ζ 1	USAF	Feb 18, 1961	VAFB	Thor-Agena B	2100
Transit 3B/Lofti 1	1961 Η 1	USN	Feb 21, 1961	AMR	Thor-Able Star	250/57
Explorer S-45	None	NASA	Feb 24, 1961	AMR	Juno II	74
Sputnik 9	1961 Θ 1	USSR	Mar 9, 1961			10,340
Sputnik 10	1961 Ι 1	USSR	Mar 25, 1961			10,330
Explorer 10	1961 Κ 1	NASA	Mar 25, 1961	AMR	Delta	79
Discoverer 22	None	USAF	Mar 30, 1961	VAFB	Thor-Agena B	2100
Discoverer 23	1961 Λ 1	USAF	Apr 8, 1961	VAFB	Thor-Agena B	2100
Vostok 1	1961 Μ 1	USSR	Apr 12, 1961	"Baikonur"		10,418
Mercury-Atlas 3	None	NASA	Apr 25, 1961	AMR	Atlas D	2000
Explorer 11	1961 Ν 1	NASA	Apr 27, 1961	AMR	Juno II	82
Explorer S-45A	None	NASA	May 24, 1961	AMR	Juno II	75
Discoverer 24	None	USAF	June 8, 1961	VAFB	Thor-Agena B	2100
Discoverer 25	1961 Ξ 1	USAF	June 16, 1961	VAFB	Thor-Agena B	2100
Transit 4A Injun 1/Greb 3	1961 Ο 1 1961 Ο 2	USN	June 29, 1961	AMR	Thor-Able Star	175 55/40
Explorer S-55	None	NASA	June 30, 1961	WI	Scout	187
Discoverer 26	1961 Π 1	USAF	July 7, 1961	VAFB	Thor-Agena B	2100
Tiros 3	1961 Ρ 1	NASA	July 12, 1961	AMR	Delta	285
Midas 3	1961 Σ 1	USAF	July 12, 1961	PA	Atlas-Agena B	3500
Discoverer 27	None	USAF	July 21, 1961	VAFB	Thor-Agena B	2100

[1]As of August 15, 1963, if applicable

TABLE 1–2 (Cont.)

INITIAL CURRENT[1] ORBITAL DATA				RESULTS
Period	Perigee	Apogee	Incl.	
93.0	128	390	82.8	Decayed 1-23-61: carried infrared experiments, no re-entry capsule
95.0 94.8	300 295	350 334	97.0 97.4	In orbit: returned micrometeorite impact data
89.8	139	204	65.0	Decayed 2-26-61: believed to be Venus probe abort
300 days 89.7	.7183AU[4] 123	1.0190AU 198	0.58[5] 65.0	In solar orbit: Venus probe, radio contact lost at 4.7M miles Decayed 2-25-61: launched Venus Probe from parking orbit
118.3 116.0	395 302	1605 1573	38.6 38.9	In orbit: optical tracking of 12-foot balloon provides density data
95.4	177	486	80.4	Decayed 7-28-62: programmer failure prevented capsule ejection
93.8	149	659	80.7	Decayed 4-20-62: carried infrared experiments, no re-entry capsule
94.5	117	511	28.4	Decayed 3-30-61: second stage and satellites failed to separate
—	—	—	—	Failed to orbit: third and fourth stage ignition malfunction
88.5	114	155	64.9	Re-entered 3-9-61: cabin with dog Chernushka recovered
88.4	111	153	64.9	Re-entered 3-26-61: cabin with dog Zvezdochka recovered after 17 orbits
112 hrs	100	145,000	33.0	Position uncertain: returned data on magnetic fields, solar wind
—	—	—	—	Failed to orbit: second stage control system malfunction
101.2	126	882	81.9	Decayed 4-16-62: capsule ejected into separate orbit, no recovery
89.1	112	203	65.0	Re-entered 4-12-61: cabin with Y. Gagarin recovered after 1 orbit
—	—	—	—	Failed to orbit: destroyed by range safety officer
108.1 107.8	304 311	1113 1096	28.8 28.8	In orbit: gamma ray counter returned data for 4 months
—	—	—	—	Failed to orbit: second stage ignition malfunction
—	—	—	—	Failed to orbit: second stage ignition malfunction
90.9	139	251	82.1	Decayed 7-12-61: capsule recovered from ocean after 33 orbits
103.7 103.8 103.8 103.8	534 551 534 551	623 616 634 618	67.0 66.8 67.0 66.8	In orbit: first nuclear power supply, transmitted until 5-63 In orbit: failed to separate, Injun provided radiation data until 4-63; Greb relayed solar X-ray data until late 1961
—	—	—	—	Failed to orbit: third stage ignition malfunction
95.0	146	503	82.9	Decayed 12-5-61: capsule recovered in mid-air on 32nd orbit
100.4 100.3	461 461	506 506	47.8 47.9	In orbit: returned 35,033 photos up to 2-27-62
160.0 161.5	1850 2106	1850 2175	91.1 91.3	In orbit: first Midas launch from Point Arguello
—	—	—	—	Failed to orbit: destroyed by range safety officer

[4]Astronomical Unit [5]Inclination from the ecliptic

TABLE 1–2 (Cont.)

NAME	INT'L DESIG.	PROJECT DIRECTION	LAUNCH DATA			WEIGHT
			Date	Site	Vehicle	
Discoverer 28	None	USAF	Aug 3, 1961	VAFB	Thor-Agena B	2100
Vostok 2	1961 Τ 1	USSR	Aug 6, 1961	"Baikonur"		10,430
Explorer 12	1961 Υ 1	NASA	Aug 15, 1961	AMR	Delta	82
Ranger 1	1961 Φ 1	NASA	Aug 23, 1961	AMR	Atlas-Agena B	675
Explorer 13	1961 Χ 1	NASA	Aug 25, 1961	WI	Scout	187
Discoverer 29	1961 Ψ 1	USAF	Aug 30, 1961	VAFB	Thor-Agena B	2100
Samos 3	None	USAF	Sept 9, 1961	PA	Atlas-Agena B	4200
Discoverer 30	1961 Ω 1	USAF	Sept 12, 1961	VAFB	Thor-Agena B	2100
Mercury-Atlas 4	1961 AA 1	NASA	Sept 13, 1961	AMR	Atlas D	2700
Discoverer 31	1961 AB 1	USAF	Sept 17, 1961	VAFB	Thor-Agena B	2100
Discoverer 32	1961 AΓ 1	USAF	Oct 13, 1961	VAFB	Thor-Agena B	2100
Midas 4	1961 AΔ 1	USAF	Oct 21, 1961	PA	Atlas-Agena B	3500
Discoverer 33	None	USAF	Oct 23, 1961	VAFB	Thor-Agena B	2100
Mercury-Scout 1	None	NASA	Nov 1, 1961	AMR	Scout	150
Discoverer 34	1961 AE 1	USAF	Nov 5, 1961	VAFB	Thor-Agena B	2100
Discoverer 35	1961 AZ 1	USAF	Nov 15, 1961	VAFB	Thor-Agena B	2100
Transit 4B Traac	1961 AH 1 1961 AH 2	USN	Nov 15, 1961	AMR	Thor-Able Star	190 240
Ranger 2	1961 AΘ 1	NASA	Nov 18, 1961	AMR	Atlas-Agena B	675
None	None	USAF	Nov 22, 1961	PA	Atlas-Agena B	
Mercury-Atlas 5	1961 AI 1	NASA	Nov 29, 1961	AMR	Atlas D	2900
Discoverer 36 Oscar 1	1961 AK 1 1961 AK 2	USAF	Dec 12, 1961	VAFB	Thor-Agena B	2100
None	1961 AΛ 1	USAF	Dec 22, 1961	PA	Atlas-Agena B	
Discoverer 37	None	USAF	Jan 13, 1962	VAFB	Thor-Agena B	2100
Composite 1	None	USN	Jan 24, 1962	AMR	Thor-Able Star	219
Ranger 3	1962 A 1	NASA	Jan 26, 1962	AMR	Atlas-Agena B	727

[1]As of August 15, 1963, if applicable

[4]Astronomical Unit

[5]Inclination from the ecliptic

TABLE 1–2 (Cont.)

INITIAL CURRENT[1] ORBITAL DATA				RESULTS
Period	Perigee	Apogee	Incl.	
—	—	—	—	Failed to orbit: second stage control system malfunction
88.6	111	160	64.9	Re-entered 8-7-61: cabin, G. Titov landed separately, 17 orbits
1585	182	48,000	33.3	In orbit: returned radiation, solar wind data until 12-6-61
91.1	105	313	32.9	Decayed 8-30-61: attempt to reach deep space orbit failed
97.3	175	606	36.4	Decayed 8-28-61: orbit lower than planned, little useful data returned
91.0	140	345	82.1	Decayed 9-10-61: capsule ejected on 33rd orbit, recovered from ocean
—	—	—	—	Failed to orbit: exploded on launch pad
92.4	154	345	82.6	Decayed 12-11-61: capsule recovered in mid-air on 33rd orbit
88.6	100	159	32.6	Re-entered 9-13-61: recovered from ocean after 1 orbit
91.0	152	255	82.7	Decayed 10-26-61: re-entry capsule separation attempt failed
90.8	147	246	81.7	Decayed 11-13-61: mid-air capsule recovery after 18 orbits
172.0 166.0	2179	2327	95.9	In orbit: West Ford cannister ejected but dipoles failed to separate
—	—	—	—	Failed to orbit: launch vehicle shut down prematurely
—	—	—	—	Failed to orbit: first stage exploded after 26 seconds
97.2	134	637	82.7	Decayed 12-7-62: system malfunction prevented capsule ejection
89.8	147	173	81.6	Decayed 12-3-61: capsule ejected on 18th orbit, recovered in mid-air
105.6 105.6 105.6 105.6	582 581 562 590	700 700 720 693	32.4 32.4 32.4 32.4	In orbit: returned geodetic data, ceased transmitting 8-62 In orbit: orientation experiment failed, returned radiation data until 8-62
88.3	98	147	33.3	Decayed 11-18-61: attempt to reach deep space orbit failed
—	—	—	—	Failed to orbit: classified payload[7]
88.5	100	148	32.5	Re-entered 11-29-61: chimp Enos recovered after 3 orbits
91.5 91.1	148 146	280 258	81.2 81.2	Decayed 3-8-62: capsule ejected on 64th orbit, recovered from ocean Decayed 1-31-62: first amateur satellite, transmitted for 18 days
94.5	145	467	89.6	Decayed 8-14-62: classified payload (initial orbital data as of 2-15-62)
—	—	—	—	Failed to orbit: malfunction following second stage ignition
—	—	—	—	Failed to orbit: insufficient second stage thrust, carried 5 satellites
406.4 days	.9839AU[4]	1.163AU	.3988[5]	In solar orbit: lunar probe, missed moon by 22,862 miles

[7]Under DOD security regulations, results of classified payload launches were not released for this and succeeding launches. Information shown here was compiled from fragmentary data released by USAF, NASA and the UN. Some assumptions were made in correlating this information, therefore accuracy of the resulting data cannot be guaranteed.

TABLE 1–2 (Cont.)

NAME	INT'L DESIG.	PROJECT DIRECTION	LAUNCH DATA			WEIGHT
			Date	Site	Vehicle	
Tiros 4	1962 B 1	NASA	Feb 8, 1962	AMR	Delta	285
Mercury-Atlas 6	1962 Γ 1	NASA	Feb 20, 1962	AMR	Atlas D	2987
None	1962 Δ 1	USAF	Feb 21, 1962	VAFB	Thor-Agena B	
Discoverer 38	1962 E 1	USAF	Feb 27, 1962	VAFB	Thor-Agena B	2100
OSO 1	1962 Z 1	NASA	Mar 7, 1962	AMR	Delta	458
None	1962 H 1	USAF	Mar 7, 1962	PA	Atlas-Agena B	
Cosmos 1	1962 Θ 1	USSR	Mar 16, 1962	Kapustin Yar		
Cosmos 2	1962 I 1	USSR	Apr 6, 1962	Kapustin Yar		
None	1962 K 1	USAF	Apr 9, 1962	PA	Atlas-Agena B	
None	1962 Λ 1	USAF	Apr 17, 1962	VAFB	Thor-Agena B	
Ranger 4	1962 M 1	NASA	Apr 23, 1962	AMR	Atlas-Agena B	730
Cosmos 3	1962 N 1	USSR	Apr 24, 1962	Kapustin Yar		
Cosmos 4	1962 Ξ 1	USSR	Apr 26, 1962	Tyuratam		
Ariel (UK 1)	1962 O 1	NASA/UK	Apr 26, 1962	AMR	Delta	132
None	None	USAF	Apr 26, 1962	PA	Blue Scout	
None	1962 Π 1	USAF	Apr 26, 1962	PA	Atlas-Agena B	
None	1962 P 1	USAF	Apr 28, 1962	VAFB	Thor-Agena B	
Anna 1A	None	USN	May 10, 1962	AMR	Thor-Able Star	355
None	1962 Σ 1	USAF	May 15, 1962	VAFB	Thor-Agena B	
None	None	USAF	May 23, 1962	PA	Blue Scout	
Mercury-Atlas 7	1962 T 1	NASA	May 24, 1962	AMR	Atlas D	2975
Cosmos 5	1962 Υ 1	USSR	May 28, 1962	Kapustin Yar		
None	1962 Φ 1	USAF	May 29, 1962	VAFB	Thor-Agena B	
None Oscar 2	1962 X 1 1962 X 2	USAF	June 2, 1962	VAFB	Thor-Agena B	10
None	1962 Ψ 1	USAF	June 17, 1962	PA	Atlas-Agena B	
None	1962 O 1	USAF	June 18, 1962	VAFB	Thor-Agena B	

TABLE 1–2 (Cont.)

INITIAL / CURRENT[1] ORBITAL DATA				RESULTS
Period	Perigee	Apogee	Incl.	
100.4 100.3	441 442	525 522	48.3 48.3	In orbit: returned total of 32,593 cloud cover photos up to 6-10-62
88.5	100	163	32.5	Re-entered 2-20-62: "Friendship 7" with J. Glenn recovered after 3 orbits
89.7	104	233	82.0	Decayed 3-4-62: classified payload (initial orbital data as of 2-27-62)
89.7	208	308	82.2	Decayed 3-21-62: capsule recovered in mid-air after 65 orbits (initial orbital data as of 3-13-62)
96.2 95.9	344 350	370 357	32.8 32.8	In orbit: transmitting real time solar flare, radiation data on 136.407 mc
93.9	147	428	90.9	Decayed 6-7-63: classified payload (initial orbital data as of 3-13-62)
96.4	135	609	49.0	Decayed 5-25-62: unannounced payload
102.0 88.7	131 98	960 170	49.0 48.9	In orbit: unannounced payload
153.0 153.0	1731 1742	2116 2107	86.7 86.7	In orbit: classified payload (initial orbital data as of 4-10-62)
91.5	98	333	73.5	Decayed 5-28-62: classified payload (initial orbital data as of 4-30-62)
Flight time: 64.0 hours				Impacted on moon: experiments inoperative due to timer failure
93.8	142	447	49.0	Decayed 10-17-62: unannounced payload
90.6	184	205	65.0	Re-entered or decayed 4-29-62: unannounced payload
100.9 100.6	242 246	754 736	53.9 53.9	In orbit: first NASA international satellite, transmitting on 136.406 mc
—	—	—	—	Failed to orbit: classified payload
				Decayed 4-28-62: classified payload
91.1	98	307	73.2	Decayed 5-26-62: classified payload (initial orbital data as of 4-30-62)
—	—	—	—	Failed to orbit: second stage ignition malfunction
94.0 91.6	180 176	401 257	82.5 82.3	In orbit: classified payload (initial orbital data as of 5-17-62)
—	—	—	—	Failed to orbit: classified payload
88.3	100	167	32.5	Re-entered 5-24-62: "Aurora 7" and S. Carpenter recovered, 3 orbits
102.8	126	994	49.1	Decayed 5-2-63: unannounced payload
89.9	121	212	74.1	Decayed 6-11-62: classified payload (initial orbital data as of 5-31-62)
90.5 90.5	131 129	241 240	74.3 74.3	Decayed 6-28-62: classified payload (initial orbital data as of 6-5-62) Decayed 6-21-62: amateur radio satellite
				Decayed 6-18-62: classified payload
92.3 90.8	234 193	244 194	82.0 82.1	In orbit: classified payload (initial orbital data as of 6-19-62)

[1]As of August 15, 1963, if applicable

TABLE 1–2 (Cont.)

NAME	INT'L DESIG.	PROJECT DIRECTION	LAUNCH DATA			WEIGHT
			Date	Site	Vehicle	
Tiros 5	1962 AA 1	NASA	June 19, 1962	AMR	Delta	286
None	1962 AB 1	USAF	June 22, 1962	VAFB	Thor-Agena B	
None	1962 AΓ 1	USAF	June 27, 1962	VAFB	Thor-Agena D	
Cosmos 6	1962 AΔ 1	USSR	June 30, 1962	Kapustin Yar		
Telstar 1	1962 AE 1	AT&T	July 10, 1962	AMR	Delta	170
None	1962 AZ 1	USAF	July 18, 1962	PA	Atlas-Agena B	
None	1962 AH 1	USAF	July 20, 1962	VAFB	Thor-Agena B	
Mariner 1	None	NASA	July 22, 1962	AMR	Atlas-Agena B	446
None	1962 AΘ 1	USAF	July 28, 1962	VAFB	Thor-Agena B	
Cosmos 7	1962 AI 1	USSR	July 28, 1962	Tyuratam		
None	1962 AK 1	USAF	Aug 1, 1962	VAFB	Thor-Agena D	
None	1962 AΛ 1	USAF	Aug 5, 1962	PA	Atlas-Agena B	
Vostok 3	1962 AM 1	USSR	Aug 11, 1962	Tyuratam		"about 10,160"
Vostok 4	1962 AN 1	USSR	Aug 12, 1962	Tyuratam		"about 10,160"
Cosmos 8	1962 AΞ 1	USSR	Aug 18, 1962	Kapustin Yar		
None	1962 AO 1	USAF	Aug 23, 1962	PA	Blue Scout	
None	1962 AΠ 1	USSR	Aug 25, 1962			
Mariner 2	1962 AP 1	NASA	Aug 26, 1962	AMR	Atlas-Agena B	447
None	1962 AΣ 1	USAF	Aug 28, 1962	VAFB	Thor-Agena D	
None	1962 AT 1	USSR	Sept 1, 1962			
None	1962 AΥ 1	USAF	Sept 1, 1962	VAFB	Thor-Agena B	
None	1962 AΦ 1	USSR	Sept 12, 1962			
None/TRS 1	1962 AX 1	USAF	Sept 17, 1962	VAFB	Thor-Agena B	/1.5
Tiros 6	1962 AΨ 1	NASA	Sept 18, 1962	AMR	Delta	281
Cosmos 9	1962 AO 1	USSR	Sept 27, 1962	Tyuratam		

[1]As of August 15, 1963, if applicable

TABLE 1–2 (Cont.)

INITIAL CURRENT[1] ORBITAL DATA				RESULTS
Period	Perigee	Apogee	Incl.	
100.5 100.4	367 377	604 593	58.1 58.1	In orbit: returned 57,857 cloud cover photos until 5-4-63
89.0	130	150	75.1	Decayed 7-7-62: classified payload (initial orbital data as of 6-30-62)
93.6	131	398	76.0	Decayed 9-14-62: classified payload (initial orbital data as of 6-30-62)
90.1	168	221	48.9	Decayed 8-8-62: unannounced payload
157.8 157.7	593 595	3503 3499	44.8 44.8	In orbit: first active repeater comsat, transmitted until 2-21-63
				Decayed 7-27-62: classified payload
90.0	122	218	70.3	Decayed 8-14-62: classified payload (initial orbital data as of 7-31-62)
—	—	—	—	Venus probe failed: destroyed by range safety officer
90.7	129	251	71.1	Decayed 8-24-62: classified payload (initial orbital data as of 7-31-62)
90.1	130	229	65.0	Re-entered or decayed 8-1-62: unannounced payload
90.2	121	227	82.3	Decayed 8-26-62: classified payload (initial orbital data as of 8-14-62)
				Decayed 8-6-62: classified payload
88.3	114	156	65.0	Re-entered 8-15-62: cabin and A. Nikoloyev landed separately after 64 orbits
88.4	112	158	65.0	Re-entered 8-15-62: cabin and P. Popovich landed separately after 48 orbits, 6 minutes after Vostok 3 re-entry; initial orbit placed Vostok 4 within 3.1 miles of Vostok 3
92.9 88.0	159 107	375 117	49.0 48.9	In orbit: unannounced payload
99.6 99.6	388 388	526 527	98.6 98.7	In orbit: classified payload (initial orbital data as of 8-31-62)
99.6	390	526	98.6	Decayed 8-28-62: probable Venus probe failure
348 days	.7046AU[4]	1.229AU	1.66[5]	In solar orbit: passed within 21,594 miles of Venus, reported surface temperature of 800°F, returned data out to 54.3M miles
90.4	114	250	65.2	Decayed 9-10-62: classified payload (initial orbital data as of 8-31-62)
				Decayed 9-6-62: probable Venus probe failure
94.4 93.2	189 178	418 352	82.8 82.8	In orbit: classified payload (initial orbital data as of 9-15-62)
				Decayed 9-14-62: probable Venus probe failure
92.8	124	383	81.9	Decayed 11-19-62: classified payload, failed to eject TRS pickaback which returned solar cell damage data (initial orbital data as of 9-30-62)
98.7 98.7	423 423	444 444	58.2 58.3	In orbit: wide angle camera operating, transmissions on 136.233, 136.922 mc
90.0	187	220	65.0	Re-entered or decayed 10-1-62: unannounced payload

[4]Astronomical Unit   [5]Inclination from the ecliptic

TABLE 1–2 (Cont.)

NAME	INT'L DESIG.	PROJECT DIRECTION	LAUNCH DATA			WEIGHT
			Date	Site	Vehicle	
Alouette	1962 BA 1	Canada	Sept 28, 1962	VAFB	Thor-Agena B	320
None	1962 BB 1	USAF	Sept 29, 1962	VAFB	Thor-Agena D	
Explorer 14	1962 BΓ 1	NASA	Oct 2, 1962	AMR	Delta	89
Mercury-Atlas 8	1962 BΔ1	NASA	Oct 3, 1962	AMR	Atlas D	3030
None	1962 BE1	USAF	Oct. 9, 1962	VAFB	Thor-Agena B	
Cosmos 10	1962 BZ 1	USSR	Oct 17, 1962	Tyuratam		
Ranger 5	1962 BH 1	NASA	Oct 18, 1962	AMR	Atlas-Agena B	755
Cosmos 11	1962 BΘ 1	USSR	Oct 20, 1962	Kapustin Yar		
None	1962 BI 1	USSR	Oct 24, 1962			
None	1962 BK 1	USAF	Oct 26, 1962	VAFB	Thor-Agena D	
Explorer 15	1962 BΛ 1	NASA	Oct 27, 1962	AMR	Delta	98
Anna 1B	1962 BM 1	USN	Oct 31, 1962	AMR	Thor-Able Star	350
Mars 1	1962 BN 3	USSR	Nov 1, 1962			1965
None	1962 BΞ 1	USSR	Nov 4, 1962			
None	1962 BO 1	USAF	Nov 5, 1962	VAFB	Thor-Agena B	
None	1962 BΠ 1	USAF	Nov 11, 1962	PA	Atlas-Agena B	
None	1962 BP 1	USAF	Nov 24, 1962	VAFB	Thor-Agena B	
None	1962 BΣ 1	USAF	Dec 4, 1962	VAFB	Thor-Agena D	
None	1962 BT 1					
Injun 3	1962 BT 1					114
None	1962 BT 3	USAF/USN	Dec 12, 1962	VAFB	Thor-Agena D	
None	1962 BT 4					
None	1962 BT 5					

[1]As of August 15, 1963, if applicable

TABLE 1–2 (Cont.)

INITIAL / CURRENT[1] ORBITAL DATA				RESULTS
Period	Perigee	Apogee	Incl.	
105.4 105.5	620 622	638 642	80.5 80.5	In orbit: ionosphere sounding satellite, transmitting on 136.591. 136.979 mc
90.3	119	241	65.4	Decayed 10-14-62: classified payload (initial orbital data as of 9-30-62)
2184 2185	174 1084	61,190 60,305	32.9 40.0	In orbit: energetic particles satellite, transmitting on 136.440 mc
89.0	100	176	32.5	Re-entered 10-3-62: "Sigma 7" with W. Schirra recovered within 5 miles of carrier after 6 orbits
90.9	103	291	81.5	Decayed 11-16-62: classified payload (initial orbital data as of 10-10-62)
90.2	130	236	65.0	Re-entered or decayed 10-21-62: unannounced payload
366 days	.9490AU[4]	1.052AU	.3901[5]	In solar orbit: lunar probe, missed moon by 450 miles
96.1 93.8	152 146	572 431	49.0 48.9	In orbit: unannounced payload
				Decayed 10-29-62: probable Mars probe failure
147.8 142.0	120 145	3452 3140	71.4 71.5	In orbit: "Starad" oayload returned radiation data until 1-18-63
312.0 314.2	194 191	10,760 10,924	18.0 18.0	In orbit: excessive spin rate degraded radiation data, transmitted until 2-63
107.8 107.8	670 672	728 733	50.1 50.2	In orbit: flash tubes operated until 1-8-63, transmitting on 162,324 mc
519 days	.9237AU[4]	1.604AU	2.683[5]	In solar orbit: Mars probe transmissions ceased after 66M miles
				Decayed 11-5-62: probable Mars probe failure
90.7	130	250	75.0	Decayed 12-3-62: classified payload (initial orbital data as of 11-7-62)
				Decayed 11-12-62: classified payload
89.8	129	202	65.2	Decayed 12-13-62: classified payload (initial orbital data as of 11-30-62)
89.2	119	175	65.0	Decayed 12-8-62: classified payload (initial orbital data as of 12-6-62)
116.0 113.8	153 147	1724 1593	70.4 70.4	In orbit: classified payload (initial orbital data as of 12-15-62)
116.3 114.9	153 152	1729 1652	70.3 70.4	In orbit: transmitting radiation data on 136.867 mc
115.6	139	1700	70.3	Decayed 7-1-63: classified pickaback
116.2 112.5	145 153	1728 1510	70.4 70.5	In orbit: classified pickaback
116.0 113.7	146 148	1718 1588	70.3 70.4	In orbit: classified pickaback

[4]Astronomical Unit [5]Inclination from the ecliptic

TABLE 1–2 (Cont.)

NAME	INT'L DESIG.	PROJECT DIRECTION	LAUNCH DATA			WEIGHT
			Date	Site	Vehicle	
Relay 1	1962 BΥ 1	NASA	Dec 12, 1962	AMR	Delta	172
None	1962 BΦ 1	USAF	Dec 14, 1962	VAFB	Thor-Agena D	
Explorer 16	1962 BX 1	NASA	Dec 16, 1962	WI	Scout	222
None	None	USAF	Dec 17, 1962	PA	Atlas-Agena B	
Transit 5A	1962 BΨ 1	USN	Dec 18, 1962	PA	Blue Scout	135
Cosmos 12	1962 BΩ 1	USSR	Dec 22, 1962	Tyuratam		
None	1963 1A[8]	USSR	Jan 4, 1963			
None	1963 2A	USAF	Jan 7, 1963	VAFB	Thor-Agena D	
None	1963 3A	USAF	Jan 16, 1963	VAFB	Thor-Agena D	
Syncom 1	1963 4A	NASA	Feb 14, 1963	AMR	Delta	86
None	1963 5A	USAF	Feb 19, 1963	PA	Blue Scout	
None	None	USAF	Feb 28, 1963	VAFB	Thor-Agena D	
None	None	USAF	Mar 18, 1963	VAFB	Thor-Agena D	
Cosmos 13	1963 6A	USSR	Mar 20, 1963	Tyuratam		
None	1963 7A	USAF	Apr 1, 1963	VAFB	Thor-Agena D	
Lunik 4	1963 8B	USSR	Apr 2, 1963			3135
Explorer 17	1963 9A	NASA	Apr 2, 1963	AMR	Delta	405
None	None	USAF	Apr 5, 1963	PA	Blue Scout	
Cosmos 14	1963 10A	USSR	Apr 13, 1963	Kapustin Yar		
Cosmos 15	1963 11A	USSR	Apr 22, 1963	Tyuratam		
None	None	USAF	Apr 26, 1963	PA	Blue Scout	
None	None	USAF	Apr 26, 1963	VAFB	Thor-Agena D	
Cosmos 16	1963 12A	USSR	Apr 28, 1963	Tyuratam		
Telstar 2	1963 13A	AT&T	May 7, 1963	AMR	Delta	175
None TRS 2 TRS 3	1963 14A 1963 14B 1963 14C	USAF	May 9, 1963	PA	Atlas-Agena B	 1.5 1.5

[8]New international designation system initiated 1-1-63

TABLE 1–2 (Cont.)

INITIAL / CURRENT[1] ORBITAL DATA				RESULTS
Period	Perigee	Apogee	Incl.	
185.9 185.0	819 815	4612 4627	47.5 47.5	In orbit: active repeater comsat, transmitting on 136.141 mc
90.5	126	241	70.0	Decayed 1-8-62: classified payload (initial orbital data as of 12-15-62)
104.4 104 3	466 459	733 741	52.0 52.0	In orbit: micrometeorite satellite, transmitting on 136.200, 136.861 mc
—	—	—	—	Failed to orbit: classified payload
99.2 99.2	432 439	455 450	90.7 90.6	In orbit: power supply malfunction after 1 day
90.5	131	252	65.0	Re-entered or decayed 12-30-62: unannounced payload
				Decayed 1-5-63: probable lunar probe failure
90.5	131	245	82.0	Decayed 1-24-63: classified payload (initial orbital data as of 1-15-63)
94.7 94.6	298 285	322 334	82.0 81.9	In orbit: classified payload (initial orbital data as of 1-31-63)
1426.5 1426.4	21,268 21,240	22,974 23,000	33.5 33.5	In orbit: communication lost at injection into synchronous orbit
97.8 97.8	305 315	498 493	100.5 100.5	In orbit: classified payload (initial orbital data as of 2-28-63)
—	—	—	—	Failed to rbit: classified payload, first use of thrust-augmented Thor
—	—	—	—	Failed to orbit: classified payload, employed thrust-augmented Thor
89.8	127	209	65.0	Re-entered or decayed 3-29-63: unannounced payload
				Decayed 4-26-63: classified payload
	55,800	434,000		In orbit: lunar probe, missed moon by 5281 miles, will enter solar orbit
96.4 96.0	158 165	570 540	57.6 57.5	In orbit: atmospheric structure satellite, transmitted until 7-63
—	—	—	—	Failed to orbit: classified payload
92.1 89.7	165 141	318 185	48.6 48.9	In orbit: unannounced payload
89.8	107	231	65.0	Re-entered or decayed 4-27-63: unannounced payload
—	—	—	—	Failed to orbit: classified payload
—	—	—	—	Failed to orbit: classified payload
90.4	129	249	65.0	Re-entered or decayed 5-8-63: unannounced payload
225.0 225.2	604 606	6713 6712	42.7 42.8	In orbit: active repeater comsat, transmitting on 136.050 mc
166.6 166.5	2254 2254	2295 2275	87.4 87.3	In orbit: classified payload, ejected 50-pound package of West Ford copper dipole needles (initial orbital data as of 5-15-63)
166.5 166.5	2246 2241	2302 2289	87.4 87.4	In orbit: radiation pickaback transmitting on 136.892 mc
166.5	2243	2287	87.3	In orbit: radiation pickaback transmitting on 136.415 mc

[1]As of August 15, 1963, if applicable

TABLE 1–2 (Cont.)

NAME	INT'L DESIG.	PROJECT DIRECTION	LAUNCH DATA			WEIGHT
			Date	Site	Vehicle	
Mercury-Atlas 9	1963 15A	NASA	May 15, 1963	AMR	Atlas D	3000
None	1963 16A	USAF	May 18, 1963	VAFB	Thor-Agena D	
Cosmos 17	1963 17A	USSR	May 22, 1963	Kapustin Yar		
Cosmos 18	1963 18A	USSR	May 24, 1963	Tyuratam		
None	None	USAF	June 12, 1963	PA	Atlas-Agena B	
None	1963 19A	USAF	June 12, 1963	VAFB	Thor-Agena D	
Vostok 5	1963 20A	USSR	June 14, 1963	Tyuratam		"about 10,160"
None Lofti 2A Greb 4 Radose None None	1963 21A 1963 21B 1963 21C 1963 21D 1963 21E 1963 21F	USAF/USN	June 15, 1963	VAFB	Thor-Agena D	
None	1963 22A	USAF	June 15, 1963	PA	Blue Scout	
Vostok 6	1963 23A	USSR	June 16, 1963	Tyuratam		"about 10,160"
Tiros 7	1963 24A	NASA	June 19, 1963	AMR	Delta	297
None Hitch-hiker 1	1963 25A 1963 25B	USAF	June 26, 1963	VAFB	Thor-Agena D	176
Geophysical Research Satellite	1963 26A	USAF	June 28, 1963	WI	Scout	
None	1963 27A	USAF	June 29, 1963	VAFB	Thor-Agena B	
None	1963 28A	USAF	July 12, 1963	PA	Atlas-Agena D	
None	1963 29A	USAF	July 18, 1963	VAFB	Thor-Agena D	
None	1963 30A	USAF	July 19, 1963	VAFB	Atlas-Agena B	
Syncom 2	1963 31A	NASA	July 26, 1963	AMR	Delta	86
None	1963 32A	USAF	July 30, 1963	VAFB	Thor-Agena D	
Cosmos 19	1963 33A	USSR	Aug 6, 1963	Kapustin Yar		

[1]As of August 15, 1963, if applicable

TABLE 1–2 (Cont.)

INITIAL CURRENT[1] ORBITAL DATA				RESULTS
Period	Perigee	Apogee	Incl.	
88.5	100	166	32.5	Re-entered 5-16-63: "Faith 7" with G. Cooper recovered after 22 orbits, within 5 miles of carrier
				Decayed 5-27-63: classified payload
94.8 94.5	162 155	490 477	49.0 48.9	In orbit: unannounced payload
89.4	130	187	65.0	Re-entered or decayed 6-2-63: unannounced payload
—	—	—	—	Failed to orbit: classified payload
90.7	120	262	81.9	Decayed 7-12-63: classified payload (initial orbital data as of 6-15-63)
88.3	109	138	65.0	Re-entered 6-19-63: cabin, V. Bykovsky landed after 82 orbits
94.5 93.7 94.1 94.1 94.0 91.7	110 114 110 103 110 102	501 449 480 487 472 341	69.9 69.9 69.8 68.9 69.9 69.8	Decayed 8-7-63: classified payload (initial orbital data as of 6-30-63) Decayed 7-18-63: VLF experiment payload Decayed 8-1-63: solar radiation payload Decayed 7-30-63: classified payload Decayed 7-27-63: classified payload Decayed 7-5-63: classified payload
99.8 99.8	450 459	475 469	90.0 90.0	In orbit: classified payload (initial orbital data as of 6-30-63)
88.4	114	145	65.0	Re-entered 6-19-63: cabin and V. Tereshkova landed separately after 49 orbits, 3 hours before Vostok 5 landing
97.4 97.4	385 382	401 409	58.2 58.2	In orbit: wide angle cameras, IR experiments operating, transmissions on 136.232, 136.922 mc
90.5 132.6 132.5	124 201 214	247 2571 2558	81.6 82.1 82.1	Decayed 7-26-63: classified payload (initial orbital data as of 6-30-63) In orbit: pickaback to measure artificial radiation decay
102.1 102.1	255 262	815 807	49.7 49.7	In orbit: no details released (initial orbital data as of 6-30-63)
94.8 94.8	308 303	319 327	82.3 82.3	In orbit: classified payload (initial orbital data as of 7-15-63)
88.3	114	116	95.4	Decayed 7-18-63: 100th Agena launched to date; classified payload (initial orbital data as of 7-15-63)
89.9	120	209	82.9	Decayed 8-13-63: classified payload (initial orbital data as of 7-31-63)
167.9 167.9	2279 2280	2321 2320	88.4 88.4	In orbit: classified payload (initial orbital data as of 7-31-63)
1408	22,132	22,823	30	In orbit: active-repeater comsat, moved to position over Brazil, early communication tests successful
90.6	94	283	74.9	Decayed 8-11-63: classified payload (initial orbital data as of 7-31-63)
92.2 92.1	168 163	322 311	49.0 49.0	In orbit: unannounced payload

1.4.4 Historical Data on Attainments

Table 1–2—tabulates the historical data on space flights to date. Early historical data on achievements by missile pioneers is available in Chapter 1 of Ehricke: *Space Flight*, Volume I.

REFERENCES TO CHAPTER 1

1. "Satellite Needs Push Reconnaissance Gains," *Aviation Week*, May 22, 1961, p. 26.
2. Morris Tepper, "Tiros Program Results," *Astronautics*, May 1961.
3. "Synoptic-Scale Radiation Processes," pp. 101-108, Contributions to Satellite Meteorology, GRD Research Notes No. 36, Geophysics Research Directorate, Air Force Cambridge Research Center, ASTIA 241835, June 1960.
4. "Novel Approaches," *Astronautics*, Oct. 1960, p. 32.
5. "Translation, Atlas of the Far Side of the Moon," U. S. Dep't of Commerce, Office of Technical Services, March 1960.
6. F. Zwicky, "Some Possible Operations on the Moon," *ARS Journal*, Vol. 30, No. 12, Dec. 1960.
7. S. F. Singer, "Some Considerations of Expected Radiation Belts of Planets Mars and Venus," Advances in the Astronautical Sciences, Vol. 6, Proceedings of the Sixth Annual Meeting of the American Astronautical Society, Jan. 1960, Macmillan Company, New York.

CHAPTER 2

SPACE ENVIRONMENT

2.1 INTRODUCTION

The purpose of this chapter is to summarize for the electronic, communications, and guidance engineer those characteristics of the space environment with which he will be directly concerned. To supplement this material where needed, the reader is urged to consult Ref. (*1*).

The chapter is organized around environmental subdivisions that develop progressively from the local area to the macroscopic: terrestrial, cislunar and lunar, solar system, galactic, and extra-galactic. Although this is the natural order of development in evolving space communication and vehicle capabilities, it should be noted that the macroscopic references may be employed as the only stable reference system for navigating among local bodies, and measuring environmental characteristics. For example, celestial coordinates requiring employment of star-tracking devices trained on galactic star sources may be the most suitable means of defining and solving an interplanetary navigational problem, especially when the target planet is eclipsed by the Sun during part of the flight.

2.2 TERRESTRIAL ENVIRONMENT

Terrestrial environment may be sub-divided into those physical characteristics which are relatively permanent in nature, those which are more transient (as weather) and those caused by external sources such as high-energy charged particles and meteors. This section is organized in this manner except that the description of the radiation belt regions of charged particles immediately follows the magnetic field description because of their close relation with one another.

2.2.1 Earth shape, size, mass, rotation, and orbit

The Earth is an oblate spheroid with an equatorial radius of 3444.058 nautical miles (6378.388 kilometers). The oblateness, ϵ, is $\frac{1}{297.00}$ which means that the polar radius is 3444.058 $(1 - \epsilon)$ nautical miles. The radius at intervening latitudes may be computed from:

$$r_\lambda = r_0\left\{1 + \left[\frac{1}{(1-\epsilon)^2} - \right]1 \sin^2 \lambda\right\}^{1/2} \tag{2-1}$$

$$= r_0\left[1 - \left(\epsilon - \frac{3}{2}\epsilon^2\right)\sin^2_\lambda + \frac{3}{2}\epsilon^2 \sin^4 \lambda \ldots\right]$$

$$= r_0\left(1 - \epsilon \sin^2 \lambda + \frac{5}{8}\epsilon^2 \sin^2 2\lambda + \ldots\right)$$

$$= 6{,}378{,}388\ (1 - 0.003670 \sin^2 \lambda + 0.0000071 \sin^2 2\lambda) \text{ in meters}$$

Topographical features and other anomalies are, of course, ignored in this approximation which is based on International Hayford spheroid computations (Ref. (*1*)).

The Earth's mass is generally taken as $m_\oplus = (5.977 \pm 0.004)\ 10^{27}$ grams (Ref. (*1*)); the density is 5.52 that of water; the gravitational constant $G = (6.668 \pm 0.005)\ 10^{-8}$ cm^3 sec^{-2} g^{-1}; and the gravitational parameter $K_\oplus = Gm_\oplus = 398{,}546$ km^3/sec^2 with limits of 399,112 km^3/sec^2 and 397,981 km^3/sec^2. As developed in Ref. (*1*), using measurements of g (gravitational acceleration) at various latitudes, a likely choice for $K_\oplus$ is $3.9858 \cdot 10^5$ km^3/sec^2. Values of g for a central force field may be taken rigorously for any field from

$$g = \frac{K}{r^2} \tag{2-2}$$

However, the oblateness of the Earth, and the inhomogeneity of the internal mass distribution make this an approximate relation. Generally, accepted values of g are:

Non-Rotating Earth

Latitude, λ	g_λ = cm/sec^2	ft/sec^2
0°	981.43	32.199
45°	983.008	32.25085
90°	983.22	32.258

Correcting for the Earth's rotation using an angular velocity of

$$\omega = 7.31958 \times 10^{-5} \text{ rad/sec,}$$

we may compute circular velocity as a function of latitude,

$$V_\lambda = r_\lambda\, \omega \cos \lambda \tag{2-3}$$

and centrifugal acceleration as a function of latitude,

$$a_\lambda = \frac{V_\lambda^2}{r_\lambda} \tag{2-4}$$

and obtain corrected values of g as a function of latitude,

$$g_{\lambda_r} = (g_\lambda - a_\lambda \cos \lambda) + i a_\lambda \sin \lambda \tag{2-5}$$

which defines both magnitude and angle off vertical in equatorial direction. Generally accepted values are (Ref. (*1*)):

	Rotating Earth	
Latitude λ	$g_{\lambda r}$ = cm/sec^2	ft/sec^2
0°	978.04	32.088
45°	980.66	32.174
90°	983.22	32.258

A more explicit formula is (Ref. (*2*)), $g_\lambda = 978.049(1 + 0.0052884 \sin^2 \lambda - 0.0000059 \sin^2 2\lambda)$.

The mean solar day is divided into 24 mean solar hours or 86,400 mean solar seconds. The Earth, when viewed from a point in space over the North Pole, is rotating counter clockwise, and travelling around the Sun in a counter clockwise direction. Thus the mean solar day is slightly longer than the actual Earth rotational period by the reciprocal of the number of mean solar days in a sidereal year. The value of the sidereal year is

$$T_\oplus = 365.2563835 \text{ mean solar days}$$

Thus the Earth rotational period is

$T_\oplus = 24\left(1 - \dfrac{1}{T_\oplus}\right)$ or 23 hours 56.068 minutes and the angular velocity is

$$\omega_\oplus = 7.31958 \times 10^{-5} \text{ rad/sec.}$$

The inclination of the Earth's equatorial plane to the ecliptic plane is (Ref. (*1*)).

$$\epsilon_\oplus = 23° \, 26' \, 59''$$

This inclination is also given (Ref. (*3*)), as

$$\epsilon_\oplus = 23° \, 27' \, 08.26'' - 0.4684 \, (t - 1900)''$$

where t is the year of interest.

The ascending node of the equatorial plane with respect to the ecliptic occurs at (Ref. (*3*))

$$l_\oplus = 173° \, 57' \, 03.6'' + 32.862 \, (t - 1900)$$

(Note: ascending nodes and all longitudes in the celestial hemisphere are taken with respect to the Vernal Equinox (about March 21) line Sun-Earth to a point in the constellation Aries; angles being positive in the counter-clockwise direction when viewed from celestial North.

The orbit of the Earth about the Sun is that of an ellipse; the semi-major axis being designated as one Astronomical Unit (A.U.). The A.U. is also the mean Earth-Sun distance (Ref. (*1*)),

$$a_{\oplus} = 1 \text{ A.U.} = 149.4 \times 10^6 \text{ km}$$

The eccentricity of the Earth's orbit is

$$e_{\oplus} = 0.0167272$$

Perihelion occurs at 0.983273 A.U. Aphelion occurs at 1.016727 A.U. The longitude of Perihelion occurs at 102° 12′ 3.2″. The longitude of the Earth at Jan. 1, 1957 was 99° 53′ 19.15″. The period of revolution about the Sun is one year or 365.2563835 mean solar days.

2.2.2 Gravitational Field

The Earth's gravitational field may be expected to follow Newton's law of gravitation. That is, the gravitational attraction between two masses, m_1 and m_2, at a separation, r, is:

$$F_g = \frac{Gm_1m_2}{r^2} \tag{2–6}$$

where $G = (6.668 \pm 0.005)\ 10^{-8}\ \text{cm}^3\ \text{sec}^{-2}\ \text{g}^{-1}$.
This attraction has been proven valid provided r is not larger than about 1 million light years (about 10^{24} cm), and not much smaller than 1 cm (Ref. (*4*)).

For practical guided or controlled space vehicles (which possess a reserve of energy, power, thrust, and maneuverability), Newton's law should be sufficient for design purposes. When planning ballistic vehicle flight trajectories, for orbits of uncontrolled satellites, further knowledge of the gravitational field will be needed. Specifically, Newton's Law requires, for exactness, that a large body (the Earth) be exactly spherical, and composed of homogeneous concentric spheric shells of uniform thickness if the Earth's mass is to be taken as acting at its center point. Thus the oblateness of the Earth, the inhomogeneity of its spheric layers (e.g. oceans versus land masses), all give rise to anomalies in the simple field which might otherwise be supposed. Herman Michielsen, based on perigee variations with latitude for various satellites, (Ref. (*5*)) has reported the Earth to be "pear-shaped", and not an oblate spheroid. Analyses of Transit IV-A satellite orbit data (Ref. (*6*)) have indicated that the Earth's equator is elliptical rather than circular. The major axis of the equator is between 11 degrees West Longitude and 169 degrees East Longitude. The minor axis of the equator is between 101 degrees West Longitude and 79 degrees East Longitude. The diameter difference is 1000 feet, thus the flattening factor is approximately $\frac{1}{40{,}000}$. The 1000-foot diameter difference may be interpreted as a 500-foot difference in sea-level (as when measured from Earth center).

The Earth's gravitational potential, U, may be developed from:

$$\frac{\delta U}{\delta x} \overset{\Delta}{=} \frac{d^2x}{dt^2} \tag{2–7}$$

(The change in potential with distance is by definition proportional to the acceleration.) For a purely central force field:

$$U = -\frac{K_\oplus}{r} \tag{2–8}$$

where $K_\oplus$ is the gravitational parameter equal to $Gm_\oplus$ and to $g_0\, r_0^2$, the values of g and r being taken at the equator.

To allow for latitude dependent harmonics, we may, following Michielsen (Ref. (*5*)) define the Earth's gravitational potential as

$$U = -\frac{K_\oplus}{r}\left[1 - \sum_{n=2}^{\infty} \mathcal{J}_r\left(\frac{r_o}{r}\right)^n P_n\,(\sin\lambda)\right] \tag{2–9}$$

where λ is the latitude and P_n is the Legendre Polynomial of order n:

$$P_n(z) = \sum_{p=0}^{m} (-1)^p \cdot \frac{(2n\text{-}2p)!\cdot z^{n-2p}}{2^n\cdot p!\cdot(n\text{-}p)!\cdot(n\text{-}2p)!}$$

where $m = \frac{1}{2}n$ or $m = \frac{1}{2}(n\text{-}1)$ whichever is an integer.

Evaluations of Vanguard, Transit, Sputnik, and other satellite orbits have led to the selection of the following $\mathcal{J}_r$ coefficient values (Ref. (*6*)):

$$\begin{aligned}
\mathcal{J}_2 &= 1082.7 \times 10^{-6}\\
\mathcal{J}_3 &= -2.5 \times 10^{-6}\\
\mathcal{J}_4 &= -1.7 \times 10^{-6}\\
\mathcal{J}_5 &= +0.3 \times 10^{-6}\\
\mathcal{J}_6 &= +0.7 \times 10^{-6}\\
\mathcal{J}_7 &= -0.6 \times 10^{-6}\\
\mathcal{J}_8 &= +0.1 \times 10^{-6}
\end{aligned}$$

Longitudinal variations in the Earth's gravitational field may be determined similarly, as has been done (Ref. (*7*)). Following W. Kaula we may define the Earth's gravitational potential as

$$U = -\frac{K_{\oplus}}{r}\left[1 - \mathcal{J}_{20}\left(\frac{r_o}{r}\right)^2 P_{20}(\sin\lambda) - \sum_{n,m} (\mathcal{J}_{nm}\cos ml + K_{nm}\sin ml)\left(\frac{r_o}{r}\right)^n P_{nm}(\sin\lambda)\right] \tag{2–10}$$

in which l is the longitude. Evaluation of the coefficients in this expression has not been completed at this writing, but is being attempted using Vanguard satellite data.

It will be noted from the foregoing that the Earth's gravitational field is basically that of a sphere, the potential function being

$$U = -\frac{K_{\oplus}}{r} \tag{2–11}$$

The most significant change is the oblateness term ($\mathcal{J}_2$ coefficient in above table) which is latitude dependent and on the order of 10^{-3} relative in magnitude to the fundamental gravitational potential. Other anomalous effects have produced measurable satellite orbit variations on the order of ±100 meters, some being latitude, and some longitude associated (Ref. (*7*)). The coefficients listed above will account for the preponderance of latitude dependent effects. Longitudinal effect coefficients can be expected to have the same order of magnitude coefficients (10^{-6} of spherical gravity potential) because of their same magnitude effect on orbit perturbations. Ocean tidal effects and underground mass inhomogeneity will be the primary causes of longitudinal variations.

For all maneuverable space craft, the basic Newtonian equations which account for Keplerian orbits will usually suffice. Inclusion of the oblateness factor and higher order harmonics terms will only be necessary for non-maneuverable vehicles whose ephemeris or destination must be known to great precision.

2.2.3 Magnetic Field

The Earth's magnetic field resembles that of a dipole whose axis is inclined about 11.4 degrees to the rotational axis. The North magnetic pole is at 78.6° N. Lat., 70.1° W. Long., and the South magnetic pole at 78.6° S. Lat. and 109.9° E. Long. The main field is entirely of internal origin. However, external influences (e.g., solar flares) cause considerable short-term variations in the field strength. Based on a magnetic dipole moment of 6.1×10^{25} gauss-cm^3 the equatorial field has been calculated as a $\frac{1}{r^3}$ function of distance from Earth's center to be (Ref. (*6*)):

r (in Earth radii)	Equatorial Magnetic Field (in milligauss)
1 (surface)	310.0
2	39.0
3	12.0
4	5.0
5	2.5

The dipole magnetic field at the Earth's surface is approximately (Ref. (*8*)).

$$H = 0.315 \cos \lambda \text{ gauss}$$
$$Z = 0.630 \sin \lambda \text{ gauss}$$

where H is the horizontal magnetic field,
Z is the vertical magnetic field, and
λ is the geomagnetic latitude.

This is a rough approximation as the vertical magnetic field, Z, at the geomagnetic North pole is 0.62 gauss, the vertical magnetic field, Z, at the geomagnetic South pole is 0.70 gauss, and the horizontal magnetic field, H, at the geomagnetic equator is 0.32 gauss. A 90° dip, or vertical field is actually experienced, not at the apparent geomagnetic poles determined from middle latitude field measurements, but at

N pole: 76° N Lat., 102° W Long
S pole: 68° S Lat., 145° E Long

Magnetic ore deposits in the Earth's crust are known to cause local anomalies in the magnetic field. Periodic variations of the magnetic field occur, and are charted and predicted on an empirical basis by the U. S. Navy Hydrographic Office. Diurnal, seasonal, and random variations can be correlated with solar influences, particularly solar flares, sunspots, and radiation effects which cause increased photoionization of air particles in the ionosphere. These effects, coupled with thermal and tide effects causing upper atmospheric winds, cause changes in the ring currents encircling the Earth, and may affect the Earth's field by as much as 1 to 4 per cent (Ref. (*9*)).

Diurnal variations of ±.01 per cent are observable on "quiet" days which may result from daily changes in the current system of the E-layer region (300,000 feet). Seasonal variations of ±.1 per cent are similarly observable. A 27-day variation corresponding to the sun's rotational period and sunspots is observable as well. Annual magnetic "storms" occur amounting to as much as 5 per cent field variation in the polar regions and 2 per cent at the equator. These "storms" are generally attributed to streams of charged particles associated with solar flare activity. These charged particle streams are thought to interact with the Earth's field to form "cavities" capable of capturing the particles in tight orbits around the magnetic lines of force. The storms occur in 27.3-day, annually, and in 11.1-year sunspot cycle periods.

In general, the magnetic field is not precisely known or charted with the exception of ship's navigation reference data which are generated empirically and reflect all known cyclic variations. The actual field effects at any distance from the Earth's surface are even less known with the possible exception of the great radiation belts of recent discovery which are described in the following section.

A survey of geomagnetic and geoelectric activity as a function of Magnetic Latitude has been made by Saunders (Ref. (*10*)). Figure 2–1 shows the geomagnetic activity in gammas (1 gamma = 10^{-5} Oersted = 10^{-5} Gauss) versus geomagnetic latitude for both electromagnetic storm and quiescent periods, including sunspot maximum and minimum effects for each period. Figure 2–2 shows the corresponding Earth-current

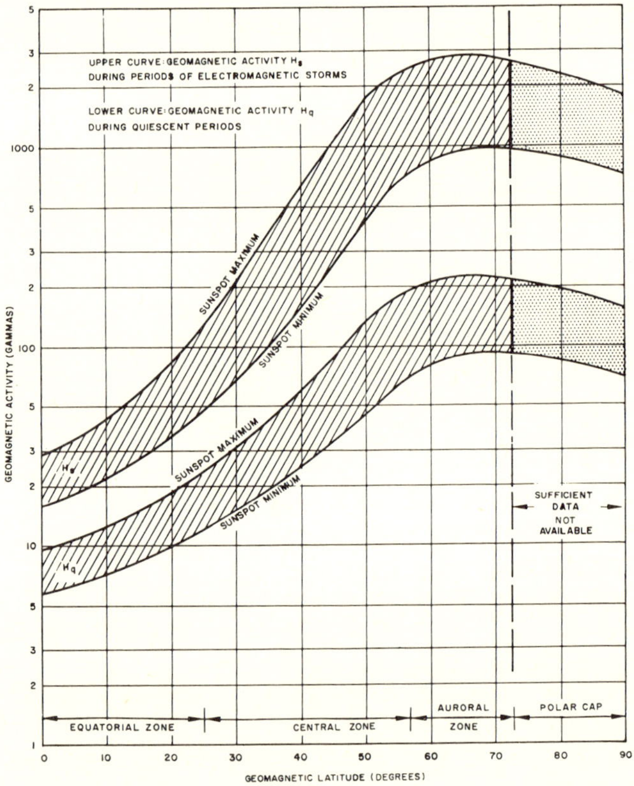

FIG. 2-1. Geomagnetic Activity as a Function of Magnetic Latitude.

activity versus geomagnetic latitude. Ignoring local anomalies in permitivity and conductivity, it was found that the prevailing Earth-current flow direction was generally E-W in the equatorial region, and N-S elsewhere.* The extension of these data radially from the Earth to present a comprehensive geomagnetic field structure model should be the subject of further research.

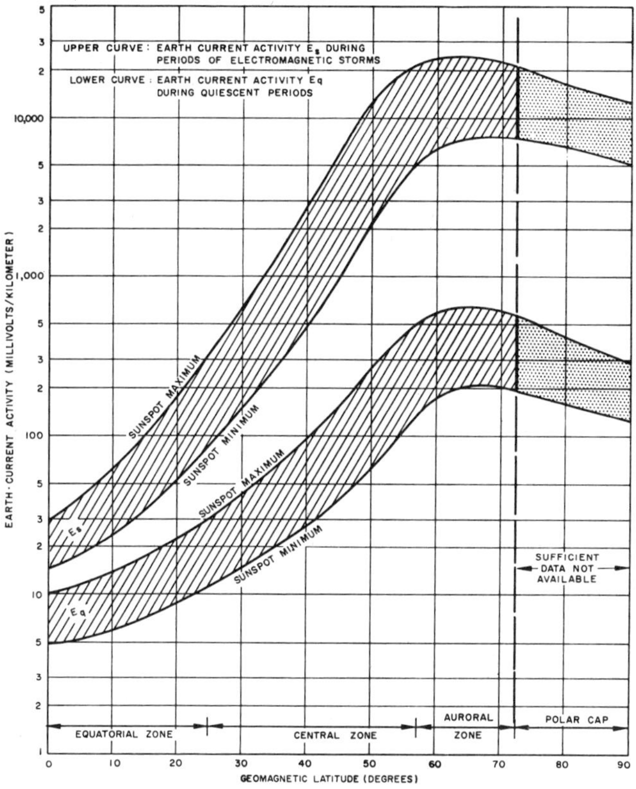

FIG. 2-2. Earth-current activity as a Function of Magnetic Latitude.

2.2.4 Great Radiation Belts

The Earth's magnetic field is known to have trapped high energy protons and electrons in two "Van Allen Belts" centered about the magnetic axis as depicted in Fig. 2–3. This figure depicts the particle counts per second (a relative measure). According to Singer (Ref. (*7*) of Chapter 1) some of the earlier work theorizing on the nature and causes of auroras, magnetic storms and ring currents in the magnetic field had predicted

*Note: Data is for Northern hemisphere only.

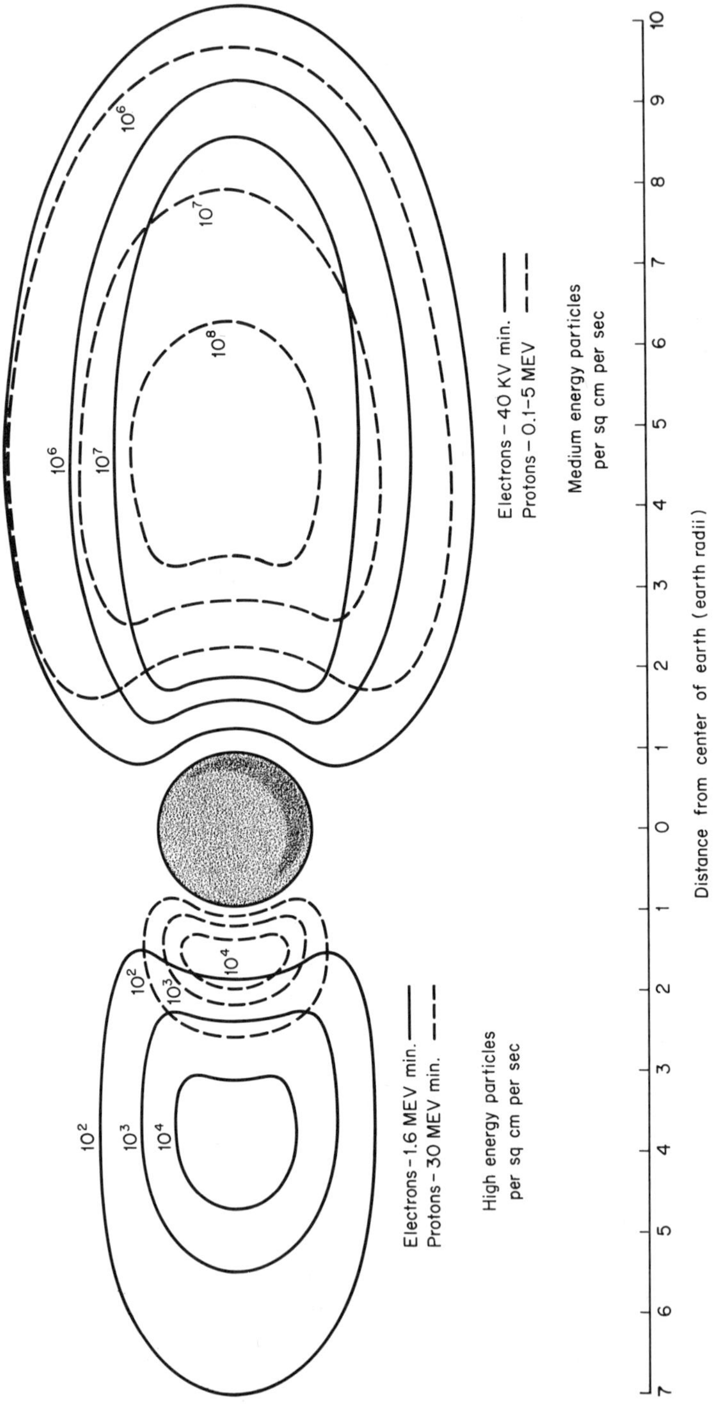

FIG. 2-3. Circumterrestrial Van Allen Belts.

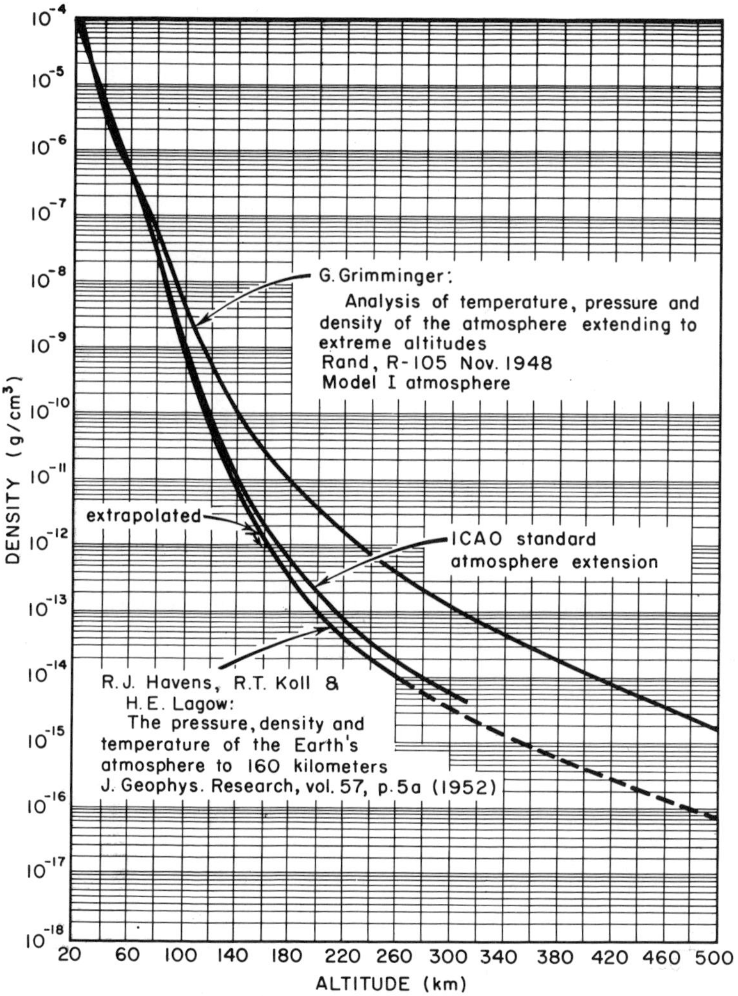

FIG. 2-4. Atmospheric Density as a Function of Altitude.

TABLE 2-1 THE EARTH'S ATMOSPHERE—

Atmospheric Shell	General Characteristics	Temperature Gradient (°C/km)	Altitude of Pause	Atmospheric Motion
Troposphere	Clouds, temperature equil. with surface	−6.5	55,000 ft (17 km) at equator; 35,000 ft (11 km) at 60°–90° lat	Turbulence; clear turbulence at pause; jet streams around pause
Stratosphere	Generally clear, dry, about constant temp. (about 217°K)	0	30 km (20 mi.) (100,000 ft) M.F.P 10^{-3} cm[1]	Lower region; jet streams upper region: calm
Chemosphere	Warm to hot at 50 km. Photochemical processes begin due to ultraviolet solar radiation. Beginning of ionization	+2 to +4 (30–50 km) −2 to −3 (50–80 km)	80 km (50 mi.) (260,000 ft) M.F.P ~0.5 cm[1]	Turbulence indicated in region of negative temperature gradient
Ionosphere	Increasing kinetic temperature.[2] Increasing ionization by solar ultraviolet; more dissociation —	+3 to +8	400 km (250 mi.) M.F.P. ~0.2−0.6 km[1]	Probably some turbulence Motion affected by magnetic field
Mesosphere	High, but decreasing kinetic temperature. Decreasing electron density, but increasing degree of ionization	Unknown	1000 km (600 mi.) (critical level for escape) M.F.P. kilometers[1]	Unknown
Exosphere	Outer atmosphere	Unknown	(1000–2000 mi.?) gradual merging with interplanetary gas	Elliptic paths and escape of neutral corpuscles

[1] M.F.P. = Mean Free Path of the molecules.

[2] Kinetic temperature is defined as the temperature a gas would have at standard pressure (1 atm) if its molecules had the same kinetic energy (i.e. translational velocity) as the rarified gas in the outer atmosphere.

NERAL CHARACTERISTICS TABLE 2–1 (Cont.)

Chemical haracteristics	Electrical Characteristics	Meteors	Biological Aspects	Technical Aspects
, O_2, 78% N_2 93% Ar, water apor, carbon oxide	Thunderstorms		40,000 ft.: ceiling for oxygen mask; 45,000 ft: ceiling for pressure breathing	Air-breathing engines
hanged; zone appears	None	Stratopause is end of trail of daylight visible meteors	45,000–80,000 ft: pressurized cabin; 80,000 ft (25 km): Max. O_3-concentration; 80,000 ft: sealed cabin required. *Physiological Space equivalence begins*	100,000 ft: approx. limit for air-breathing engines. Rocket propulsion remains sole means of controlled locomotion
nning of O_2-ssociation; pper region-frared night glow of OH; 50 km: 18% ; 82% N_2	D-layer (daytime) producing radio blackout upon flares in solar chromosphere	Upper region: beginning of disappearance of ordinary meteors	Space equivalence. Heating problem due to flight speed. Cosmic radiation	Slip flow region. Chemopause is about upper region for glide velocities of the order of 24,000 ft/sec
ompletely ssociated; dissocia- n	Atmosphere is electric: current sheets flowing 100–300 km: E, F_1, F_2-layers (daytime) E, F-layers (night time) Aurorae; Night air glow	100 km: upper ends of typical meteor trails	Space conditions	Free molecule flow region. Temporary satellites; thermal and electrical rocket propulsion
N_2 disso- tion. Re- n in which 2000 A is sorbed	Region of highest aurorae (sunlight aurorae). Beginning of lower Van Allen radiation belt	Penetrated by meteors	Space conditions. Insignificant absorption of electromagnetic radiation	Temporary to permanent satellites; departure orbits for lunar and interplanetary flights
nown	Ions will not escape but follow magnetic force lines Van Allen radiation belt	Space conditions	Space conditions. No longer any absorption of electromagnetic radiation	Permanent terrestrial satellites

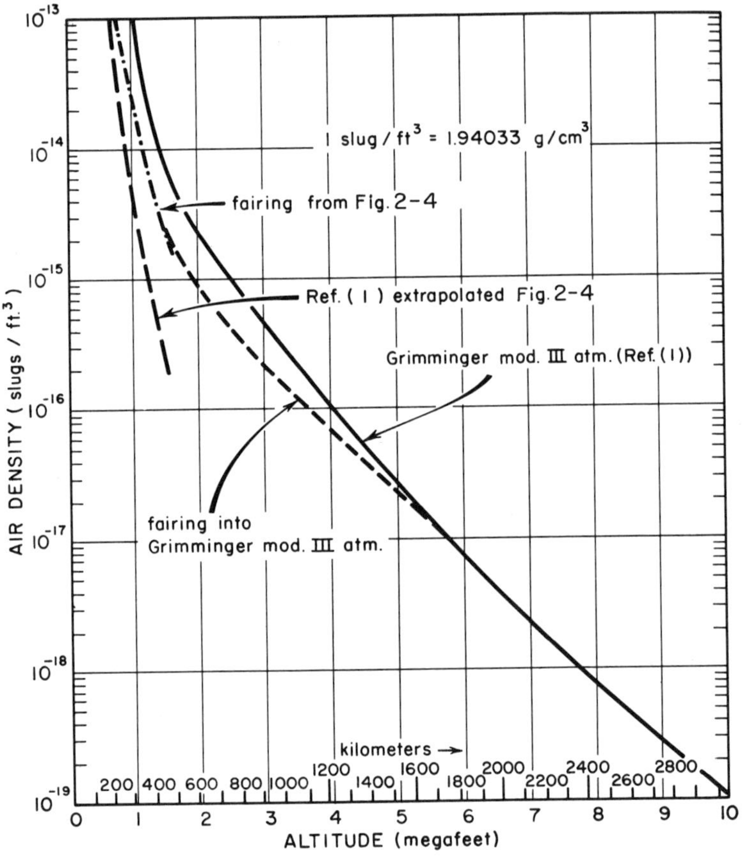

FIG. 2-5. Air Density at Terrestrial Satellite Altitude.

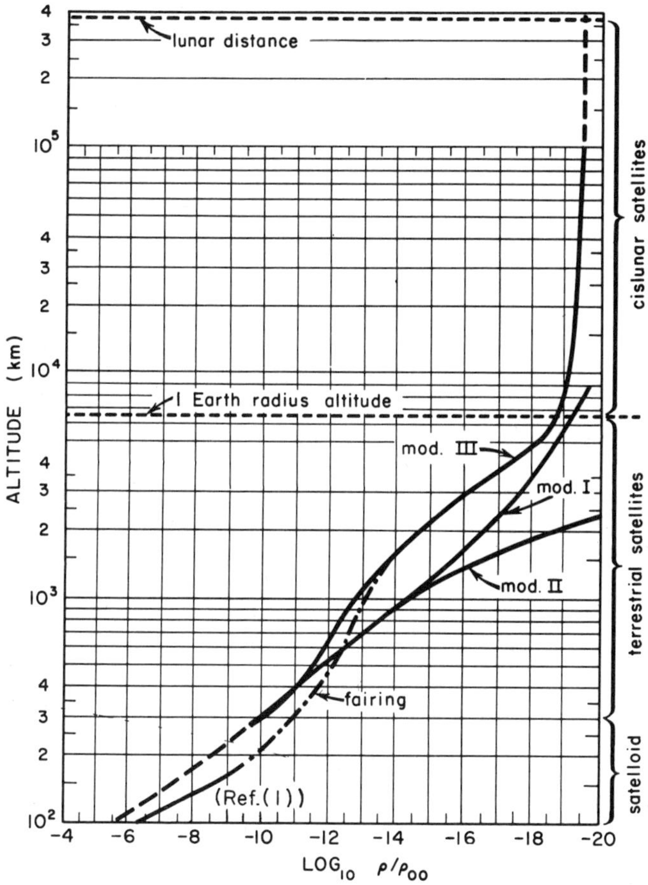

FIG. 2-6. Possible Variation of Density at Extreme Altitudes.

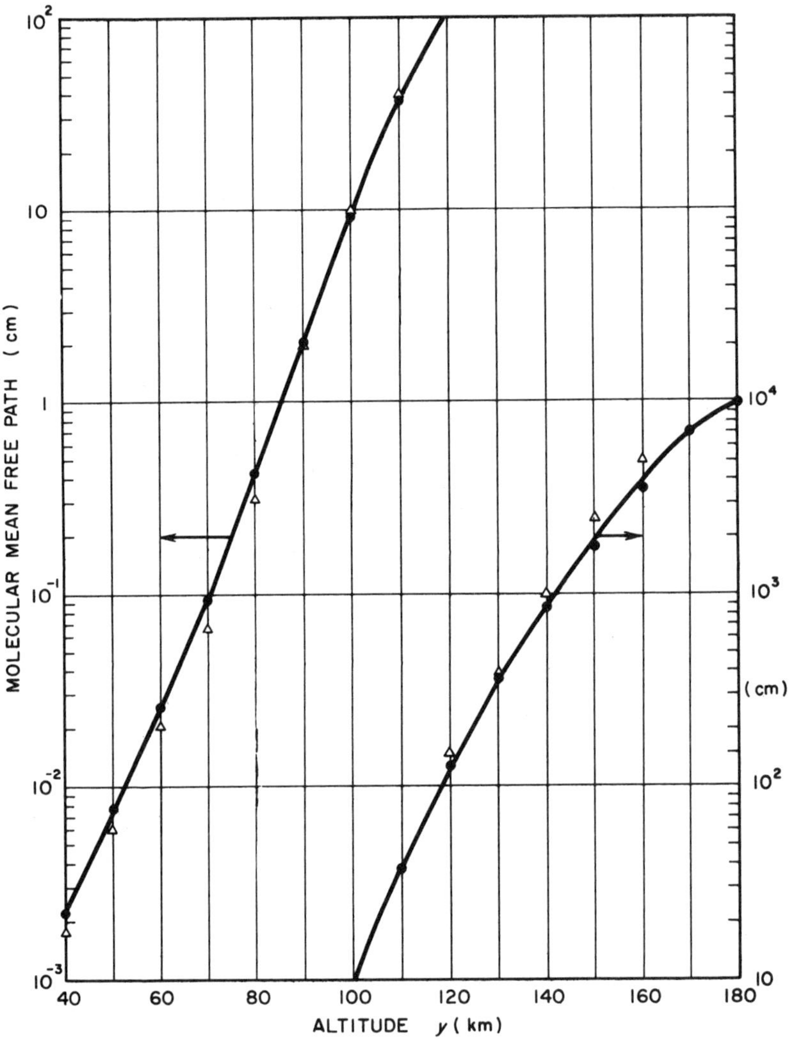

FIG. 2-7. Molecular Mean Free Paths as a Function of Altitude.

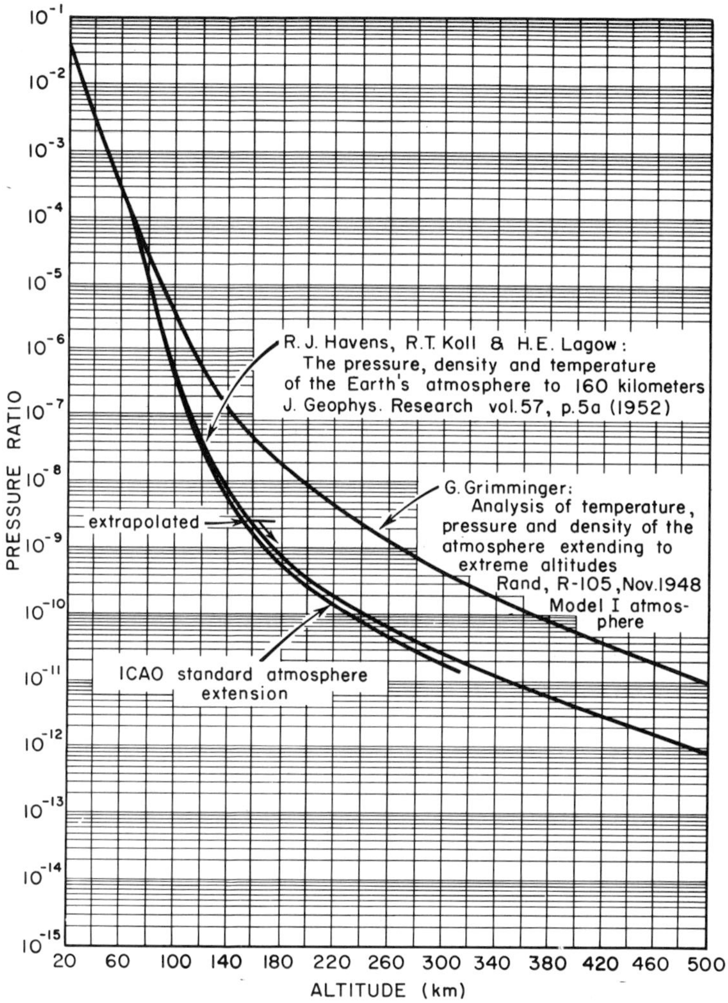

FIG. 2-8. Atmosphere Pressure Ratios as a Function of Altitude.

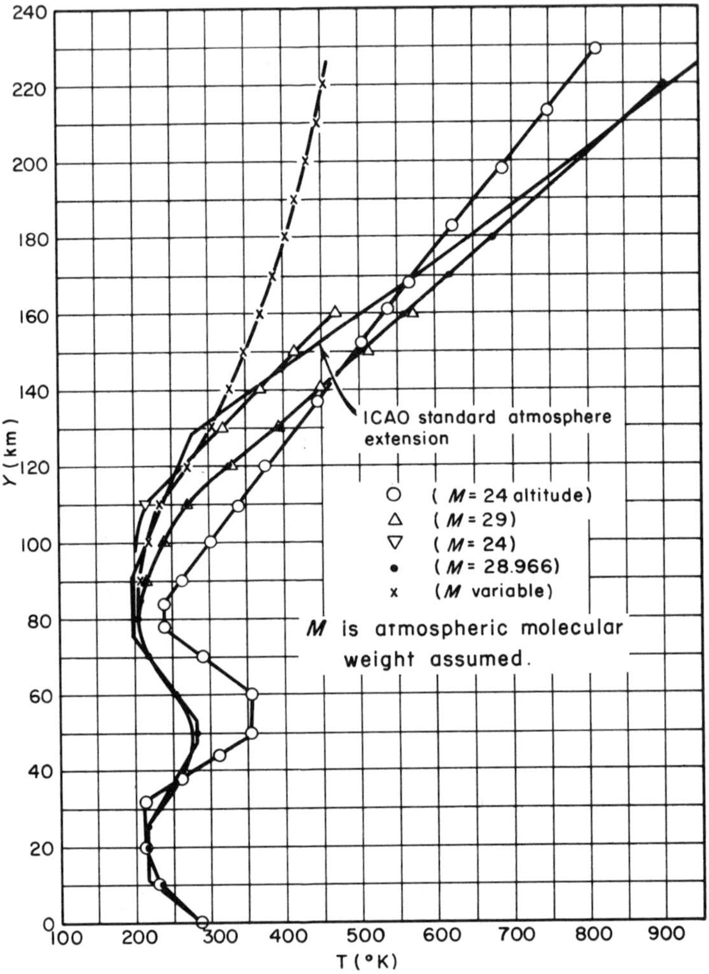

FIG. 2-9. Temperature Variation with Altitude.

that such a zone could exist, but that it was inaccessible unless some external disturbance temporarily distorted the Earth's magnetic field (e.g. solar flare action).

The inner belt is theorized to be predominantly composed of high-energy protons possessing energies in excess of 40 Mev (million electron volts). The outer belt is believed to be predominantly lower-energy electrons, 10–50 Kev (thousand electron volts). Thus the inner belt is known as the "hard" belt, and the outer as the "soft".

It is inevitable that space-craft must traverse through these belts. Satellites may be placed in orbit above, below, or even between, but exploration vehicles to the moon and beyond will generally pass through these belts because the energy requirements to maneuver through a polar exit and then into the ecliptic plane will likely be at least as prohibitive weight-wise as to shield passengers and instruments against effects of these particles. Of particular future interest to the electronic engineer will be the effect of high energy proton bombardment on structures such as antennas, including deformation and thermal noise effects, and on various circuit components and materials.

2.2.5 Atmospheric Characteristics

The Earth's atmosphere may be described as composed of concentric, oblate spheroid layers which are successively called troposphere, stratosphere, chemosphere, ionosphere, mesosphere, and exosphere. Table 2–1 provides general characteristic information on these layers. Physical Data, Relevant Equations, and Composition are provided in Tables 2–2, 2–3, and 2–4. Figures 2–4 through 2–12, show altitude functions of atmospheric density, molecular mean free path, pressure, temperature, wind velocity, and atmospheric constituents. Diurnal, seasonal and longer term periodic variations exist, but are usually considered as deviations from the mean values presented here as design reference material. Further details on these tables and figures and their derivation are available in Ref. (*1*) and the references contained therein.

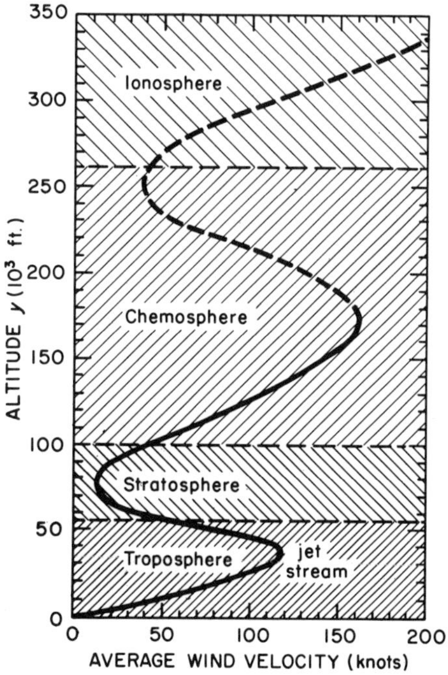

FIG. 2-10. Average Wind Velocities in the Atmosphere.

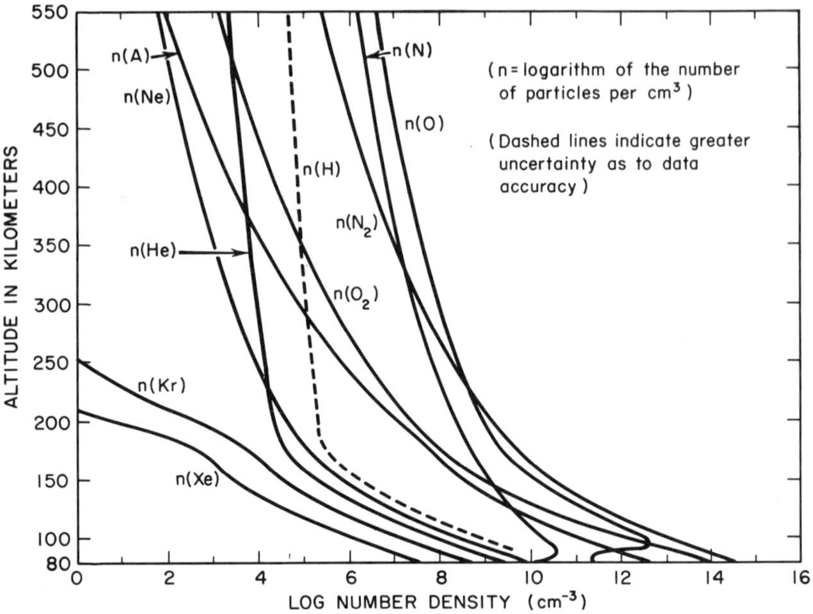

Fig. 2-11. Altitude Distribution of Atmospheric Constituents.

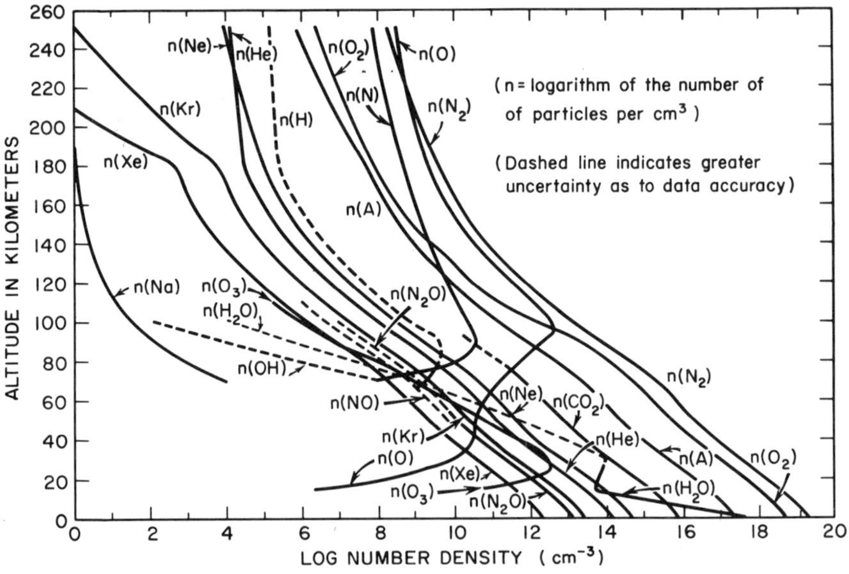

Fig. 2-12. Altitude Distribution of Atmospheric Constituents.

TABLE 2–2 THE EARTH'S ATMOSPHERE—PHYSICAL DATA

Normal temperature	0°C = 273.16°K = 32°F
Normal pressure	760 mm Hg
	29.921 in. Hg
	1013.246 millibar (mb)
	1.013 kg/cm^2
	14.7 psi
Normal density	0.0012928 g/cm^3
	0.00250790 $slug/ft^3$
Molecular weight	28.970
Specific heats	$c_p = 0.2403$ cal/g °K
	$c_v = 0.1715$ cal/g °K
Ratio of specific heats	$c_p/c_v = \gamma = 1.401$
Speed of sound at N.P.T.[1]	331 m/sec
	1086 ft/sec
Molecules per cm^3 (N.P.T.)[1]	$2.688 \cdot 10^{19}$
Molecular diameter	$3.46 \cdot 10^{-8}$ cm
Mean free path (N.P.T.)[1]	$6.98 \cdot 10^{-6}$ cm
Coefficient of viscosity	$1.72 \cdot 10^{-4}$ poise
Thermal conductivity	$5.6 \cdot 10^{-5}$ cal/cm sec °K
Scale height of atmosphere (height for pressure change by 10^{-1} with altitude)	$0.0293 \cdot T$ (°K) km
Height of homogeneous atmosphere (N.P.T.)[1]	7.991 km $\sim$ 800,000 cm
Mass per cm^2 of 1 atm-cm (= 1 cm N.P.T.)[1] of gas of molecular weight M	$4.461 \cdot 10^{-5} \cdot M$ g/cm^2

[1] N.P.T. = normal pressure and temperature, as defined above.

The absorptivity of the atmosphere over a wide spectrum of wavelengths (x-rays to short microwaves) is portrayed in Fig. 2–13. The absorption of microwave radio energy by oxygen and water vapor in a fully saturated sea level atmosphere is depicted as a function of distance in Fig. 2–14. These curves have been combined by Perlman *et al.* (Ref. (*11*)) to obtain a combined surface attenuation rate, Fig. 2–15. In combination with an equivalent atmosphere thickness curve for inclined paths, Fig. 2–16, Perlman *et al.* have also obtained Earth-to-Space absorption charts for oxygen and water vapor as functions of both frequency (or wavelength) and inclination of ray path, Fig. 2–17 and 2–18. Examination of these figures will indicate that one should stay below the K band of microwaves for Earth-to-Space communications if at all possible, but that useful "holes" do exist in the vicinity of 30 and 100 kmc if feasible antenna aperture size is tightly constrained. This is particularly significant when operating from high altitudes as indicated particularly by Fig. 2–17.

Detailed information on infra-red, ultraviolet, and visual wavelength absorption by atmospheric gases (and ions) is available in the literature (particularly Ref. (*12*)). These data confirm the absorption chart above, Fig. 2–13.

TABLE 2–3 THE EARTH'S ATMOSPHERE—RELEVANT EQUATIONS[1]

* Ideal gas equation, $pV = \frac{R_0}{M} T = \frac{p}{\rho}$

* Number of molecules per cm³ (Loschmidt number), $N_L = \frac{N_A}{V} = N_A \frac{p}{T} \frac{M}{R_0}$

* Mean free path of molecules, $\lambda = \frac{1}{N_L \pi \sigma^2 \sqrt{2}}$

* Molecular root-mean-square velocity, $\sqrt{\overline{V^2}} = \sqrt{\frac{3R_0T}{M}}$

*Arithmetic mean of all velocities in a gas, $\overline{V} = 0.92\sqrt{\overline{V^2}}$

* Speed of sound, $v_s = \sqrt{\frac{\gamma p}{\rho}} = \sqrt{\frac{\gamma R_0 T}{M}}$

* Hydrostatic pressure relation, $\log \frac{p}{p_1} = -\int_{y_1}^{y} \frac{1}{H}\, dy = \int_{y_1}^{y} \frac{Mg'}{R_0T}\, dy$

* Gas density, $\rho = \frac{N_L}{N_A} M$

Refractive index, $n = 1 + 2.876 \cdot 10^{-4} + 1.629 \cdot 10^{-6}\lambda^{-2} + 1.36 \cdot 10^{-8}\lambda^{-4}$

(λ = wavelength in 10^{-4} cm (μ))

Rayleigh scattering, $\sigma_s = 1.043 \frac{32\pi^3(n-1)^2}{3N_L\lambda^4}$ (λ in μ)

p = pressure

V = volume of 1 mole gas

R_0 = universal gas constant[2]

ρ = density

T = absolute gas temperature

σ = mean diameter of molecules making up the gas (cf. Table 3–7b)

M = mean molecular weight of gas mixture[2]

γ = c_p/c_v = ratio of specific weights

y = altitude

g' = apparent gravity

N_A = Avogadro's constant = $(6.02472 \pm 0.00036) \cdot 10^{23}$(g-mol)$^{-1}$

[1] For additional gas kinetic equations cf. *Space Flight*, Vol. III, *Space Operations*.

Equations marked by an * apply also to the atmospheres of other planets, taking their gas composition into account. Note that numerical values must be in consistent units.

[2] Cf. *Space Flight*, Vol. III, *Space Operations*.

Refraction indices of the atmosphere for radio waves may be computed (Ref. (*13*)) as,

$$(n_{t,p,f} - 1)\, 10^6 = 288 \frac{p}{760} \times \frac{1}{1 + 0.00366t} + \frac{6.7f}{(1 + 0.00366t)^2} \quad (2\text{–}12)$$

TABLE 2–4 COMPOSITION OF THE DRY ATMOSPHERE
(M = 28.93)

Gas	Wt. %	Vol. %	Thickness (cm N.T.P.)
Nitrogen	78.09	75.51	624,600
Oxygen	20.95	23.15	167,600
Argon	0.93	1.28	7440
Carbon dioxide	3.10^{-2}	$4.6 \cdot 10^{-2}$	220
Neon	$1.8 \cdot 10^{-3}$	$1.25 \cdot 10^{-3}$	14
Helium	$5.2 \cdot 10^{-4}$	$7.2 \cdot 6^{-5}$	4.2
Methane	$1.5 \cdot 10^{-4}$	$9.4 \cdot 10^{-5}$	1.2
Krypton	$1 \cdot 10^{-4}$	$2.9 \cdot 10^{-4}$	0.8
Nitrous oxide	$5 \cdot 10^{-5}$	$8 \cdot 10^{-5}$	0.4
Hydrogen	$5 \cdot 10^{-5}$	$3.5 \cdot 10^{-6}$	0.4
Ozone	$4 \cdot 10^{-5}$	$7 \cdot 10^{-6}$	0.3
Xenon	$8 \cdot 10^{-6}$	$3.6 \cdot 10^{-5}$	$6 \cdot 10^{-2}$
Nitrogen dioxide	$1 \cdot 10^{-7}$	$2 \cdot 10^{-7}$	$8 \cdot 10^{-4}$

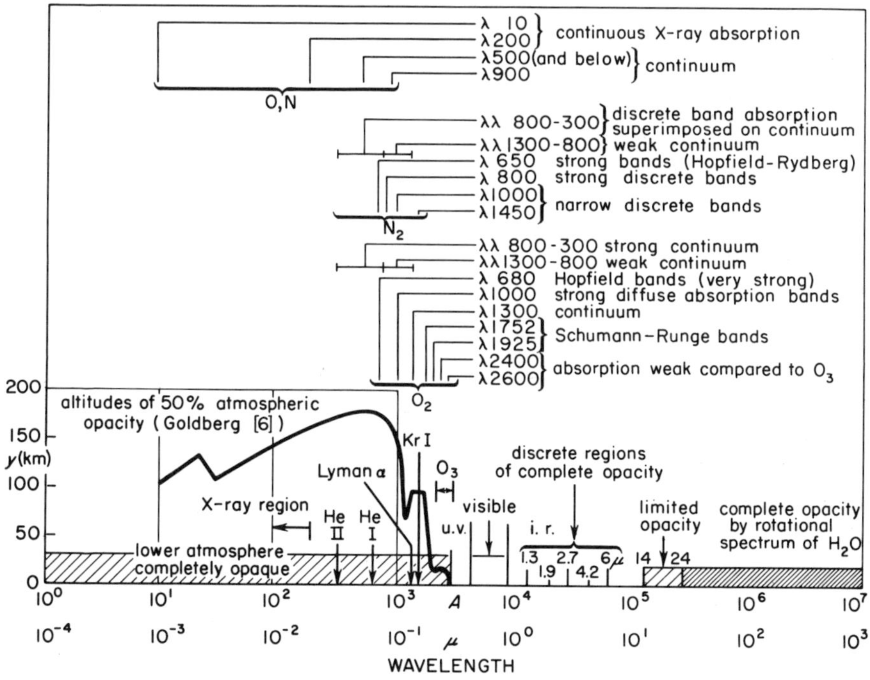

FIG. 2-13. Absorptivity of the Atmosphere.

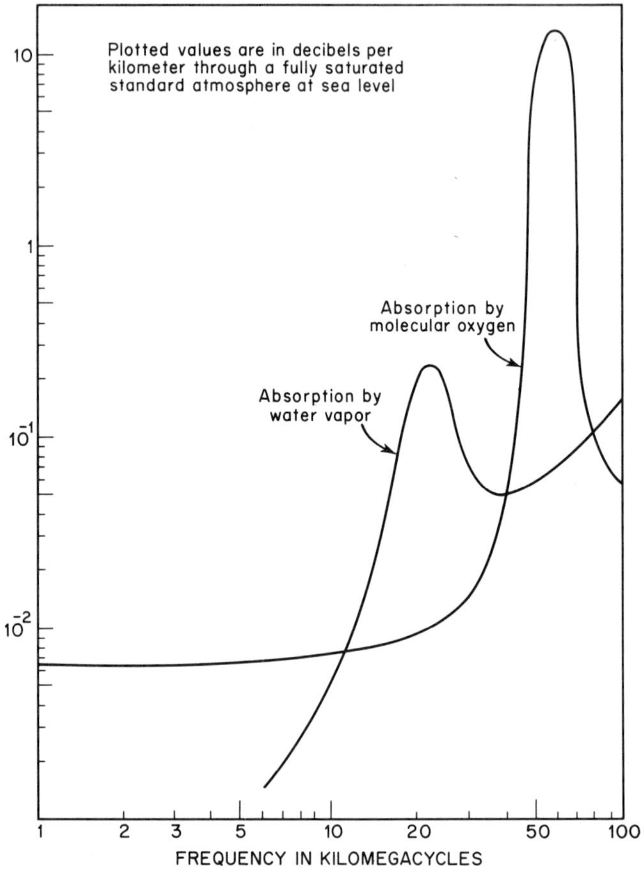

FIG. 2-14. Absorption Rates for Microwave Radio Energy by Oxygen and Water Vapor in Air.

where n is the refraction index as a function of temperature, t, in degrees centrigrade; air pressure, p, in mm Hg; and water vapor pressure, f, in mm Hg. Note the absence of wavelength in radio-wave refraction, but the dependence on temperature, water-vapor and air pressure which indicates the refractive index will vary as a function of altitude and time.

The ionosphere layers, D, E, F_1 and F_2, are concentric shells that exhibit pronounced effects on radio propagation. These shells are generally believed to be caused by the ionization of constituent gases by extraterrestrial (primarily solar) radiation. Table 2–5 defines these layers by altitude of maximum electron concentration. The D-layer ionization is believed to result from O_2 and NO absorption at $\lambda < 1019\text{Å}$ and $\lambda < 1300\text{Å}$ respectively, the NO absorption occurring just below the ionosphere in the chemosphere. D-layer effects have been observed to be strongly

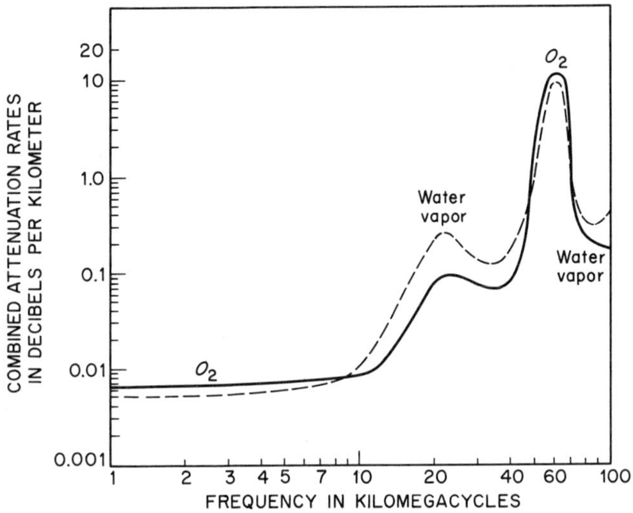

FIG. 2-15. Combined Surface Attentuation Rate for Microwave Radio Energy by Oxygen and Water Vapor.

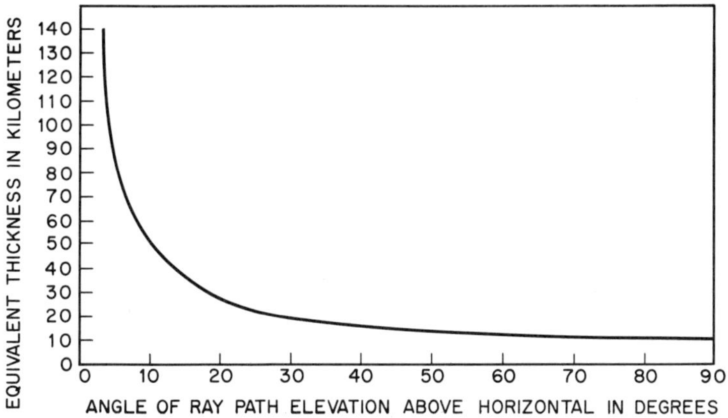

FIG. 2-16. Equivalent Atmosphere Thickness for Inclined Ray Paths.

responsive to solar flare occurrences which produce intense ultraviolet (predominantly Lyman α) radiation. E-layer ionization is believed to result from O_2 absorption between λ910Å and λ1000Å. The electron density jump provided by the E-layer has provided a particularly useful reflection shell for radio waves below 3 mcs. F_1-layer ionization is believed to result from ionization of atomic oxygen by solar light quanta of $\lambda < 910$Å. The F_1-layer has reflected vertical incidence microwaves of 5–10 mcs. F_2-layer ionization results from ionization of both atomic oxygen and nitrogen by solar light quanta of $\lambda < 850$Å. The ionosphere electron density profile is depicted (as best known currently in Ref. (*14*))

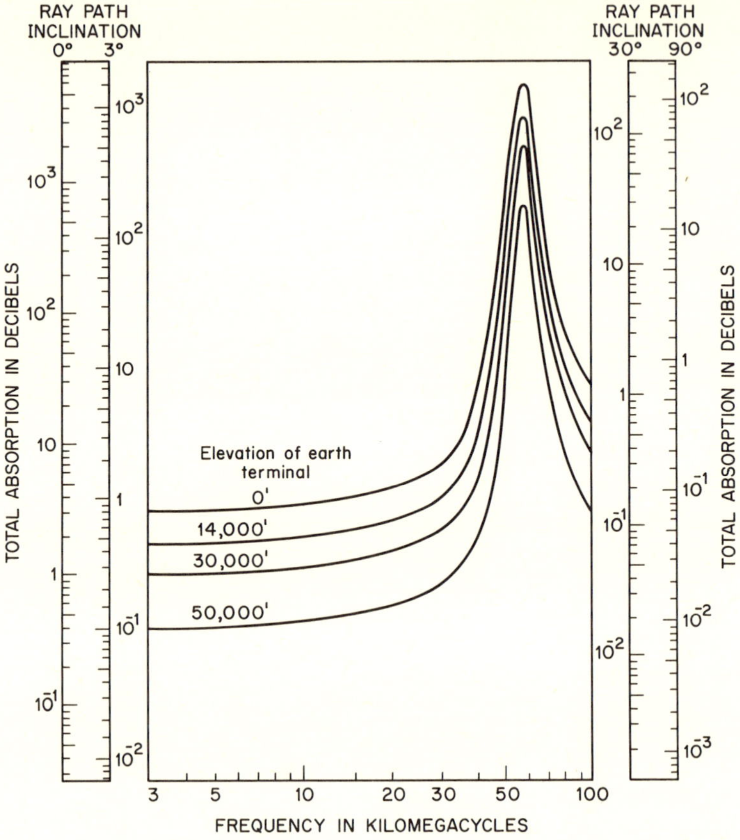

FIG. 2-17. Earth-to-Space Absorption by Oxygen.

in Fig. 2–19. Attempts are being made to ascertain the nature of the ionosphere by radio "soundings" of its top side using rocket and satellite vehicles (Ref. (*15*)), and by transmitting a group of harmonically related frequencies through it (Ref. (*14*), (*16*)). Effects to be measured will include: (1) Faraday effect *m* the rotation of the plane of polarization which is dependent on frequency, magnetic field strength, direction of propagation and columnar electron density (total number of electrons along the path), (2) Doppler effect *m* the differential Doppler resultant from passing harmonically related waves through an electron plasma, and determining the electron effect on the wavelengths (a variable with wavelength), (3) Angle of Arrival *m* a measure of the refractive effect of the electrons in the ionosphere. Diurnal, annual and sunspot variations in the ionosphere are considerable, and significant data are available (Ref. (*17*), (*18*), (*19*), (*20*), (*21*)). Ionospheric interactions with inter-

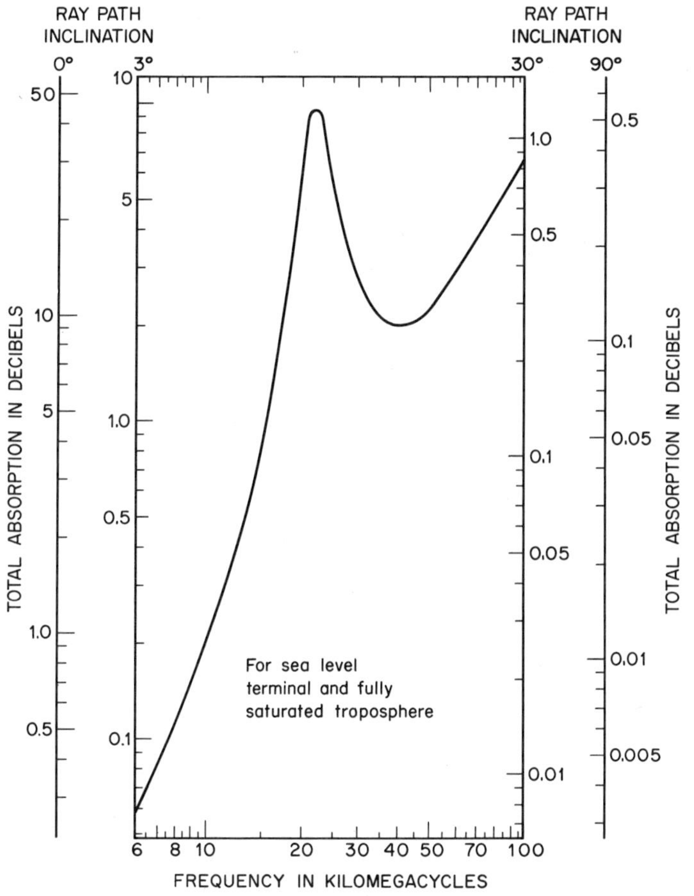

Fig. 2-18. Earth-to-Space Absorption by Water Vapor.

planetary gas, thermal radiation and convection, geomagnetically bound currents, solar and lunar tidal effects are all believed to exist (Ref. (*22*)), but none have been explored sufficiently well to be predictable on a theoretical basis.

The aurora and air glow phenomena are visible night displays whose occurrences are associated with solar eruptions. These phenomena occur at about 100 to 1000 km altitude, and are believed to result from protons travelling from the sun (perhaps aided by magnetic field focusing). Basic data are needed to understand these phenomena better (Ref. (*1*), (*23*), (*24*)). Chapman (Ref. (*25*)) has depicted the different altitude bands for Aurora discovered by Störmer, the Norwegian mathematician and naturalist, as is shown in Fig. 2–20. This figure shows that pre-sunrise and post-sunset

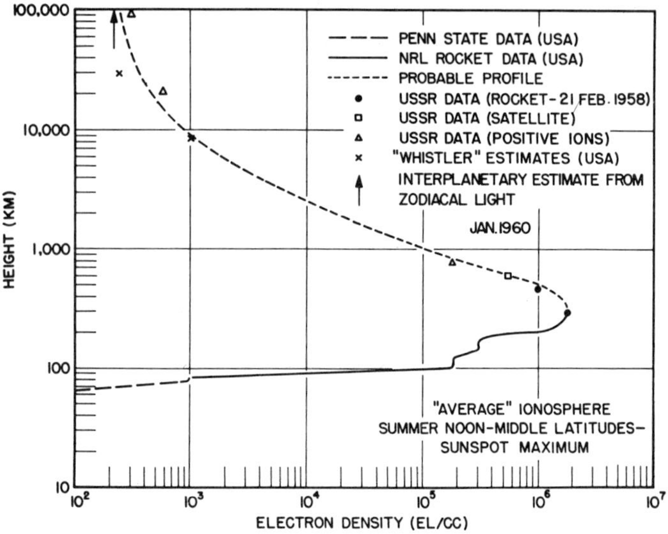

FIG. 2-19. Ionosphere Electron Density Profile.

TABLE 2–5 THE EARTH'S ATMOSPHERE—IONOSPHERE LAYERS

Ionospheric Layer	D	E	F_1	F_2
Altitude of maximum e^--concentration (km)	70	120	170	250
Particles per cm³	$2\cdot10^{15}$	$2\cdot10^{12}$	$2\cdot10^{10}$	10^9
Atmospheric composition (in order of abundance)	N_2, O_2	N_2, O_2, O	N_2, O, N	O, N_2, N
For Sun in zenith:				
e^-_{max} /cm³ at no sunspots	300?	$1.34\cdot10^5$	$2.4\cdot10^5$	$5.9\cdot10^5$
e^-_{max} /cm³ at 100 sunspots	?	$1.88\cdot10^5$	$3.59\cdot10^5$	$1.77\cdot10^6$
Rate of e^--production per cm² per sec for Sun in zenith:				
at zero sunspots	?	$5\cdot10^8$	$1.9\cdot10^9$	$2.3\cdot10^9$
at 100 sunspots	?	10^9	$4.2\cdot10^9$	$6.9\cdot10^9$

Height of base of exosphere = 500 km.
Particles per cm³ at base of exosphere = 10^7.

auroras may be observed which actually occur in sunlit atmosphere but are observable from territory in the Earth's night shadow. Note that auroras occurring in sunlight are generally at 300 to 1000 km altitude, while those occurring at night are generally at 100 to 300 km altitude. The

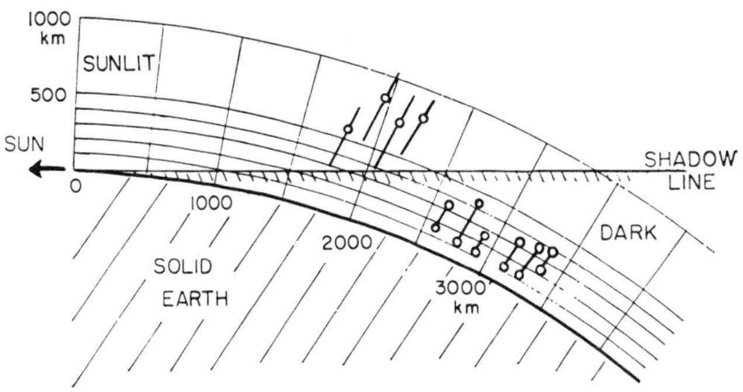

FIG. 2-20. Heights of Typical Auroras.

frequency of occurrence of the auroras has been charted as isochasm lines which clearly orient aurora occurrence frequency with the North and South magnetic poles. Current theories indicate that solar radiations, consisting predominantly of hydrogen protons and electrons, are deflected by the Earth's magnetic field so as to impinge on the atmosphere in the high magnetic latitudes. Particles approaching at equatorial latitudes are deflected at about 10 Earth radii by the magnetic field, protons to the West and electrons to the East. Some particles continue on past the Earth while some become trapped permanently in what are now known as the Van Allen belts. Thus the same solar radiation which results in Van Allen belts (charged particle regions) at equatorial latitudes is believed to provide the protons which penetrate to the atmosphere along the magnetic field lines in the high latitudes to produce aurora and air glow. Maximum aurora occurrence is in zones 20° to 25° from the geomagnetic axis (geomagnetic latitude of 65 to 70 degrees). Radio noise at medium and high frequencies is generated with the auroras, and VHF and UHF radio waves experience reflection and scattering up to 144 mc (Ref. (*26*)), thus the auroras represent an obstacle to both visual and radio frequency communications. The best known data associated with auroral phenomena (Ref. (*27*)) are presented in Table 2–6.

2.2.6 Ionospheric Weather and Electrical Noise

Weather comprises those dynamic, transient, and turbulent effects experienced in the space environment in an essentially unpredictable fashion. That is, we can predict diurnal and season variations in temperature, seasonal variations in precipitation, etc., but we cannot predict much in advance the specific path and magnitude of a storm. It is of course hoped, as discussed in Chapter 1, that satellite geophysical observations will provide an easily interpretable gross picture of atmospheric weather. Figure 2–10, page 77, has indicated average wind velocities in

TABLE 2–6. AURORAL PHENOMENA

Solar streams and clouds

	Protons	Electrons
velocity	up to 2×10^8 cm/sec	up to 2×10^8 cm/sec
energy	up to 20 Kev[a]	$1 \approx 2$ ev[b]
number density[c]	$100/cm^3$	$100/cm^3$
flux	up to $2 \times 10^{10}/cm^2$ sec	up to $2 \times 10^{10}/cm^2$ sec

DCF field

	Location of the stream front[d]			
	4a	5a	6a	∞
intensity[e]				
at the front of the stream	500 γ	260 γ	150 γ	0
at Earth's surface				
dayside	93 γ	44 γ	24 γ	0
nightside	44 γ	24 γ	15 γ	0

DR field

Intensity at Earth's surface: up to 400 γ. (The induction effect is subtracted.)

Intensity of the *DR* field at Earth's surface for the production of a neutral line at 6 Earth radii must be about 20 γ.

Trapped particles: illustrative data at 6 Earth radii

	Protons	Electrons
energy	500 kev	30 kev
speed	10^9 cm/sec	10^{10} cm/sec
maximum E-W[f] drift speed	W 250 km/sec	E 16 km/sec
oscillation time[f,g]	23 sec	2 sec
time of revolution round the Earth[f]	24 min	6 hr
Oscillations per revolution[f]	63	10,100

Radiation belts and particles

	Protons	Electrons
	Observed	
	V1, inner belt, at 1.5*a*	
energy	40 mev	20 kev
flux	$2 \times 10^4/cm^2$ sec	$2 \times 10^9/cm^2$ sec
	V2 belt, at 3.5*a*	
energy	60 mev	20 kev
flux	$10^2/cm^2$ sec	$10^{11}/cm^2$ sec
	Inferred	
	V3 belt, ring current belt, beyond 6*a*, in outer part of V2	
energy	500 kev	20 kev
flux	(?)	$10^9/cm^2$ sec

Auroral zone

	Location	Corresponding location in the equatorial plane connected by lines of force
center of the auroral zone[h]	66°30′ *gm* lat	6.3*a*
low latitude aurora[i]	down to about 55°33′	3.1*a*

Auroral particles in the auroral zone

	Protons	Electrons
diffuse form		
energy	150 kev	39 kev
flux[j]	$5 \times 10^5/cm^2$ sec	$10^7/cm^2$ sec
active form		
energy	150 kev	30 kev ∼ 250 kev
flux[j]	$5 \times 10^3/cm^2$ sec	$10^{11}/cm^2$ sec
lines and bands produced	Hα	5577 (OI), 3914 (N_2^+), etc.

TABLE 2–6 (Cont.)

Auroral particles in the equatorial plane; at 6 Earth radii

	Protons (100 kev)	Electrons (30 kev)
radius of gyration	180 km	1.7 km
period of gyration	0.26 sec	1.5×10^{-4} sec

Auroral curtains

length, at least 2000 km
thickness

	Proton layer (Hα layer)	Electron layer
diffuse form	30 km	3 km
	(700 km)[k]	(70 km)
active form	30 km (?)	250 m
	(often invisible)	(6 km)

irregularities

	Characteristic length	Cause
large scale irregularity (folded structure)	$10 \sim 500$ km	instability of the neutral line
small scale irregularity (ray structure)	$1 \sim 4$ km	instability of the thin electron beam-sheet

lifetime

diffuse form	$1 \sim 5$ days (intense display)	
active form	$1 \sim 2$ hrs (intermittent)	

Auroral ionosphere: Auroral electrojet

auroral electrojet

	Westward jet	Eastward jet
sense of the emf	Equatorward	poleward
emf	up to 60 v/km	20 v/km
current intensity	up to 3×10^6 amp	$<10^6$ amp
magnetic disturbing force at ground level	up to 1500 γ	$< 500\ \gamma$
electron density in the aurora		
diffuse aurora	up to $5 \times 10^5/cm^3$	
activity form	up to $7 \times 10^7/cm^3$	

conductivity of the ionosphere
up to 1.2×10^{13} esu

emf along the neutral line
10^{-5} v/cm at about 6 Earth radii.

---

[a] On the basis of geomagnetic evidence, we think that the stream may at times contain protons of energy 500 kev or more confined by transported magnetic fields, perhaps tangled.

[b] Arnold et al. (Ref. (*27*)) observed a flux $10^7/cm^2$ sec of 50 kev electrons in a solar stream.

[c] Possibly this may attain $1000/cm^3$.

[d] The distance from Earth's center; unit: $a = 6380$ km.

[e] Calculated using the image dipole approximation.

[f] The ratio of velocity along to that perpendicular to the lines of force in the equatorial plane is assumed to be about 30.

[g] Time for one complete oscillation between the two mirror points.

[h] The auroral zone is located between 65 and 70 deg *gm* latitude. More than 5 auroral curtains are often seen simultaneously in the belt.

[i] The curtain structure of auroras often appears there during great magnetic storms.

[j] Flux in the aurora: to obtain the corresponding flux at 6 Earth radii in the equatorial plane multiply the given number by 2.5×10^{-2}.

[k] The material in parentheses shows the corresponding thickness in the equatorial plane.

the atmosphere. The "jet stream" occurs at about 40,000 feet altitude in the upper troposphere. This air current is known to reach velocities of 250 miles per hour, and follows a winding path around the Earth travelling from west to east. Portions of this stream become detached, forming whirlpools which become intense storms in the lower troposphere. Very little is known about "weather" conditions in the outer atmospheric layers except that magnetic storms are experienced at times of intense solar activity, and that "whistlers" are generated in the ionosphere by streams of highly ionized particles precipitating into the dense electron layers in the presence of the Earth's magnetic field (Ref. (*28*) and (*29*)). These whistlers are in addition to those generated by lightning discharges in the troposphere.

Whistlers propagate as VLF signals along the Earth's magnetic field lines, so that a whistler heard at a particular latitude in the Northern hemisphere will also be heard at the same geomagnetic latitude in the Southern hemisphere on the same geomagnetic longitude but with a slight time delay. Figure 2–21 illustrates a typical propagation path between observing stations at Annapolis and Cape Horn. Figures 2–22 and 2–23 illustrate the type of whistler echo trains that might be heard at opposite ends of a magnetic field line when the causative impulse is a lightning bolt at one end. Note that at the causative pulse end, 2-hop, 4-hop and 6-hop echoes may be heard (called a "long whistler"), while at the opposite end, 1-hop, 3-hop, 5-hop, and 7-hop "short whistler" echoes may be heard. In addition, the longer wavelength components of the whistler are delayed more with each hop. This wavelength dependent delay has been interpreted to indicate a peak electron density in the exosphere of 500/cc by Story in 1953 (Ref. (*29*)).

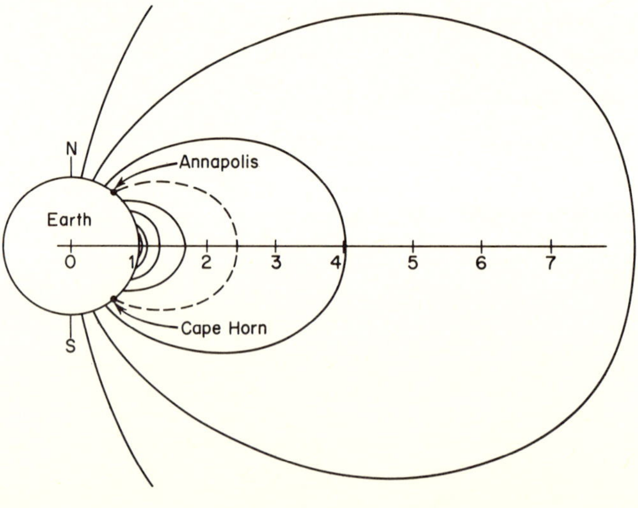

FIG. 2-21. Whistler Propagation Path Between Annapolis and Cape Horn.

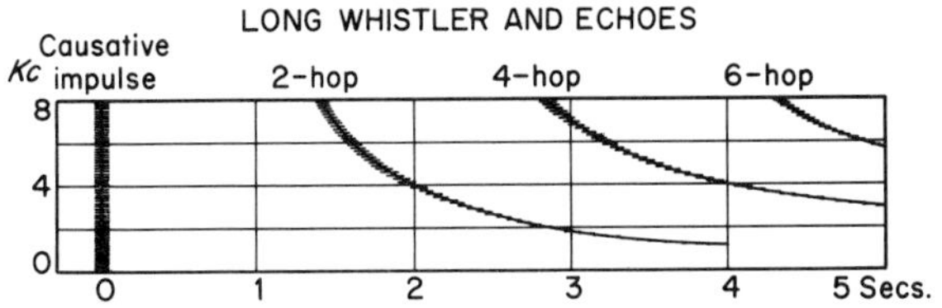

FIG. 2-22. Idealized Whistler Echo Train at Causative End of a Magnetic Field Line.

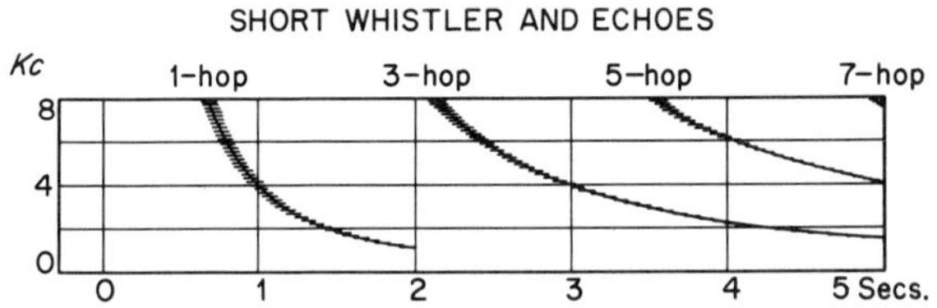

FIG. 2-23. Idealized Whistler Echo Train at Opposite End of a Magnetic Line Field.

The Appleton-Hartree formula for the square of the refractive index assuming zero losses is

$$n^2 = 1 + \frac{f_p^{\,2}}{f(f_H - f)} \tag{2–13}$$

where f_p = plasma frequency $= 10^4\sqrt{\dfrac{N}{1.24}}\ cps$,

N = number of electrons per cm³,

f = wave frequency in cps,

f_H = electronic gyro-frequency $= 2.8 \times 10^6\ B$ cps, and

B = flux density of earth's field in gauss.

Group velocity is related to refractive index by

$$V_g = c\left(n + f\frac{dn}{df}\right)^{-1} \tag{2–14}$$

These two equations have been solved and simplified to yield (Ref. (*29*)):

$$V_g = 2c\frac{f^{1/2}\,(f_H - f)^{3/2}}{f_H\, f_p} \tag{2–15}$$

which reaches a maximum at

$$f = f_H/4 \tag{2–16}$$

and yields $V_g = 0$ at $f = 0$ and $f = f_H$.

Since the minimum electron gyro frequency determines the upper "cut-off frequency for a whistler, and this frequency is dependent on magnetic field strength, we may plot upper cut-off frequency as a function of geomagnetic latitude, Fig. 2–24. This figure shows the upper frequency which may be observed at any geomagnetic latitude. For low latitudes and frequencies below 10 kc, we may take $f \ll f_H$ which yields

$$V_g = 2c \frac{\sqrt{f(f_H)}}{f_p} \tag{2–17}$$

which has been used by some research workers (Ref. (*29*)).

Using this expression we may integrate over the path length to obtain the one-hop time difference

$$T = \int_{\text{path}} \frac{ds}{V_g} = \frac{1}{2c} \int_{\text{path}} \frac{f_p}{\sqrt{f(f_H)}}\, ds = \frac{D}{\sqrt{f}} \tag{2–18}$$

where $D = \dfrac{1}{2c} \displaystyle\int_{\text{path}} \frac{f_p}{\sqrt{f_H}}\, ds.$

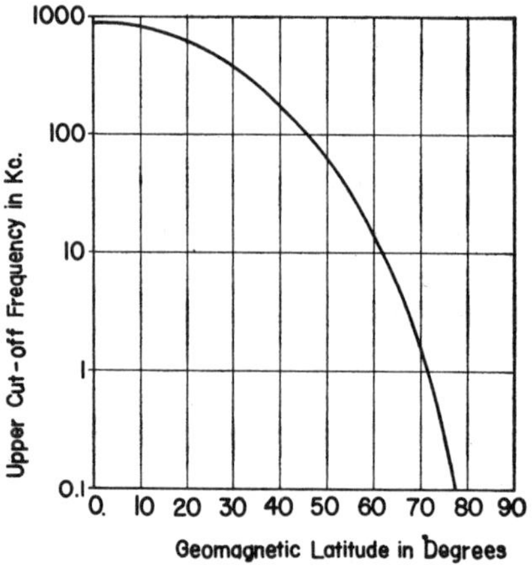

FIG. 2-24. Upper Cut-off Frequency as a Function of Geomagnetic Latitude.

This relation is being employed on an iterative and frequency basis (Ref. (*29*)) in an attempt to determine the actual propagation path, gyro-frequency variation, electron distribution and density over the path length. The relation is particularly sensitive to regions near the top of the path; thus, low latitude whistlers may be employed to explore the ionosphere F_2 region.

Whistlers generated in the exosphere are called "nose" whistlers. That is, the impinging stream of protons interacts with the electron cloud to generate frequencies on both sides of the $f = f_H/4$ maximum value, and two frequencies are received simultaneously at a ground station yielding a nose shape.

Diurnal variations in whistler occurrence are very pronounced. Studies have shown peak occurrences of as high as nine per minute at some listening stations during the hours around sunrise and sunset with a minimum of zero at noon being typical, and only slight dips near midnight. Day-today variations are considerable, as well, and are quite likely associated with terrestrial and solar storm activity. Gallet (Ref. (*30*)) has classified VLF noises further as indicated in Tables 2–7 and 2–8, and has deduced that the "noises" not associated with lightning discharges are excited in the exosphere by streams and bunches of high-speed ionized particles

TABLE 2–7. SYSTEMATIC CLASSIFICATION OF OBSERVED VLF NOISES

2 Large Groups:

I. *WHISTLERS*—Cause: Lightning Discharges.
All frequencies emitted at once; shape due to dispersion along the path.

II. *VLF EMISSIONS*—Not caused by lightning.
Strongly associated to magnetic perturbations.

TWO PRINCIPAL TYPES:

1. *Continuous* in both time and frequency.
Steady state situation. *Hiss.*
2. *Discrete*, but often with repetition tendency.
Transient situation. Many classes recognized.

A third group more complex and rarer exists:

III. Interactions between Whistlers and VLF emissions.
The interaction involves either the continuous or the discrete VLF emissions.

precipitating into the ionized gases and electrons of the exosphere in the presence of the Earth's magnetic field. The excitation mechanism is considered to be similar to traveling wave tube operation. The frequencies generated are dependent on local electron density in the exosphere, N; local magnetic field intensity, B; and particle velocity, V; and may be expressed,

$$f = \frac{f_H}{2}\left[1 \pm \left\{1 \pm \left(\frac{f_p}{f_H}\frac{V}{c}\right)^2\right\}^{1/2}\right] \tag{2–19}$$

For middle latitudes the lower frequency solution of this expression simplifies to

$$f_1 = \left(\frac{V}{c}\right)^2 \frac{f_p^2}{f_H} \tag{2–20}$$

which explains many types of "noise" assuming a particle velocity, V, of the order of 10,000 km per second. The second frequencv reduces to (given the same assumptions):

$$f_2 = f_H\left[1 - \left(\frac{V}{c}\frac{f_p}{f_H}\right)^2\right] \tag{2–21}$$

which becomes very large.

By postulating an exosphere model such that N/B stays constant along a magnetic line of force (note that B varies approximately as r^{-3}, where r is the distance from geomagnetic center), Gallet explained the various

TABLE 2–8

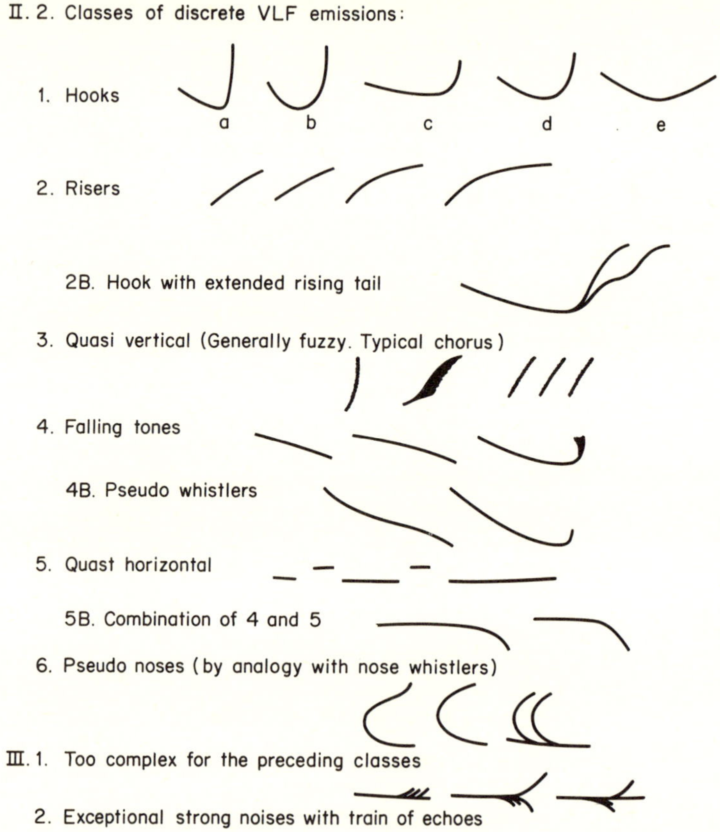

"noises" classified in Table 2–8 and determined an electron density curve, Fig. 2–25. By assuming various particle velocity magnitudes, Gallet has also shown that a particle velocity of the order of 10,000 km per second is necessary to explain the various "noise shapes" recorded at terrestrial observation stations. These velocities are an order of magnitude higher than solar-earth transit time deductions of particle velocities (as in solar flare impingement-time studies); however, these velocities agree with the velocity requirements for protons to penetrate to 100 km altitudes in the generation of auroras. It may be that 10,000 km per second velocities result from hydromagnetic interaction between the geomagnetic field and the large interplanetary solar clouds with velocities of 1000 km per second magnitude. The fast clouds postulated possess particle kinetic energies between one and two orders of magnitude less than current estimates of electron and proton energies in the Van Allen belts.

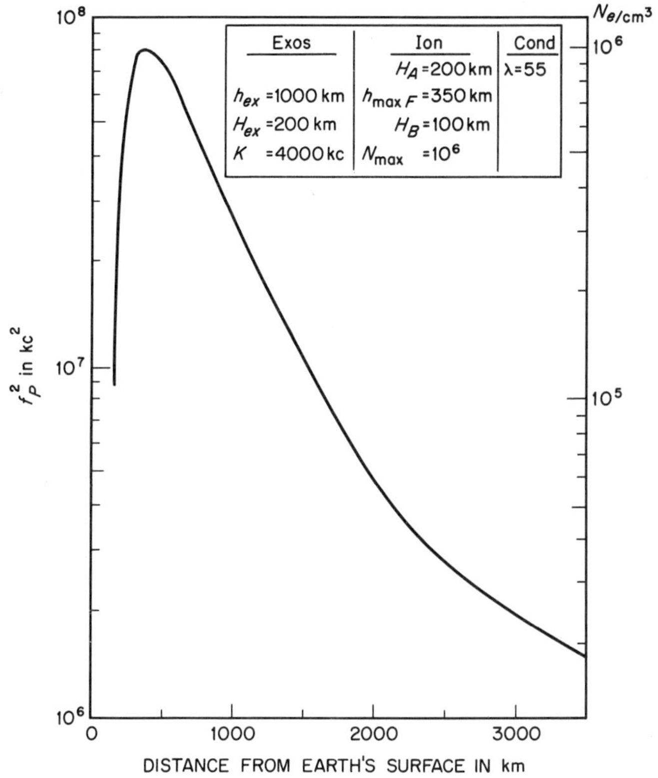

FIG. 2-25. VLF Emission Theory Model Electron Density Distribution.

Considerable effort is being expended at this writing in attempts to "sound" the ionosphere from above using probes and satellites, and by assessing the Faraday (polarization rotation), Doppler, and refraction effects on frequency spectrums transmitted through the ionosphere. Such data should eventually provide even more than the mean values and diurnal (and other) variations in ionosphere columnar electron density, magnetic field flux, and solar radiation flux. Continuous observations should permit an assessment of the true nature of ionosphere-exosphere "weather" through measurement of cause and effect relations, and the establishment of "weather" predictions with their impact on space communications and travel.

2.2.7 Meteoric Activity and Dust

Direct measurements are being attempted of meteoric materials and dust in the various outer atmospheric belts. Heavy "dust" bands have been reported (Ref. (*31*)), but no quantitative data are available as yet.

O'Keefe (Ref. (*32*) and (*33*)) has studied the extremely dense meteor showers of Feb. 9, 1913, and related these to the tektite meteors located in an orbit plane related band. Speculations on the tektites (glassy silica meteors with low refraction indices), has indicated that each shower produces a particular group of tektites, and that each shower results from the collision of a parent body which has been orbiting (established from entry velocity) with high eccentricity about the Earth and finally penetrates the atmosphere. The true density of such natural satellites has not been established.

Meteoric activity, defined as particle impingement from sources beyond the Earth gravitational field predominance, is another phenomena which has yet to be measured precisely. Nazarova (Ref. (*34*)) has reported on attempts to measure numbers of meteoric impacts on a terrestrial satellite. Impact density varied from zero to eleven per square meter per second with instruments sensitive to particles as small as 8×10^{-9} gm with 40 km per second velocity. Daily and altitude variations are not well known; at least they are not predictable with any certainty, and much further study and measurement will be needed.

Meteor penetration of the atmosphere has produced some useful estimates of meteoric flux. Generally accepted estimates based on visual magnitude have been tabulated (Ref. (*35*)) and are given in Table 2–9. The origin of meteors is unknown. Correlation has been established between some meteoric showers and some comet occurrences, but other comets pass by without corresponding showers, and other showers occur without corresponding comet passings. Studies of meteor trajectories indicate they travel in elliptic orbits; this sets an upper limit on their velocity of the parabolic escape velocity required to leave the solar system. About 90 per cent of meteors are stony with a density of 6.6 slugs/ft^3 while the remaining meteors are iron and nickel with a density of approximately 15 slugs/ft^3. Ehricke (Ref. (*1*)) has tabulated orbital data for meteor streams which is reproduced here as Table 2–10.

Table 2–9 Meteoric flux

Visual Magnitude	Adjusted Number per Day	Mass (slug)	Kinetic Energy (ft lb)	Average Flux over the Hemisphere (ft^{-2} sec^{-1})
0	4.5×10^{5}	8.1×10^{-5}	7.1×10^{5}	9.5×10^{-16}
5	4.5×10^{7}	8.1×10^{-7}	7.1×10^{3}	9.5×10^{-14}
10	4.5×10^{9}	8.1×10^{-9}	7.1×10^{1}	9.5×10^{-12}
15	4.5×10^{11}	8.1×10^{-11}	7.1×10^{-1}	9.5×10^{-10}
20	4.5×10^{13}	8.1×10^{-13}	7.1×10^{-3}	9.5×10^{-8}
25	4.5×10^{15}	8.1×10^{-15}	7.1×10^{-5}	9.5×10^{-6}
30	4.5×10^{17}	8.1×10^{-17}	7.1×10^{-7}	9.5×10^{-4}

METEOR STREAMS—ORBITAL DATA[1]

Name	Maximum	Geocentric Velocity (km/sec)	Semi-Major axis, a (A.U.)	Eccentricity, e	Perihelion, R_P (A.U.)	Aphelion, R_A (A.U.)	Longitude of Ascending Node ☊ (deg)	Longitude of Perihelion ω (deg)	Orbital Inclination to Ecliptic, i	Associated Comet
Quadrantids	Jan. 3	46	$\rightarrow \infty$	$\rightarrow 1.0$	0.9	$\rightarrow \infty$	282		85	
Lyrids	Apr. 21	51		0.985	0.92		30	213	80	1861
Eta Aquarids	May 5	66		0.967	0.595		45	103	162	Halley
Arietids	June 7	37		0.94	0.10		77	29	18	
Zeta Perseids	June 7	27		0.79	0.35		77	61	4	
Beta Taurids	July 1	29		0.85	0.34		278	244	5	
Delta Aquarids	July 28	50	$\rightarrow \infty$	$\rightarrow 1.0$	0.04	$\rightarrow \infty$	308		56	
α Piscid Australids	July 29	46	90	0.99	0.9		282		85	
Perseids	Aug. 12	61		0.96	0.963		139	155	116	1862 III
Giacobinids	Oct. 9	22		0.715	1.0		196	174	31	1933 III
Orionids	Oct. 20	67		0.97	0.57		28	143	161	Halley
Taurids	Nov. 7	27		0.85	0.33				3	Encke
Leonids	Nov. 15	72		0.905	0.98		233	177	163	1866 I
Andromedids	Nov. 23	16		0.756	0.859		245	224	13	Biela
Geminide	Dec. 12	36.2		0.90	0.14		259	326	34	

[1] Note the high geocentric velocity of the meteor streams in retrograde orbits ($i > 90°$).

TABLE 2–10

The basic data available on meteor density come from visual radiations resulting from atmosphere impacts. Figure 2–26 is an attempt to relate visual magnitude with mass, while Fig. 2–27 attempts to relate visual magnitude with particle diameter. Visual or stellar magnitude is a measure of the brightness or light intensity of a star. The ratio of two luminosities l_1/l_2 is related to magnitudes m_1 and m_2 by the equation

$$0.4\,(m_2 - m_1) = \log\,(l_1/l_2) \qquad (2\text{–}22)$$

That is, luminosity ratios vary by a factor of

$$100^{1/5} = 2.512$$

and:

$$\log 2.512 = 0.4$$

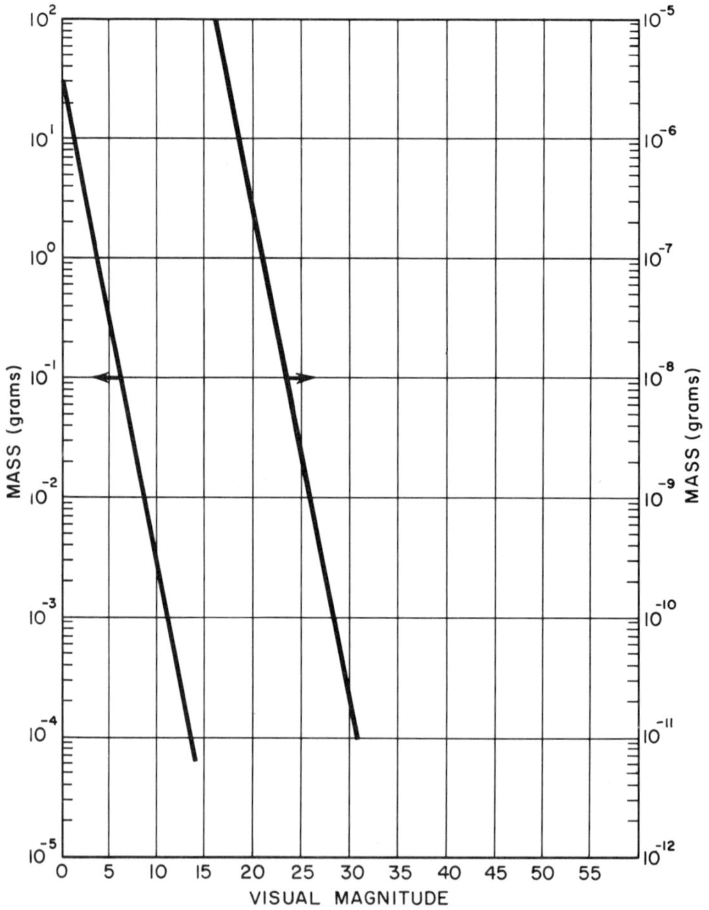

FIG. 2-26. Relation Between Mass and Visual Magnitude According to Whipple (*36*).

If the luminosity of stars of 1st magnitude is set = 100 = 2.512^5, we may say:

Magnitude	*Luminosity*
1	$2.512^5 = 100$
2	$2.512^4 = 39.81$
3	$2.512^3 = 15.85$
4	$2.512^2 = 6.31$
5	$2.512^1 = 2.512$
6	$2.512^0 = 1.0$

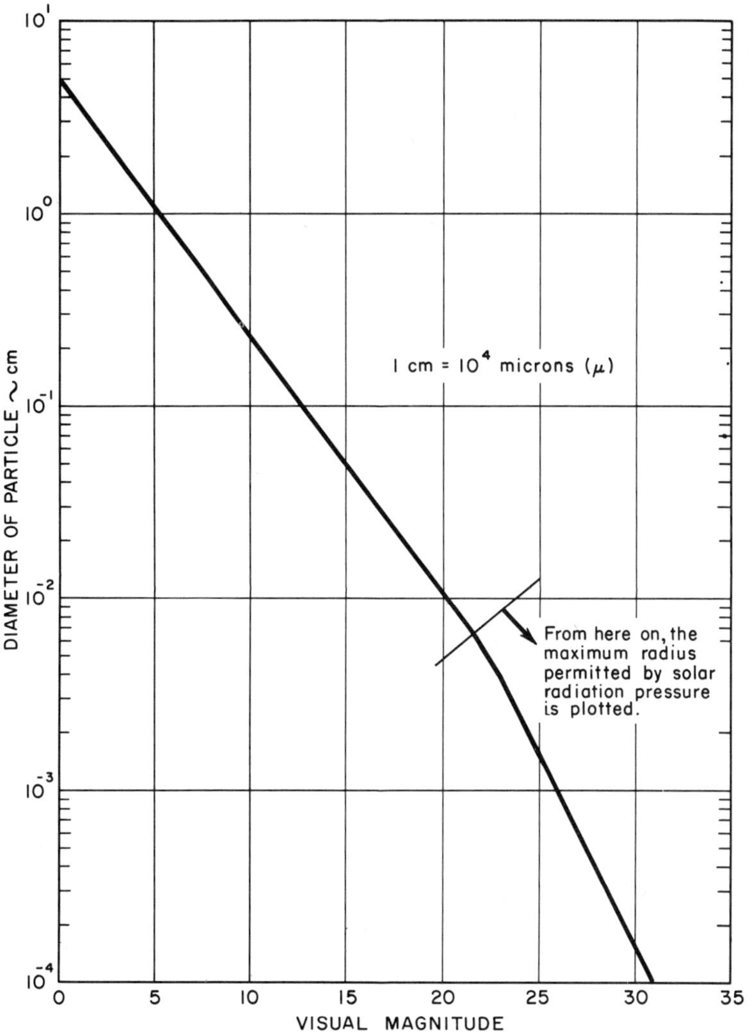

FIG. 2-27. Relation Between Visual Magnitude and Diameter of Particle According to Whipple (*36*).

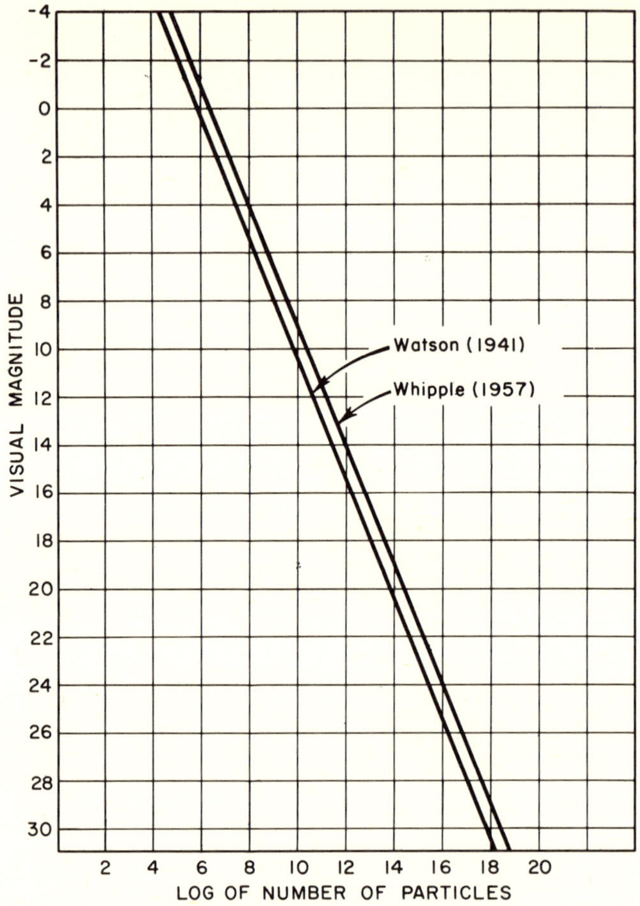

Fig. 2-28. Number of Meteoric Particles Encountered by the Earth in 24 hours According to Watson (*37*) and Whipple (*36*).

The number of meteoric particle encounters by the Earth per day has been plotted (Ref. (*1*)), and is shown here as Fig. 2–28. From these data, an attempt has been made to estimate the number of hits on a 3-meter diameter sphere in orbit around the Earth and in orbit around the sun at 1 A.U. These data are shown in Fig. 2–29. Note that a sphere orbiting the Earth encounters one-half the number of particles as does a sphere in solar orbit away from the Earth. This effect results from the Earth's "sweeping" of particles, thus shadowing its satellite for one-half of its orbit.

Actual meteoric particle count, size, velocities, and distribution should eventually be determined in a direct empirical manner. That is, by sampling from rockets and satellites. The erosion probabilities and effects

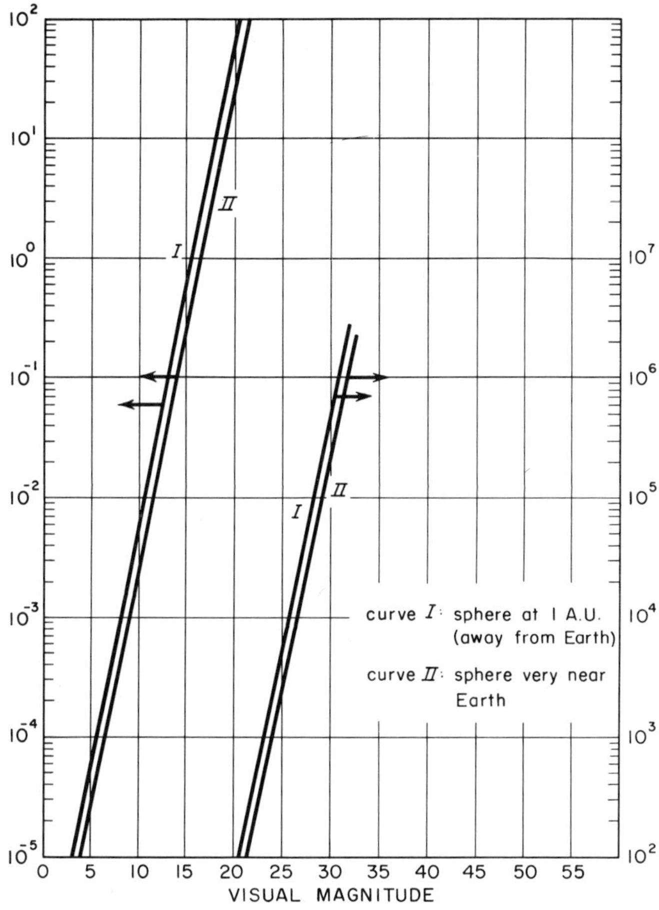

FIG. 2-29. Number of Hits on a Sphere of 3-meter Diameter in 24 Hours According to Whipple (*36*).

need to be studied much more thoroughly in space laboratories. Of particular importance will be the erosion effects on atmospheric entry bodies because the subsequent aerodynamic heating may be accelerated causing further excessive erosion and metal oxidation. Whipple has charted meteoric velocity and penetration into aluminum versus meteor visual magnitude as illustrated in Fig. 2–30 (Ref. (*36*)). From his data Ehricke (Ref. (*1*)) has estimated that meteors of the order of 1000 micron diameter (0.1 cm) would penetrate (puncture) a 0.4 cm-thick aluminum skin about once every 50 days for a space ship in orbit.*

*Note: Ship size is unspecified; meteoric impact rate may be taken as proportional to cross-section area fronting on meteor stream, the puncture or wall thickness will vary with shape contour.

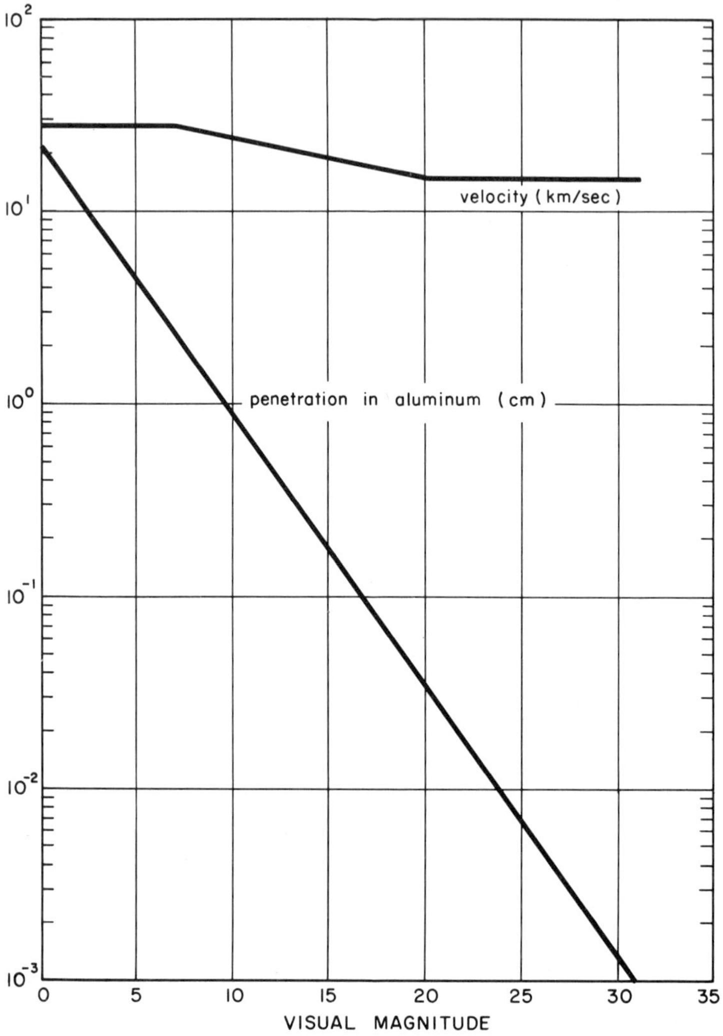

Fig. 2-30. Velocity and Penetration in Aluminum of Meteorite and Matter According to Whipple (*36*).

2.3 CISLUNAR AND LUNAR ENVIRONMENT.

Cislunar environment is the environment encountered between the Earth and Moon. Lunar environment, is associated directly with the Moon itself. Although the Moon is not the largest mass satellite in the solar system, it is the largest in proportion to its parent planet. In fact, the mass ratio between the Moon and Earth is approximately 0.226 which is sufficient to cause the Earth-Moon system to act as a double "star" with both bodies orbiting about a common center of mass approximately 5000 km (2800 nautical miles) from the center of the Earth. Although this barycenter is within the Earth, it is closer to the Earth's surface than center, and the Earth's center describes an elliptic orbit of one month period with respect to the barycenter. The Sun's attraction on the Earth-Moon system is so strong that significant perturbations are produced on the motion of the two-body system. The Moon produces ocean tides on the Earth. Tidal forces on the Moon have slowed it down to where its periods of rotation and revolution coincide so that the same side always faces the Earth. Longitudinal, latitudinal and parallactic librations of the lunar sub-terrestrial point about the Earth-Moon axis result in 59 per cent of the Lunar surface being visible from the Earth over an extended period of time.

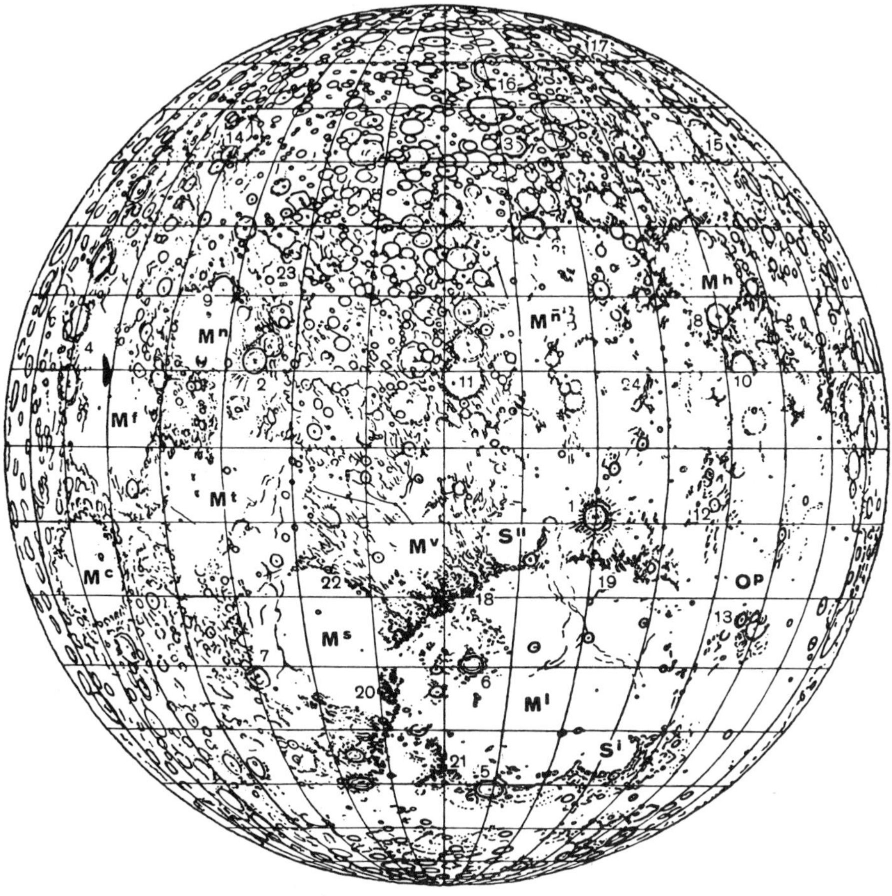

Fig. 2-31. Map of the Moon.

TABLE 2–11 ADDITIONAL LUNAR DATA

Distance from Earth	Mean	Max.	Min.
(km.)	384,400	406,700	356,400
(mi.)	238,900	252,700	221,500
(n. mi.)	207,600	219,500	192,500
(Earth radii)	60.267	63.800	55.900
Angular diameter	31′7″.2	29′19″.8	33′24″
Base at lunar distance	1.85 km	1.97 km	1.73 km

Mean equatorial parallax: 57′2″.7

Flattening: polar 1/2000; equator ellipse 1/1667

Dynamic flattening: $\frac{C - A}{C} = 0.0006269 \pm 0.0000027$

(equatorial) $\frac{B - A}{C} = 0.000118 \pm 0.000057$

Density: 0.643 of Earth or 3.34 specific gravity

Gravitational acceleration: 162 cm/sec^2 = 4.94 ft/sec^2 (about 1/6 of Earth)

Surface area: 0.074 of Earth

Volume: 0.02 of Earth

Albedo: 0.07

Inclination equator to orbital plane: 1° 31′ 21″.1

Inclination of lunar orbit to ecliptic: mean 5° 8′ 43″, max. 5° 20′, min. 4° 57′

Eccentricity of orbit: 0.043 to 0.067: mean: 0.0549

Lunar libration longitude: 7°.9; latitude 6°. 85

Parallactic libration: 1° 1′ 35″ (max.)

Visible surface area: 0.59 of total

1° selenocentric = 16″.6 geocentric = 30.33 km (mean)

Period of rotation: 27.32 days (sidereal month)

One Earth year = 12.36827 lunations

Barycenter of Earth-Moon system:

at mean distance	1645 km (beneath Earth surface)
at max. distance	1390 km
at min. distance	2010 km

At mean distance: point of gravitational "neutrality": 346,000 km from Earth 38,400 km from Moon

Mean daily motion of Moon with respect to inertial space: 13° 10′ 35″.03

Precession period of apsidal line: 8.85 years = 3232.58 days

Yearly precession of apsidal line: 40° 40′ 31″

Yearly regression of nodes: 19° 21′ 19″

Period of nodal precession: 18.6 years

Length of lunar shadow cone: 372,650 km to 374,450 km

Surface temperature (sunshine): 100°C (212°F)

(at beginning of total eclipse; i.e., in penumbra): −70°C (−94°F)

(at termination of eclipse): −110°C (−166°F)

Surface material: probably porous material (ash, pumice)

Monthly temperature curve for visible Moon: 200–270°K

(4-day lag with respect to received solar heat)

Mean sub-surface temperature: 230°K (−45°F)

Atmospheric constituents: possible traces of CO_2, SO_2, Kr

Atmospheric pressure: $< 10^{-5}$ atm

Lunar magnetic field: ?

2.3.1 Physical Parameters

Lunar physical data are tabulated in Table 2–11. The zero point of the lunar coordinate system is taken as the average position of the center

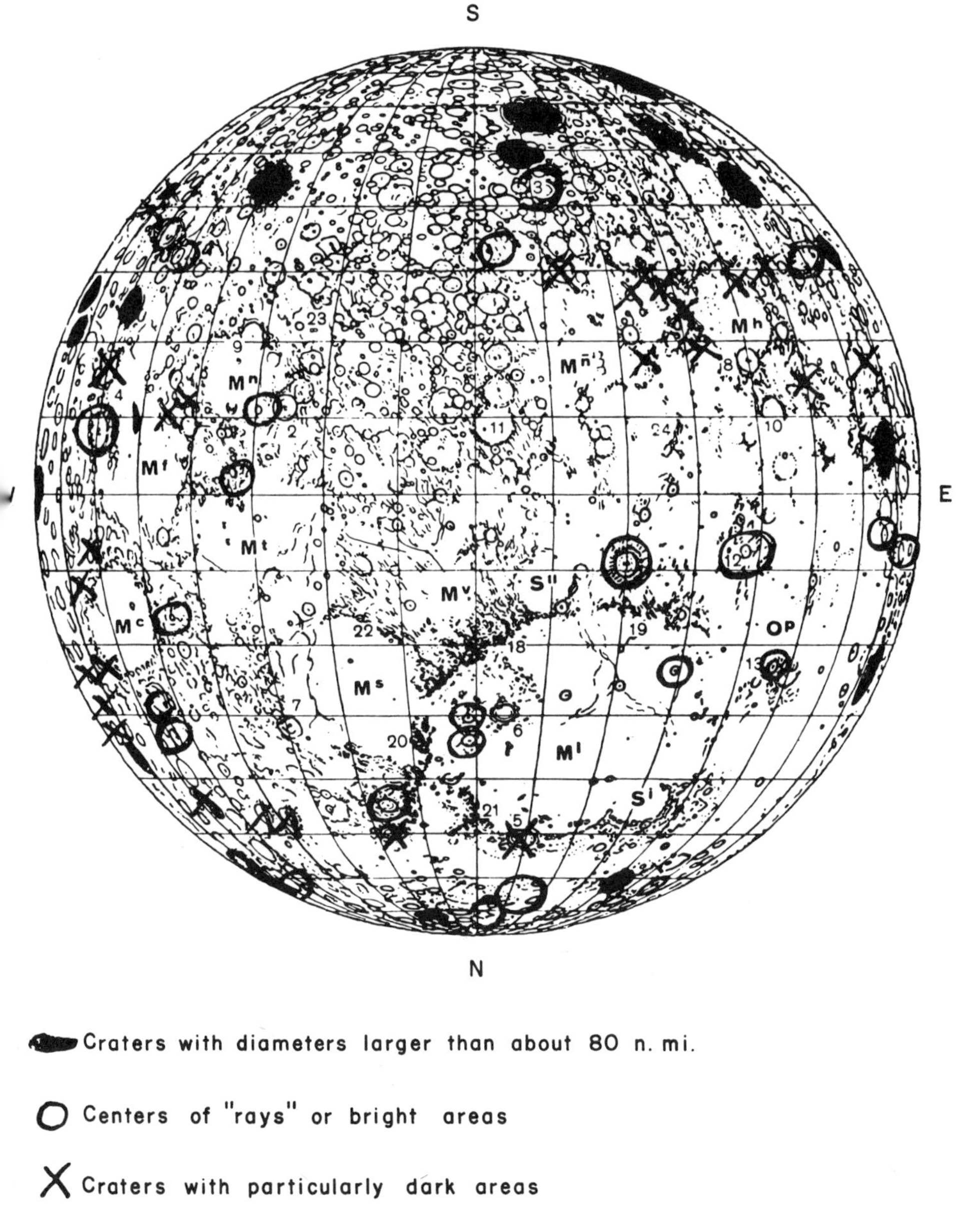

FIG. 2-32. Distribution of Prominent Lunar Surface Features.

of the lunar disc. Longitudes, latitudes, and librations are measured relative to this point. A map of the Moon is portrayed in Fig. 2–31 which indicates the significant features with respect to latitude and longitude lines.

The Sun's strong attraction on the Earth-Moon system causes the Earth-Moon distance to vary with a monthly period and amplitude of greater than ± 5 per cent (specific data included in Table 2–11). This distance variation causes a variation in orbital velocity which results in the longitudinal and latitudinal librations of the subterrestrial point. Parallactic librations occur as a result of shifts in the Earth observer's position with the Earth's rotation.

Figure 2–32 indicates the distribution of prominent lunar surface features. Altitude measurements of lunar craters and mountain ranges have indicated 15,000 to 20,000 foot altitude differences. Some craters are notably darker or brighter than the surrounding terrain, and much speculation has occurred as to the minerals or physical reasons that cause these differences.

Great cracks, or at least, radial lines are observed to cross the lunar surface which may indicate a parallel to the Earth's fault lines which shift with earthquakes. Whether the parallel is direct or not has not been ascertained.

Televised photographs of the far side of the Moon have been obtained using a circumlunar probe (Ref. (*38*)). Figure 2–33 is a map of the Moon's far side. The most striking difference between the near and far sides of the Moon is that the great plains on the near side are not duplicated on the far side. Rather, there seems to be a great tableland with numerous craters and rings.

The lack of sufficient gravity to hold an Earth-type atmosphere has indicated that whatever atmosphere is present (as from surface evaporation from the solid rocks or dust) must be heavy gases. Much speculation has occurred as to the presence of dust coatings and their thickness on the lunar surface, and it is hoped that the arguments will be settled or at least constrained by lunar sampling probes in the near future. Arguments for considerable dust layers can be developed from the extreme surface temperature variations ($+212°$F to $-166°$F) which would tend to pulverise most rocks on the surface. The lack of much atmospheric pressure ($<10^{-5}$ atm) would indicate an absence of wind to "blow" the dust about; thus a thin dust layer may exist on the lunar surface which thermally insulates the interior rock structure from the extreme surface temperature variations. The extreme temperature variations may produce "freezing" of the atmospheric constituents on the dark side and generate an equivalent to wind motion (although not very akin to terrestrial winds) which may stir up the dust layers at lunar sunset. The light gravity effects will require steep slopes for dust to drift or settle downhill appreciably, so that dust-free surfaces are likely only around crater edges, on mountain peaks, and in crevice walls. The dust may have some electrostatic charge, and may be acted upon by sunlight pressure and solar proton bombardment as well as by electrostatic interaction and lunar gravity. This indicates

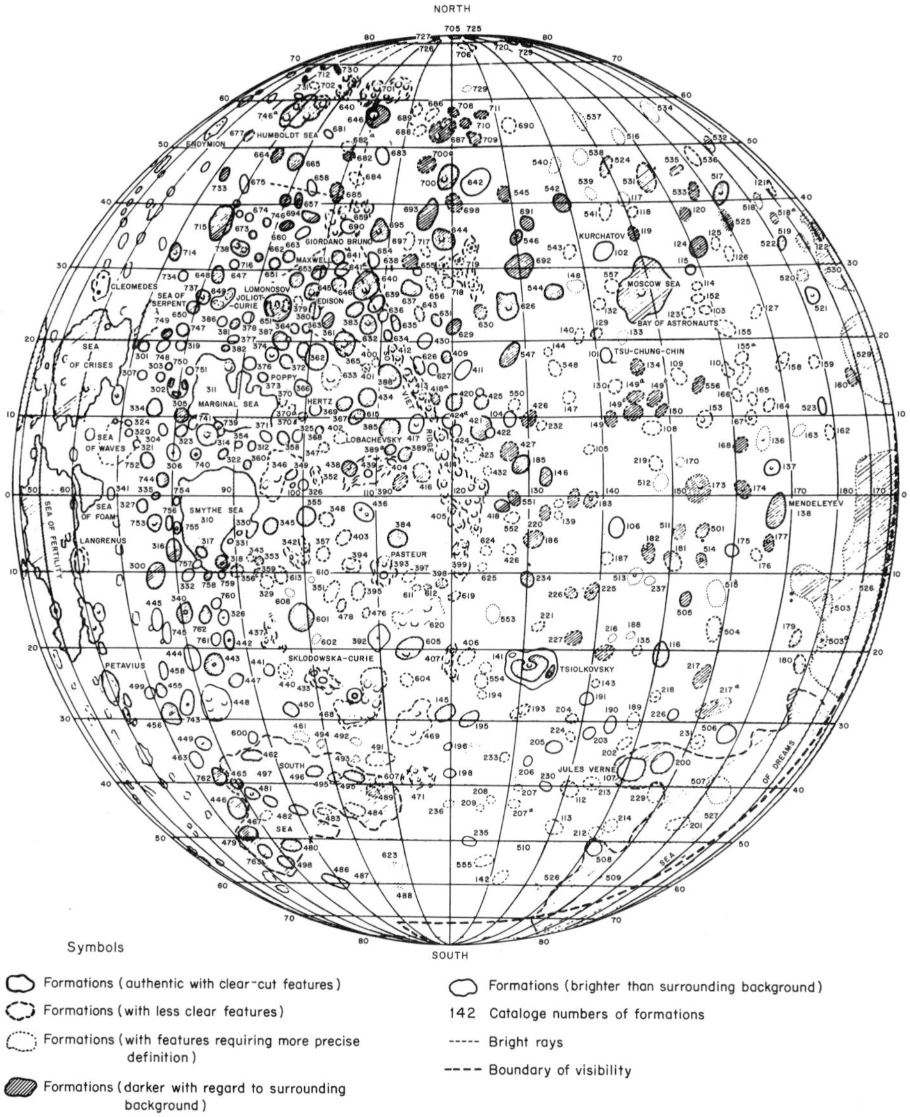

Fig. 2-33. Map of the Far Side of the Moon.

that the dust may be relatively thick only where compacted in the crevices, and that a light layer of active dust may be encountered on direct lunar contact.

The absence of a terrestrial-type atmosphere, particularly troposphere and ionosphere, which would permit communication by scatter or reflection techniques, has caused some study of lunar surface propagation (Ref. (*39*)). Taking into account the Moon's closer horizon and increased cosmic noise effects, but assuming a smooth homogeneous spherical shape, it has been predicted that a carrier frequency of one megacycle will permit communication by surface wave propagation over approximately 140 km with 1 kw radiated power and a receiver bandwidth of 20 kc. Further details of this calculation are contained in Ref. (*39*).

2.3.2 Gravitational Field

The gravitational field of the Moon is generally considered to be an r^{-1} function as is the central force field of a homogeneous sphere. The Moon is actually an oblate spheroid with an elliptical equator. The polar flattening factor is 1/2000; the equatorial, 1/1667. These factors are much smaller than the Earth's polar flattening factor of 1/297 which produces an r^{-3} field function with a coefficient on the order of 1×10^{-3} of the primary (r^{-1}) field component. The Central Force field concept thus applies more rigorously to the Moon than to the Earth. The terrestrial field strength is significant as one leaves the Moon's immediate vicinity however. Instead of a pure central force field we experience a complex, double-centered force field with unequal gravitational parameters associated with the two centers, and in which one force center (the Earth) may have significant higher order terms, (r^{-3}, r^{-5}).

By assuming a small mass relative to the lunar mass, and assuming the Earth and Moon are each central force fields which may be super-

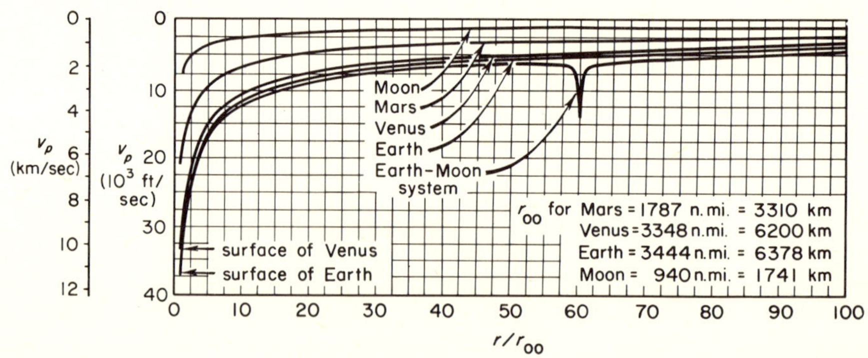

FIG. 2-34. Gravitational Field of Earth, Moon and Earth Moon System in Terms of Parabolic Velocity.

imposed, we may write the gravitation potential function as the vector summation of the Earth and Moon potential functions. Thus,

$$U = U_{\oplus} + U_{☾} \tag{2–23}$$

$$U = \frac{-K_{\oplus}}{r_{\oplus}} - \frac{K_{☾}}{r_{☾}} \tag{2–24}$$

where $K_{\oplus}$ and $K_{☾}$ are the terrestrial and lunar gravitational parameters, and $r_{\oplus}$ and $r_{☾}$ are the distances from the terrestrial and lunar centers of gravity. The parabolic (or escape) velocity may be used conveniently to provide a graphic picture of the nature of this field, Fig. 2–34. The parabolic velocity, V_p, is the velocity at which a vehicle follows a parabolic flight path, and is related to the circular velocity, V_c, by,

$$V_p = V_c\sqrt{2} \tag{2–25}$$

where the circular velocity is related to the field potential, U, and the gravitational parameter by,

$$V_c\sqrt{|U|} = \sqrt{\frac{K}{r}} \tag{2–26}$$

If we imagine Fig. 2–34 as being rotated about the ordinate axis ($r = 0$) so that the plotted curves form funnel-shaped figures, we may observe that the normal operation in such a funnel-shaped field requires velocities that are inverse functions of the distance from field-center; something like spinning marbles in a funnel. The important observation to make here is that the Earth's "funnel" has a funnel-shaped indentation on one side (or in one direction), and that this indentation rotates about the center funnel. This illustrates the relative magnitude of lunar "focusing" of lunar and translunar probes, and the difficult "fire control" problem of aiming a free flying projectile to hit or be deflected by the lunar "funnel" as it rotates about the much larger Earth-centered funnel.

Much analysis of velocity and orbit requirements in this type of field has been done under the heading of Perturbation Theory. Erhicke (Ref. (*1*)) has a comprehensive coverage of the LaGrangian, Hamiltonian, and Jacobian solutions to the problem. Of particular interest is the Jacobian, which assumes (1) that one of the masses in the 3-body problem is negligibly small, and (2) a constant system angular rotation about the barycenter. This problem will be discussed more fully in the succeeding chapter on dynamics.

2.3.3 Magnetic Field

Whether the Moon possesses a magnetic field is an unresolved question at the time of this writing. Attempts to measure the lunar magnetic field strength have indicated nothing discernible at the Earth's distance (Ref. (*40*)). Estimates of the linear magnetic field strength predict values on the order of 10 gamma at the lunar surface. Knowledge of the lunar magnetic field strength and its orientation would have a bearing on the validity of the several theories on lunar construction, composition, and formation. Should the lunar field strength be significant, it would interact with the Earth's magnetic field and tend to reorient the Moon. Future explorations of this lunar characteristic will have significant value also in predicting magnetic effects on instruments operating in the lunar environment.

2.3.4 Radiation

Radiation in lunar and cislunar space includes both the electromagnetic and energetic particle type. Electromagnetic radiation will be experienced throughout the radio to X-ray wave-length bands. Particle radiation may be divided into two main groupings: particles originating beyond the solar system, and particles from within the solar system. The dividing line between these two particle types is taken as the kinetic energy required to inject particles into intergalactic space (or, sometimes, solar escape velocity). Primary cosmic radiation, i.e., radiation possessing energy in excess of one Bev is assumed to be of extra-galactic or extra-solar origin. Radiation of less energy, being trapped by the solar central force field, is adjudged to originate from within the solar system, with the Sun as the principal source. Further discussion of radiation environment is relegated to subsequent sections on solar system, galactic, and extra-galactic environment.

2.3.5 Meteoric Activity and Dust

The meteoric activity and dust to be encountered in cislunar space is unknown, in the sense that no knowledge is available as to how it might differ from that experienced in immediate terrestrial space. Lunar "sweeping" may be expected to clear meteors from the lunar "wake," just as Earth "sweeping." Thus meteors will generally be (encountered largely on the forward travelling sides of both bodies, with only higher velocity meteors "catching up" from the lee side. The gravitational fields may be expected to have some focusing (or at least deflecting) effect on such particles. Dust, as a surface feature on the Moon, has been discussed previously under lunar physical parameters.

2.4 THE SOLAR SYSTEM

The solar system is defined as that portion of space and the material therein in which the Sun is the dominant gravitational sink and energy source. It includes all bodies whose kinetic energy is insufficient to enable a parabolic or hyperbolic escape trajectory from the Sun's influence.

Alternatively, it includes all bodies and material in elliptic (including circular) orbit about the Sun, the Sun itself and radiation emanating from or impinging on these bodies.

The dominant body in the solar system is the Sun. Orbiting about the Sun are nine planets which, in turn, have a number of satellites in subsidiary orbits. Also in orbit about the Sun are other smaller bodies called asteroids, comets, meteors, dust and gas. The principal features of the Sun which cause it to dominate these bodies are (1) its size—the Sun is a sphere of nearly 10 times the diameter of the next largest body in the system; (2) its mass—the Sun represents approximately 99.2 per cent of the total mass of the Solar System; (3) its gravitational field—a consequence of the huge mass, this field is the agent by which all the parasite bodies are retained in orbit about the Sun; (4) its electromagnetic radiation, the primary source of visual light throughout the System, and the primary source of radio, microwave, infrared, ultraviolet and X-ray spectra; (5) its thermal radiation, the primary temperature determining agent throughout the System; (6) its corpuscular or energetic particle radiation, the source of considerable concern to man-in-space enthusiasts as a hazard to prolonged space flight, and a prime agent in forming ionosphere layers in planetary atmospheres; (7) its "atmosphere," the photosphere, chromosphere, and corona, which radiate the visible and electromagnetic radiations observable from the Earth; (8) its magnetic field, probably the strongest magnetic field in the Solar System, and (9) its prominences, visually identifiable "solar storms" which result in radiation and magnetic storms in the Earth's upper atmosphere.

2.4.1 Physical Data

Primary physical data, for the various bodies in the solar system, are presented in Tables 2–12 through 2–17. The primary physical data of the Sun and the nine planets are presented in Tables 2–12 and 2–13 as tabulated by Ehricke (Ref. (*1*)). These data include angular diameter as viewed from the Earth, radii in terms of Earth radius and standard units, oblateness, density, mass, gravitational parameter, activity sphere (maximum radial distance from planet at which its gravity field predominates over the Sun's field in attracting a small particle) and period of rotation in sidereal time. Planetary surface physical characteristics, including atmospheric constituents, are presented in Table 2–14. Table 2–15 shows the physical data for the 31 known satellites of the planets. Table 2–16 presents physical data on 13 major asteroids. (There are more than 1600 cataloged, and probably 10 times as many more.) Table 2–17 presents data on the major comets and Table 2–18 presents data on the principal meteor streams. Subsequent sections will discuss the significance of the data presented in Tables 2–12 through 2–18.

TABLE 2–12 THE

Planet	Symbol	No. of Satellites	Major Semi-axis (A.U.) [a]	Eccen-tricity [e]	Perihelion (A.U.) [R_P]	Aphelion (A.U.) [R_A]	Mean
							Of Ascending Node [☊]
1	2	3	4	5	6	7	8
							° ′ ″
Mercury	☿		0.387099	0.2056259	0.307501	0.466697	47 49 17.6
Venus	♀		0.723332	0.0067935	0.718418	0.728246	76 17 33.7
Earth	⊕	1	1.000000	0.0167272	0.983273	1.016727	—
Mars	♂	2	1.523691	0.0933654	1.381431	1.665951	49 13 33.2
Jupiter	♃	12	5.202803	0.0484305	4.950829	5.454777	100 0 50.7
Saturn	♄	9	9.538843	0.0556922	9.007604	10.070082	113 16 52.6
Uranus	♅	5	19.181973	0.0472012	18.276561	20.087385	73 43 51.4
Neptune	♆	3	30.057707	0.0085724	29.800040	30.315374	131 18 24.2
Pluto	♇		39.51774	0.2486438	29.691899	49.343581	109 38 0.2

Cols. 4, 5, 8–11, and 13 *The American Ephemeris and Nautical Alamanac*, 1957. The mean distance is equal to the major axis of the planet's orbit.
Cols. 6 and 7 computed from 4 and 5.
Col. 12 Russell, Dugan and Stewart, *Astronomy*, Vol. 1 (1945).

2.4.2 Orbits, Rotations and the Ecliptic Plane

Tables 2–12, 2–13, 2–15, 2–16, 2–17 and 2–18 have provided basic orbital and rotational data on the planets, their satellites, asteroids, and meteor streams. To provide a clear picture of the meaning of some of the terms used, a few definitions and illustrations are required.

The standard reference direction in space is the Sun-Earth line at the vernal equinox. This line is directed towards a point in the constellation of Aries, and is designated by the Greek symbol ♈. The celestial sphere is divided into Northern and Southern hemispheres by the *ecliptic plane* which is merely the plane of Earth's motion about the Sun. The Northern celestial hemisphere is designated as the one containing the Earth's North pole. When viewing motions from the North pole of the ecliptic plane, planets, satellites and other bodies moving in a counter-clockwise direction are said to be in *direct orbits*, and those bodies moving in a clockwise direction are in *retrograde orbits*. All the nine planets are in direct orbits; a few satellites and other bodies are in retrograde orbits. Orbit inclinations are measured at the *ascending node*, i.e. where the orbit crosses the ecliptic

PLANETS—ORBITAL DATA

longitude		Inclination		Sidereal Mean Daily Motion (Sec) [μ]	Mean Orbital Velocity (⊕=1) $[V_c]_a$	Period of Revolution	
Of Perihelion [ω]	Of Planet at Epoch Jan. 1, 1957 [L]	Orbital to Ecliptic [i]	Equatorial to Orbit [ε]			In Earth Years	y d h
9	10	11	12	13	14	15	16
° ′ ″	° ′ ″	° ′ ″	° ′ ″				
76 47 11.1	57 22 35.13	7 0 14.2	—	14,732.420	1.607271	0.2411	0 88
130 57 57.9	218 19 2.64	3 23 39.1	—	5767.670	1.175794	0.6156	0 224 17
102 12 3.2	99 53 19.15	—	23 26 59	3548.193	1.0	1.0	1 0 0
335 16 2.9	44 23 7.04	1 50 50.8	25 12	1886.519	0.8068546	1.8822	1 322 0
13 37 47.5	168 43 16.58	1 18 19.9	3 6.9	299.128	0.4384109	11.86	11 313 21
92 12 20.4	243 57 27.62	2 29 42.2	26 44.7	120.455	0.3237816	29.46	29 167 22
169 57 44	128 24 17.04	0 46 22.9	97 59	42.235	0.2283249	84.0	84 0 0
44 14 51.9	210 20 37.36	1 46 26.5	29	21.532	0.1823988	164.8	164 292 0
223 10 30.2	137 38 8	17 8 38.4	—	14.283	0.7590757	247.7	247 255 12

Col. 14 Computed from $K_{\odot}=1.324948\times10^{11}$ km³/sec² according to mean orbital velocity $=(1/a)^{1/2}$ the value of $K_{\odot}$ was found from the equation $K_{\odot}=a_{\oplus}[V_c]^2_{a,\oplus}$, where $[V_c]_{a,\oplus}$ and $a_{\oplus}$ were taken as $[V_c]_{a,\oplus}=29.77$ km/sec and $a_{\oplus}=1$ A.U. $=149.4\times10^6$ km.

going from South to North. An inclination in excess of 90 degrees indicates a retrograde orbit. The longitude of the ascending node is measured on the ecliptic in a counter-clockwise direction (as viewed from North pole) with the Vernal Equinox direction as 0 degrees longitude. Orbit eccentricity may vary from 0 (for a circle) to 1 (for a parabola). All elliptic orbits may thus be defined by their eccentricity. In elliptic orbits, the Sun will be at one of the foci; the point of nearest approach to the Sun is designated the *perihelion;* the point farthest away, the *aphelion* (as with Earth satellites we have perigee and apogee). The line connecting the perihelion and aphelion is the *major axis* of the elliptic orbit. The longitude of perihelion is measured in the ecliptic with reference to the Vernal Equinox. Gravitational inter-actions between the planets and other orbiting bodies have caused shifts in the longitude of perihelion for some bodies, so the time at which a longitude of perihelion is valid is usually given, and is called an *epoch.*

The Sun rotates with its axis inclined 83 degrees to the ecliptic plane. The period of rotation varies with latitude being 25 days at the equator,

TABLE 2–13 THE

Planet	Diameter (app. angular from ⊕) (sec. of arc)	Radius (Equator)			Ob-lateness [ϵ]	Density		Volume	Mass	
		⊕=1	n. mi.	km		⊕=1	Water =1	⊕=1	$\frac{M_{\odot}}{M_{planet}}$	⊕=1
1	2	3	4	5	6	7	8	9	10	11
Sun	1919.62	109.2	375,557	696,000	—	0.2554	1.41	1.302×10^{6}	1.0	3.3253×10^{5}
Mercury	4.7–12.7	0.39	1350	2500	0	0.9058	5.00	0.06	6.12×10^{6} $\pm 43{,}000$	0.0543
Venus	9.9–64.5	0.973	3348	6200	0	0.8877	4.90	0.92	4.08645×10^{5} ± 208	0.8136
Earth	—	1.0	3440	6378.39	$\frac{1}{297}$	1.0	5.52	1.0	3.32488×10^{5}	1.0
Mars	3.5–25.1	0.52	1787	3310	$\frac{1}{192}$	0.7609	4.20	0.153	3.0880×10^{8} ± 3000	0.1077
Jupiter	30.8–50.0	10.97	37,732	69,880	$\frac{1}{15.4}$	0.2409	1.33	1318	1.0474×10^{3} ± 0.03	318.35
Saturn	14.9–20.6	9.03	31,075	57,550	$\frac{1}{9.5}$	0.1286	0.71	736	3.49764×10^{3} ± 0.27	95.3
Uranus	3.4–4.2	4.0	13,769	25,500	$\frac{1}{14}$	0.2283	1.26	64	2.2869×10^{4} ± 300	14.58
Neptune	2.2–2.4	3.9	13,499	25,000	$\frac{1}{45}$	0.2917	1.61	60	1.9314×10^{4} ± 300	17.26
Pluto	0.4(?)–0.6(?)	—	—	—	—	—	—	1.0(?)	4.0×10^{5} $\pm 30{,}000$	0.8312

Cols. 3, 5, 7 and 8 H. C. Urey, *The Planets* (1952), except for Sun.
Col. 4 computed from 5.
Cols. 2, 8, 9 and 19 Russell, Dugan and Stewart, *Astronomy I* (1945).
Col. 10 W. Rabe, *Astronautical Journal*, Vol. 55, pp. 112–126, May 1950. $M_{\odot}/(m_{\oplus}+m_{☾}) = 329{,}000 \pm 3000$.
Col. 11 computed from 10.
Col. 12 computed from $K_{pl} = K_{\oplus} m_{pl}/m_{\oplus}$, except for Earth which was computed as described in par. 3-1. The value for the combined Earth-Moon mass is $K_{\oplus + ☾} = 403{,}395$ km³/sec² $= 1.424567 \times 10^{16}$ ft³/sec² $= 63{,}505$ n. mi.³/sec² $= 1.20728 \times 10^{-19}$ A.U.³/sec².

27.5 days at ± 45 degrees and 33 days at ± 80 degrees. The Solar North Pole points about half-way between Polaris and Vega to celestial sphere coordinates of right ascension α = 19h 04m and declination $\delta = +64°$.

The standard length unit in the solar system is the Astronomical Unit (A.U.), the mean solar distance of the Earth. The value of the A.U. in Earth terms has been quoted by American scientists as 92,956,000 statute miles and by Russian scientists as 149,457,000 kilometers (92,868,000 statute miles). The difference in these measurements taken during the same time period has not been explained (Ref. (*41*)), and some further work is needed in this area. The technique employed by both groups was a radar echo time measurement at the inferior conjunction (closest passing)

PLANETS—PHYSICAL DATA

Gravitational Parameter, K				Activity Sphere (A.U.)	Period of Rotation, ω	
$\frac{km^3}{sec^2}$	$\frac{n.\ mi.^3}{sec^2}$	$\frac{ft^3}{sec^2}$	$\frac{A.U.^3}{sec^2}$		h. m.	rad/sec
12	13	14	15	16	17	18
1.324948×10^{11}	2.085813×10^{10}	4.67898×10^{21}	3.965293×10^{-14}	—	$24^d.2$ (Equ.) $33^d.5$ (75°λ)	—
2.16494×10^{4}	3408	7.645392×10^{14}	6.479236×10^{-21}	0.000746	—	1.98333×10^{-5}
6.2423×10^{5}	51,041	1.14499×10^{16}	9.703408×10^{-20}	0.00412	—	7.93013×10^{-5}
3.9858×10^{5}	62,746.8	1.407528×10^{16}	1.192867×10^{-19}	0.00618	23 56.068	7.31958×10^{-5}
4.2906×10^{4}	6754	1.51520×10^{15}	1.284088×10^{-20}	0.00378	24 37.38	7.08821×10^{-5}
1.26498×10^{8}	1.991419×10^{7}	4.46723×10^{18}	3.785844×10^{-17}	0.3216	9 50.5	1.77341×10^{-4}
3.78811×10^{7}	5.96348×10^{6}	1.33776×10^{18}	1.133704×10^{-17}	0.3246	10 2	1.73953×10^{-4}
5.79364×10^{6}	912,070	2.04599×10^{17}	1.733911×10^{-18}	0.346	10 48	1.61605×10^{-4}
6.86004×10^{6}	1.07995×10^{6}	2.42259×10^{17}	2.053066×10^{-18}	0.5805	15 48	1.10464×10^{-4}
3.31237×10^{5}	52,146	1.16975×10^{16}	9.913239×10^{-20}	0.2366	—	—

Cols. 13–15 derived from 12.

Col. 16 the activity sphere describes the space in which the planet, rather than the Sun, should be regarded as center body in the perturbation analysis of hyperbolic encounter between the planet and a particle of negligible mass (cf. par. 4-32). The activity sphere, therefore, may be used to define the circumplanetary space (e.g. circumterrestrial space). Values were computed, using the values from Col. 4 (Table 3-1a) and the mass ratios Sun/planet in Col. 10 of this table (cf. Chap. 6).

of Venus between March and May 1961, the frequencies employed being 2388 mc (U.S.) and something in the neighborhood of 1000 mc (U.S.S.R.). By measuring all solar distances in A.U., all that is needed is a precise measurement of a single solar system distance in Earth units (feet, meters, etc.) to convert all solar system distances. Figure 2–35 depicts the orbital inclinations of the planets and major satellites and asteroids (minor planets). Note that, with the exception of the asteroids and the planet Pluto, all the major bodies orbit close to the ecliptic plane. Satellites, it will be observed, orbit in a great variety of inclinations while accompanying their parent planets about the Sun. Shown also is the *invariable plane* (I.P.) which is defined by the condition that the angular momentum of

TABLE 2–14 THE

Planet	Mass Gravitational Acceleration at 45° Latitude			Centrif. Relief at 45° Latitude	Apparent Gravit. Acceleration at 45° Latitude		Circular Velocity at Equat. Radius Distance		
	g_{00} units	cm/sec²	ft/sec²	cm/sec²	cm/sec²	ft/sec²	km/sec	ft/sec	A.U./sec
1	2	3	4	5	6	7	8	9	10
Mercury	0.35238	346.3904	11.36449	—	—	—	2.94276	9.6547×10^{3}	1.57737×10^{-10}
Venus	0.85805	843.4703	27.67285	—	—	—	7.231538	2.37258×10^{4}	3.15795×10^{-10}
Earth	1.00000	983.008	32.25085	1.70667	981.30133	32.19486	7.90959	2.59497×10^{4}	3.45381×10^{-10}
Mars	0.40894	401.9938	13.18875	0.83152	401.16228	13.16148	3.60036	1.18122×10^{4}	1.57143×10^{-10}
Jupiter	2.81509	2767.252	90.78892	109.885	2657.367	8.71838	42.5467	1.395886×10^{5}	1.54755×10^{-9}
Saturn	1.29640	1274.368	41.80986	87.071	1187.297	38.95323	25.6560	8.41733×10^{4}	1.12049×10^{-9}
Uranus	0.97479	958.2298	31.43792	33.29815	924.93165	30.34546	15.0733	5.14215×10^{4}	6.58391×10^{-10}
Neptune	1.14183	1122.431	36.82509	15.25288	1107.17812	36.32466	16.5651	5.4347×10^{4}	2.2946×10^{-4}
Pluto	0.40096	394.151	12.93144	—	—	—	7.43214	2.43836×10^{4}	5.0307×10^{-8}

Cols. 2–7: Mean gravitational acceleration was computed from $g = K/r_{45}^2$, using K values in Table 3-1b for a radial distance at 45° latitude, because of oblateness of most planets. This radius was obtained from $r_{45} = r_0.(2-\epsilon)/2$ where r_0 is given in Table 3-1b. The centrifugal relief, due to planet rotation, is given also for 45°. The equatorial value is roughly twice as much. The apparent gravitational acceleration is the difference between col. 3 and col. 5.

Cols. 8–10: Circular velocity computed from $V_c = \sqrt{K/r_0}$ using the values for K and radius given in Table 3-1b. (For Pluto $r_{♇} = r_{\oplus}(m_{♇}/m_{\oplus})^{1/3} = 5996.69$ km was used assuming equal density of both planets.)

Cols. 11–13: Parabolic velocity computed from $V_p = V_c\sqrt{2}$.
Col. 14: S = solid, C = clouds.

PLANETS—SURFACE DATA

Parabolic Velocity at Equat. Radius Distance					Temperature (°K)			Atmosphere	
km/sec	ft/sec	A.U./sec	Condition of Visible Surface	Albedo	Black Spher.	Max. (Day-side)	Av. Temp.	Composition	Abundance N.P.T. (cm)
11	12	13	14	15	16	17	18	19	20
4.16652	1.36538×10^{4}	1.82467×10^{-10}	S	0.058	450	625	—	Traces of A^{40}, Kr, X(?)	—
10.72620	3.35528×10^{4}	4.46602×10^{-9}	C	0.76	329	324	229	CO_2 CO N_2O CH_4 C_2H_4 C_2H_6 NH_3	100,000 <100 <100 <20 <3 <1 <4
11.1859	3.6645×10^{4}	4.8865×10^{-10}	S	0.39	279	349	246	N_2 O_2 CO_2 CH_4 N_2O O_3	625,000 168,000 220 1.2 0.4 0,3
5.091667	1.67049×10^{4}	2.22234×10^{-10}	S	0.148	225	307	217	CO_2 SO_2 O_3 N_2O CH_4 C_2H_2 C_2H_6 NH_3 [Mostly N_2(?) A(2200?)]	440 <0.003 <0.05 <200 <10 <2 <1 <2
60.1902	6.2426×10^{5}	2.6272×10^{-9}	C	0.51	122	145	102	CH_4 NH_3	15,000 700
36.2831	1.19039×10^{5}	1.58460×10^{-9}	C	0.50	90	107	76	CH_4 NH_3 O_3 SO_2	35,000 <250 <0.1 <0,01
321.3168	6.9937×10^{4}	9.3111×10^{-9}	C	0.66	64	69	49	CH_4 O_3 SO_2	220,000 <0,1 <0.01
23.4268	7.68592×10^{4}	1.02609×10^{-9}	C	0.62	51	56	40	CH_4	370,000
10.51067	3.44837×10^{4}	7.11451×10^{-8}	?	0.16	45	60	42	—	—

Col. 16: Perfect conductor.

Cols. 15, 17–20: G. P. Kuiper, *The Atmospheres of the Earth and the Planets* (1952) Chapter XII, Table 1 and Table 9. N.P.T. (cm): equivalent height of atmosphere component at normal pressure and temperature. The Earth's atmosphere is equivalent to about 800,000 cm (N.P.T.). Data given are provisional. Notes for Mercury added by author. Bracketed note for Mars according to G. de Vaucouleurs, *The Physics of the Planet Mars* (Ref. (*19*)). Maximum temperature given by $T_{max} = 394° \, (1\text{-Albedo})^{1/4}/\sqrt{\text{mean orb. dist.}}$ Average temperature given by $T_{av} = T_{max}/\sqrt{2}$. The low temperature values of Venus are due to the planet's high albedo.

TABLE 2–15 THE SATELLITES

Planet	Satellite	Stellar Magni- tude	App. Distance as seen from ☉ (Sec. of Arc)	Distance from Planet		Period of Revolution		Eccen- tricity, (*e*)	Inclina	
				Planet Radius =1	km	d h m	Days		Satellite Orbit to its Proper Plane (or the Planet's Equator)	Mean inc. of Proper Plane to Planet's Orbit
1	2	3	4	5	6	7	8	9	10	11
Earth	Moon	−12.5	529.75	60.267	384,400	27 7 43.19	27.321661	0.05490	28° 35′ to 18° 19′	5° 8.55′
Mars	Phobos	11.5	8.46	2.79	9350	0 7 39.23	0.3189103	0.021	1 8	25° 11′
	Deimos	13	21.25	6.93	23,500	1 6 17.92	1.2631354	0.003	1 46	24 16
Jupiter	V	13	48.06	2.540	181,500	0 11 57.38	0.49817938	0.0028	27.3	3 6.9
	I 10	5.5	111.78	5.905	422,000	1 18 27.56	1.76913783	0.0	1.6	3 6.7
	II Europa	5.7	177.86	9.401	671,400	3 13 13.70	3.55188113	0.0003	28.1	3 5.8
	III Ganymede	5.1	283.7	14.995	1,071,000	7 3 42.56	7.15455265	0.0015	11.0	3 2.3
	IV Callisto	6.3	498.99	26.379	1,884,000	16 16 32.19	16.68901862	0.0075	15.2	3 42.7
	VI	13.7	3037	161	11,500,000	240.7	250.7	0.155		
	VII	16	3113	165	11,750,000	260	260	0.207		
	X	17.8	3116	165	11,750,000	260	260	0.08		
	XI	16	5990	315	22,500,000	692	692	0.21		
	VIII	17.6	6240	330	23,500,000	739	713–768	0.29–0.45		
	IX	17.4	6360	332	23,700,000	758	758	0.1–0.4		
	XII									
Saturn	Mimas	12.1	26.83	3.11	185,700	0 22 37.09	0.9424219	0.019	1 31	26 44.7
	Enceladus	11.6	34.42	3.99	238,200	1 8 53.11	1.3702178	0.0001	1.4	26 44.7
	Tethys	10.5	42.6	4.94	294,800	1 21 18.44	1.8878025	0.0	1 5.6	26 44.7
	Dione	10.7	54.57	6.33	337,700	2 17 41.16	2.7369159	0.002	0 0	26 44.7
	Rhea	10	76.20	8.84	527,500	4 12 25.20	4,5175026	0.0009	20.5	26 41.9
	Titan	8.3	176.67	20.48	1,223,000	15 22 41.41	15.9454201	0.0289	18.3	26 7.1
	Hyperion	13	214.13	24.82	1,484,000	21 6 38.40	21.2766667	0.1043	1 31	26 44.7
	Japetus	10.1 to 11.9	514.73	59.68	3,563,000	79 7 55.54	79.330234	0.0284	13 51.8	16 18.1
	Phoebe	14.5	1870.4	216.8	12,950,000	550 10.6	550.4416	0.1659	(148.9)	(174.7)
Uranus	Ariel	15.2	13.78	7.35	191,800	2 12 29.35	2.5203796	0.007	(0)	—
	Umbriel	15.8	19.2	10.2	267,300	4 3 27.61	4.1441748	0.008	(0)	—
	Titania	14	31.5	16.8	438,700	8 16 56.45	8.705865	0.023	(0)	—
	Oberon	14.2	42.12	22.4	586,600	13 11 7.06	13.463235	0.01	(0)	—
	V									
Neptune	Triton	13.6	16.23	14.2	354,000	5 21 2.66	5.8768437	0.0	(160)	
	Nereid									

Cols.: 3–7, 10–13, 15 and 17 Russell, Dugan and Stewart, *Astronomy* I (1945).
Col. 9 Russell, Dugan and Stewart, *Astronomy* I (1945), W. T. Skilling and R. S. Richardson, *Astronomy* (1947) show some values with considerable deviation: Phobos, 0.017; eccentricities for Jupiter I through Jupiter V are designated as variable; Jupiter X, 0.14051; Enceladus, 0.0046; Hyperion, 0.119.

OF THE SOLAR SYSTEM

on											
lean inc. of atellite's Orbit to Planet's Orbit	Longitude of Ascending Node	Direction of Rotation	Diameter (km)	Mass		Gravitational Parameter		Escape Velocity		Surface g	Atmosphere Component (cm) N.P.T.
				☿ = 1	☾ = 1	$\frac{km^3}{sec^2}$	$\frac{ft^3}{sec^2}$	$\frac{km}{sec}$	$\frac{ft}{sec}$	⊕ = 1	
12	13	14	15	16	17	18	19	20	21	22	23
		Direct	3476	0.2259	1	4890	1.7270×10^{14}	2.372	7873	0.1650	SO_2(<0.0003) O_3 (<0.005) Kr (?)
26° 19′		Direct	~15								
26 2		Direct	~8								
3 34.2		Direct	75–150								
3 8.3		Direct	3730	0.2236	0.99	4841	1.7097×10^{14}	2.279	7476	0.1337	CH_4(<200) NH_3(<40)
3 33.9		Direct	3150	0.1446	0.64	3130	1.1053×10^{14}	1.994	6541	0.1286	CH_4(<200) NH_3(<40)
3 13.3		Direct	5150	0.4766	2.11	10,319	3.6440×10^{14}	2.831	9288	0.1586	CH_4(<200) NH_3(<40)
3 57.9		Direct	5180	0.2982	1.32	6455	2.2797×10^{14}	2.233	7325	0.0981	CH_4(<200) NH_3(<40)
28 45	181°	Direct	120?								
27 58	243	Direct	50?								
28	82	Direct	20?								
48 1	208	Retrograde	25?								
57	61	Retrograde	50?								
53	232	Retrograde	22?								
26 25.4		Direct	450?	0.000117	0.000516	2.523	8.911×10^{10}	0.1498	491	0.00508	
27 2.4		Direct	500?	0.000264	0.00117	5.722	2.0206×10^{11}	0.2139	702	0.00933	
30 9.9		Direct	1100?	0.00206	0.0088	43.035	1.5198×10^{12}	0.3956	1298	0.01450	
26 44.7		Direct	1100?	0.00316	0.014	68.465	2.4178×10^{12}	0.4990	1637	0.02306	
27 50.3		Direct	1600?	0.00678	0.03?	146.67	5.1796×10^{14}	0.6055	1987	0.02335	
26 46.1		Direct	4200	0.4337	1.92	9390	3.3159×10^{14}	2.990	9811	0.21694	CN_4(20,000) NH_3(<300)
28 15.7		Direct	400?	<0.000136	$<6 \times 10^{-4}$	<2.934	$<1.0362 \times 10^{11}$	<0.1713	<562	<0.0075	
26 17 to 26 56		Direct	1300?	0.00429	0.019	92.17	3.2549×10^{12}	0.5325	1747	0.0223	
5 18.7		Retrograde	300?								
7 59		Retrograde	500?								
7 59		Retrograde	400?								
7 59		Retrograde	1000?								
7 59		Retrograde	900?								
0 (1900)		Retrograde	4500?	0.4066	1.8	8803	3.1086×10^{14}	2.7972	9177	0.1772	

Col. 16 Computed from $m_{☿}/m_{☾} = 4.427$

Col. 18 computed from the mass in terms of Moon's mass given in 17. Moon's mass derived from Earth's mass is 32859 × 10^{25} g.

Cols. 20–22 computed from gravitational parameter 18.

Col. 23 G. Kuiper, *The Atmosphere of the Earth and the Planets* (1952), p. 308; cf. Cols. 22 and 23 in Table 3-1b.

TABLE 2–16 ASTEROIDS[2]

No. 1	Name 2	t_0 3	M_0 4	ω 5	Ω 6	i 7	$e=\sin\varphi$ 8
			°	°	°	°	
1	Ceres	1957 June 11	279.880	71.853	80.514	10.607	0.0759
2	Pallas	1957 June 11	271.815	309.759	172.975	34.798	0.2340
3	Juno	1957 June 11	329.336	246.133	170.438	12.993	0.2585
4	Vesta	1957 June 11	79.667	149.134	104.102	7.132	0.0889
7	Iris	1949 Mar. 25	135.449	144.093	259.829	5.509	0.2315
15	Eunomia	1900 Jan.	13.53	94.98	294.42	11.76	0.1870
116	Sirona	1929 Apr. 22	29.764	91.551	64.842	3.580	0.1404
345	Tercidina	1925 Jan. 1	352.891	228.308	213.246	9.728	0.0613
433	Eros	1931 Jan. 18	0.586	177.930	304.071	10.831	0.2398
434	Hungaria	1925 Jan. 1	300.916	123.100	175.196	22.504	0.0737
719	Albert	1911 Oct. 2	7.929	151.940	186.094	10.825	0.5406
887	Alinda	1942 Jan. 31	358.049	348.119	111.029	9.024	0.5398
944	Hidalgo	1948 Feb. 29	335.824	57.506	21.273	42.529	0.6557
1019	Strackea	1924 Mar. 14	259.187	121.198	144.244	26.972	0.0716
1025	Riema	1943 Dec. 22	294.188	347.568	163.100	26.871	0.0389
1139	Atami	1941 June 19	96.784	205.517	213.215	13.102	0.2552
1172	Aneas	1948 Aug. 7	271.438	45.617	246.784	16.687	0.1029
1173	Anchises	1948 Aug. 7	264.330	30.710	284.132	6.980	0.1376
1208	Troilus	1948 Aug. 7	233.295	292.565	47.985	33.694	0.0924
1221	Amor	1948 June 3	22.165	25.549	171.202	11.924	0.4346
1235	Schorria	1931 Dec. 8	359.694	43.005	12.767	25.020	0.1537
1404	Ajax	1948 Aug. 7	325.207	57.197	332.320	18.132	0.1118
1437	Diomedes	1948 Aug. 7	230.321	126.771	315.330	20.563	0.0436
1509	Esclangona	1939 Jan. 1	276.797	267.061	283.286	22.320	0.0329
1566	Icarus	1950 Aug. 7	53.499	30.912	87.746	22.979	0.8266
1580	Betulia	1952 Dec. 14	291.862	158.888	61.874	52.037	0.4928
1583	Antilochus	1950 Nov. 15	336.205	186.229	221.067	28.303	0.0542
1600	1947 UC	1948 Feb. 19	322.366	49.765	60.314	21.177	0.0376
1620	Geographos	1954 Dec. 24	195.077	276.211	336.999	13.325	0.3352
	1948 Oa	1948 Oct. 7	0.000	126.314	274.191	9.397	0.4360
	Apollo	1932 Apr. 25	319.984	284.878	36.077	6.422	0.5663
	Adonis	1936 Feb. 25	22.086	39.537	352.538	1.480	0.7792
	Hermes	1937 Nov. 7	327.038	90.687	35.367	4.685	0.4746

[1] Ref. Ephemerides of minor planets by the Institute of Theoretical Astronomy at Leningrad (in Russia), except for Geographos and 1948 Oa where elements were computed and kindly made available by Prof. S. Herrick, University of California.

μ sec	a A.U.	R_P A.U.	R_A A.U.	Closest Approach to Earth	Diameter Miles	Period of Rotation h m	Albedo
9	10	11	12	13	14	15	16
770.695	2.7675	2.5574	2.9776		480		0.06
768.884	2.7718	2.1232	3.4204		304		0.07
814.046	2.6683	1.9785	3.3581		120		0.12
977.646	2.3617	2.1517	2.5717		240		0.26
963.378	2.3849	1.8328	2.9370			6 12	
825.347	2.6439	2.1495	3.1383			3 2	
770.820	2.7672	2.3787	3.1557			9 40	
1000.550	2.3255	2.1829	2.4681			8 47	
2015.293	1.4581	1.1084	1.8078	0.1742 (1931)	15	5 16	
1308.770	1.9443	1.8010	2.0876				
853.665	2.5852	1.1876	3.9828		2–3		
886.351	2.5212	1.1602	3.8822				
254.384	5.7944	1.9950	9.5938		15–30		
1342.468	1.9116	1.7747	2.0485				
1274.509	1.9790	1.9020	2.0560				
1305.814	1.9472	1.4503	2.4441				
300.254	5.1881	4.6542	5.7220				
308.455	5.0958	4.3946	5.7970				
302.757	5.1595	4.6828	5.6362				
1334.746	1.9190	1.0850	2.7530	0.1075 (1932)			
1343.831	1.9103	1.6167	2.2039				
302.437	5.1631	4.5859	5.7403				
304.209	5.1431	4.9189	5.3673				
1391.117	1.8668	1.8054	1.9282				
3171.432	1.0777	0.1869	1.9685	0.038 (1968)			
1090.753	2.1954	1.1135	3.2773				
292.703	5.2770	4.9910	5.5630				
1411.228	1.8490	1.7795	1.9185				
42.612	1.2441	0.8271	1.6611	0.032 (1969)	1		
36.960	1.3679	0.7715	1.9643				
1958.60	1.4861	0.6445	2.3277	0.0699 (1932)	1–2		
1284.03	1.9692	0.4348	3.5036	0.0161 (1936)	1–2		
2420.68	1.2904	0.6780	1.9028	0.0052 (1937)	<1		

Explanation of symbols: t_0 = Epoch, M_0 = Mean anomaly at epoch t_0, ω = Longitude of the perihelion, ☊ = Longitude of the ascending node, i = inclination of orbital plane to ecliptic, e = Eccentricity, μ = Mean daily motion, a = Major semi-axis, R_P, R_A = Perihelion and aphelion distance. Unit of col. 13 is A.U.

TABLE 2–17 SOME PERIODIC COMETS (FALL 1932)

Family	Name	Discovery	R_P[1] (A.U.)	R_A[2] (A.U.)	i[3] (eg)	T[4] (yr)
♃	Encke	1818	0.3	4.1	12.5	3.3
	Grigg-Skjellernp	1902	0.9	4.9	17.5	5.0
	Tempel II	1873	1.3	4.7	12.8	5.2
	Neujmin II	1916	1.3	4.8	10.6	5.4
	Brorsen I	1846	0.6	5.6	29.4	5.5
	Tempel III L. Swift	1869	1.2	5.2	5.4	5.7
	de Vico—E. Swift	1844	1.4	5.1	2.1	5.9
	Tempel I	1867	1.8	4.8	9.8	6.0
	Pons—Winnecke	1858	1.0	5.6	18.9	6.0
	Perrine I	1896	1.2	5.8	15.7	6.5
	Kopff	1906	1.7	5.3	8.7	6.6
	Giacobini II—Zinner	1900	1.0	6.0	30.7	6.6
	Biela	1826	0.9	6.2	12.6	6.6
	d'Arrest	1851	1.4	5.7	18.1	6.6
	Finlay	1886	1.1	6.2	3.4	6.9
	Holmes	1892	2.1	5.1	20.8	6.9
	Borelli	1905	1.4	5.8	30.5	6.9
	Brooks II	1889	1.9	5.4	5.6	6.9
	Faye	1843	1.6	5.9	10.6	7.3
	Schaumasse	1911	1.2	6.8	14.7	8.0
	Wolf I	1884	2.4	5.8	27.3	8.3
♄	Tuttle I	1858	1.0	10.3	55.0	13.5
	Neujmin I	1913	1.5	12.0	15.2	17.7
♅	Pons-Forbes	1818	0.7	18.0	28.9	28.7
♆	Westphal	1852	1.3	30.0	40.9	61.7
	Brorsen II—Metcalf	1847	0.5	33.2	19.2	69.1
	Pons-Brooks	1812	0.8	33.7	74.0	71.6
	Olbers	1815	1.2	33.6	44.6	72.7
	Halley	—[5]	0.6	35.5	102.2	76.0

[1] Perihelion distance; [2] aphelion distance; [3] orbital inclination to ecliptic; [4] period of revolution; [5] last seen in 1910; discovery unknown.

the solar system about an axis in this plane is zero, and is maximum about an axis normal to this plane. The ecliptic is inclined about 1°43′ with respect to the invariable plane. Figures 2–36 and 2–37 depict (on different scales as may be noted from the common Martian orbit) the Inner and Outer solar systems showing the orientation of the longitudes of the ascending modes. The Jovian satellite system is like a small solar system in itself, and presents some interesting studies. Figure 2–38 depicts the orbit relations of the twelve jovian satellites at epoch 1951. This system may be observed to have a dense inner system and a vast outer system

METEOR STREAMS—ORBITAL DATA[1]

Name	Maximum	Geocentric Velocity (km/sec)	Semi-Major axis, a (A.U.)	Eccentricity, e	Perihelion, R_P (A.U.)	Aphelion, R_A (A.U.)	Longitude of Ascending Node Ω (deg)	Longitude of Perihelion ω (deg)	Orbital Inclination to Ecliptic, i	Associated Comet
Quadrantids	Jan. 3	46	$\to \infty$	$\to 1.0$	0.9	$\to \infty$	282		85	
Lyrids	Apr. 21	51		0.985	0.92		30	213	80	1861
Eta Aquarids	May 5	66		0.967	0.595		45	103	162	Halley
Arietids	June 7	37		0.94	0.10		77	29	18	
Zeta Perseids	June 7	27		0.79	0.35		77	61	4	
Beta Taurids	July 1	29		0.85	0.34		278	244	5	
Delta Aquarids	July 28	50	$\to \infty$	$\to 1.0$	0.04	$\to \infty$	308		56	
α Piscid Australids	July 29	46	90	0.99	0.9		282		85	
Perseids	Aug. 12	61		0.96	0.963		139	155	116	1862 III
Giacobinids	Oct. 9	22		0.715	1.0		196	174	31	1933.III
Orionids	Oct. 20	67		0.97	0.57		28	143	161	Halley
Taurids	Nov. 7	27		0.85	0.33				3	Encke
Leonids	Nov. 15	72		0.905	0.98		233	177	163	1866 I
Andromedids	Nov. 23	16		0.756	0.859		245	224	13	Biela
Geminide	Dec. 12	36.2		0.90	0.14		259	326	34	

[1] Note the high geocentric velocity of the meteor streams in retrograde orbits ($i > 90°$).

TABLE 2–18

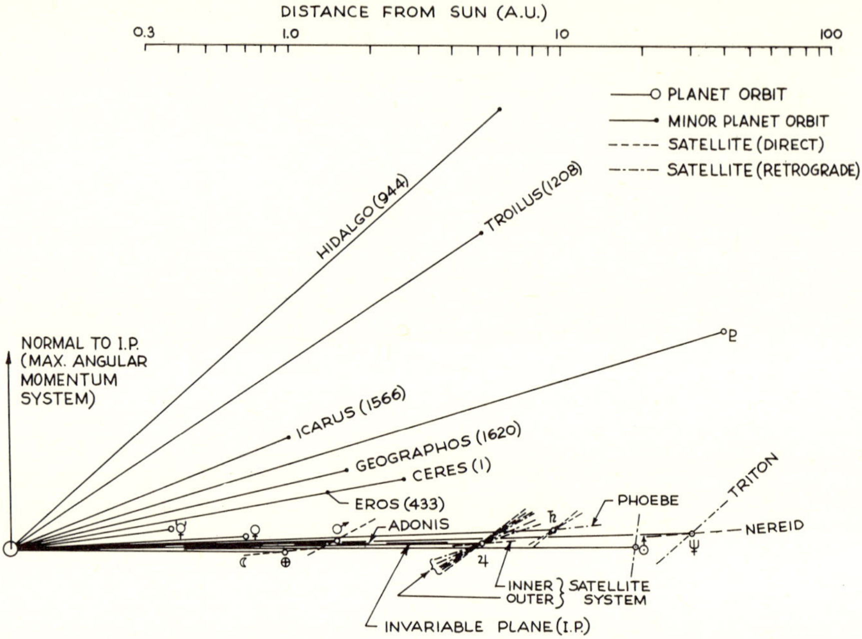

Fig. 2-35. Orbital Inclinations of Planets and Major Satellites and Asteroids (Minor Planets).

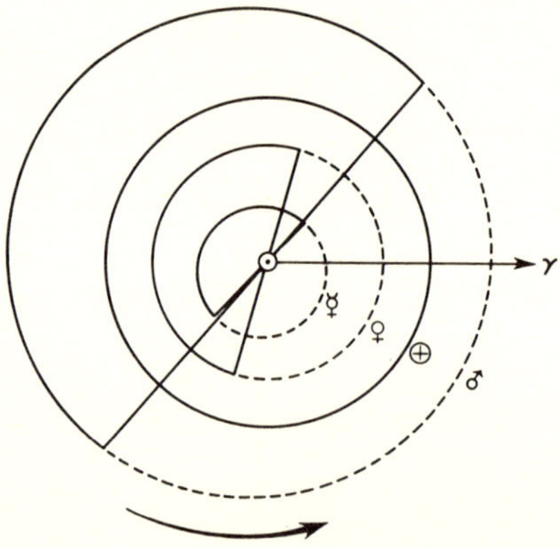

Fig. 2-36. Orbits and Ascending Nodes of the Inner Solar System Planets.

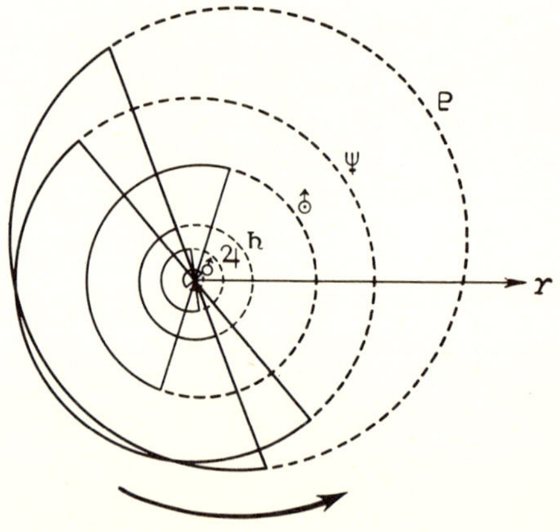

Fig. 2-37. Orbits and Ascending Nodes of the Outer Solar System Planets.

similar to the solar system subdivision. The four innermost moons are called the Galilean satellites after their discoverer, Galileo. The five inner satellites revolve in almost circular, direct orbits. Moons VI, VII, and X of the outer system follow direct orbits. Moons VIII, IX, XI, and XII follow retrograde orbits. The orbits of VIII and IX are greatly disturbed, not even closing on a repetitive orbit. All of the twelve moons undergo considerable inter-action so that their orbits vary to the extent that their sequence in order of distance from Jupiter may vary. More specific data on the orbits and physical characteristics of these satellites and others are contained in Table 2–15. The satellites of Saturn, Uranus, Neptune, and Mars present other interesting "miniature solar systems," although without the great variety of conditions found in the Jovian system. The satellites of Uranus and Neptune are all in retrograde orbits.

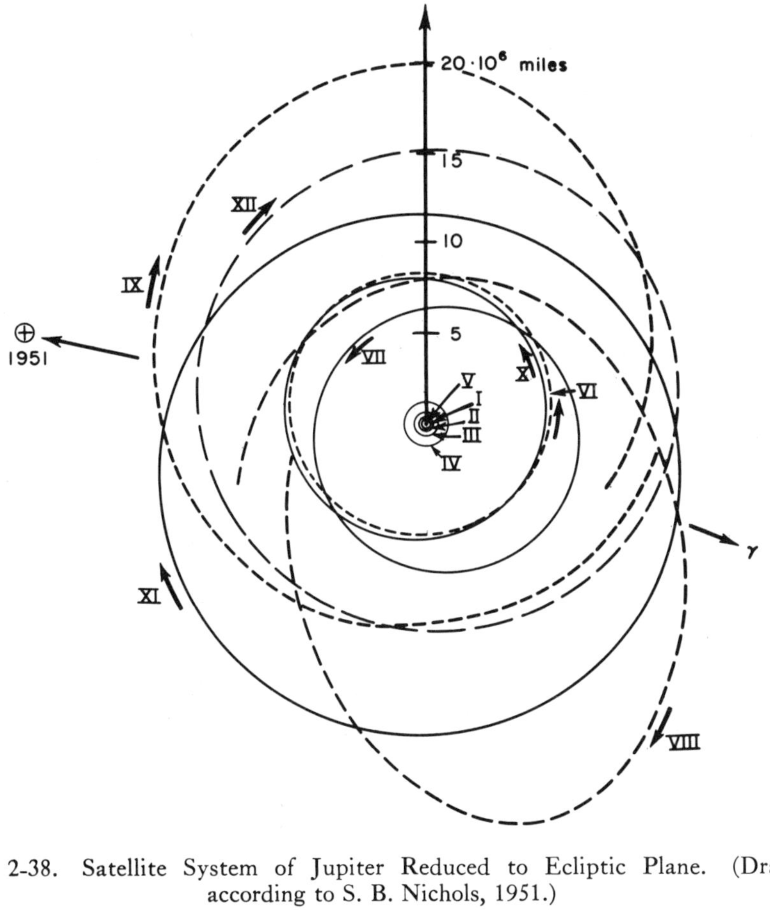

FIG. 2-38. Satellite System of Jupiter Reduced to Ecliptic Plane. (Drawn according to S. B. Nichols, 1951.)

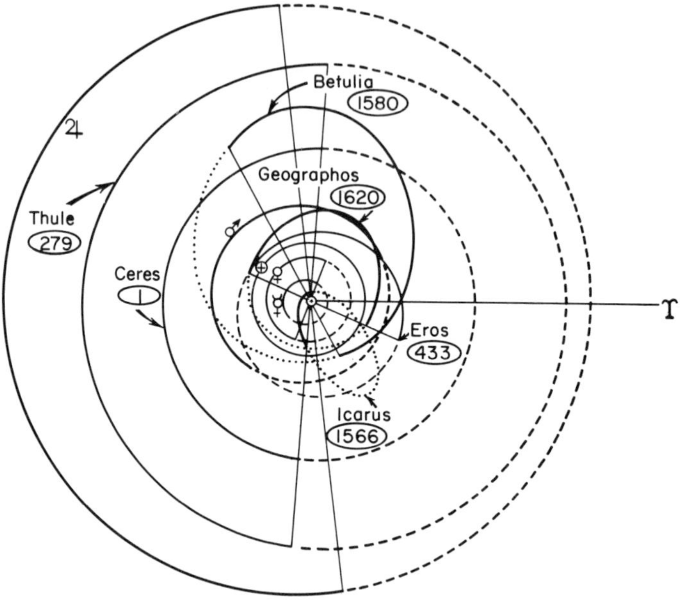

Fig. 2-39. Orbits of Major Asteroids in the Inner Solar System.

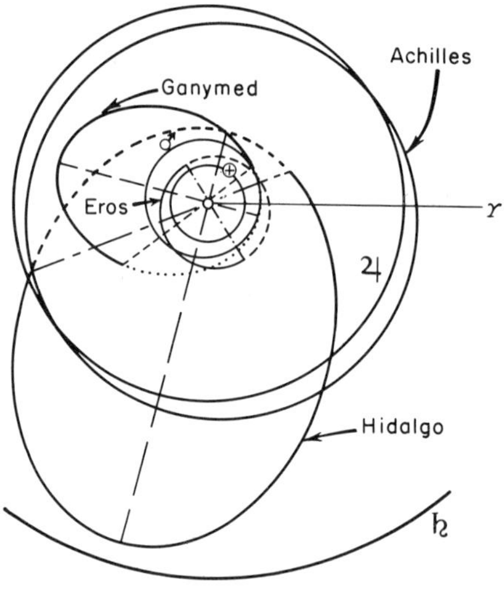

Fig. 2-40. Orbits of Major Asteroids in the Jovian Belt.

Orbits of some of the major asteroids are depicted in Figures 2–39 and 2–40 along with planet orbits in their proximity. Note the high eccentricity of some. Icarus crosses both the Mercurial and Martian orbits, and Betulia, Geographos, and Hidalgo are others with notable high eccentricity orbits. In general, the asteroids fill a "gap" between Mars and Jupiter, and may be the result of the explosion of some intermediate planet, or the failure of some planet to coagulate properly (perhaps under the gravitational influence of its sizeable neighbor, Jupiter). Two interesting groups of Asteroids are known as the Trojans. These bodies are co-orbital with Jupiter about the Sun, and lie at the two equilateral libration points. The Trojans follow periodic oscillations about the libration points in three axes which may be a physical corroboration of the inherent stability of orbiting in equilateral relation with two other bodies of unequal mass, and hence point the way to naturally stable space station locations throughout the solar system as well as the Earth-Moon system.

Comets exhibit some extremely eccentric orbits as will be noted in Table 2–17; an extreme example is Halley's comet with perihelion distance of 0.6 A.U., aphelion distance of 35.5 A.U., an orbit inclination of 102.2 degrees, and a period of 76 years. As the orbit eccentricity approaches unity, the orbit approaches an escape parabola, so that there may be some extremely long-period comets with near-parabolic orbits which would be observed only once in centuries and would not be identifiable on successive returns as the same comet. Short-period comets tend to orbit within 18 degrees inclination from the ecliptic, with only a very few (five known) with inclinations less than 6 degrees. A second group of short period comets orbits with approximately 90 degrees inclination. Long-period comets seem to be concentrated in the 80 to 150 degree orbit inclination band. Periodic comets can be correlated by their aphelion distances with the orbits of the four largest planets as is done in Table 2–17. Figure 2–41 illustrates a process by which a high eccentricity comet may be captured by a major planet to become one of that planet's comet system. As the comet approaches the planet and the Sun, the planet's gravitational pull on the comet acts along the lines connecting the instantaneous positions of the planet and the comet. Initially, as may be seen easily by resolving the instantaneous position line or force direction vector, the comet's eccentricity is increased, and it is pulled closer to the planet. However,

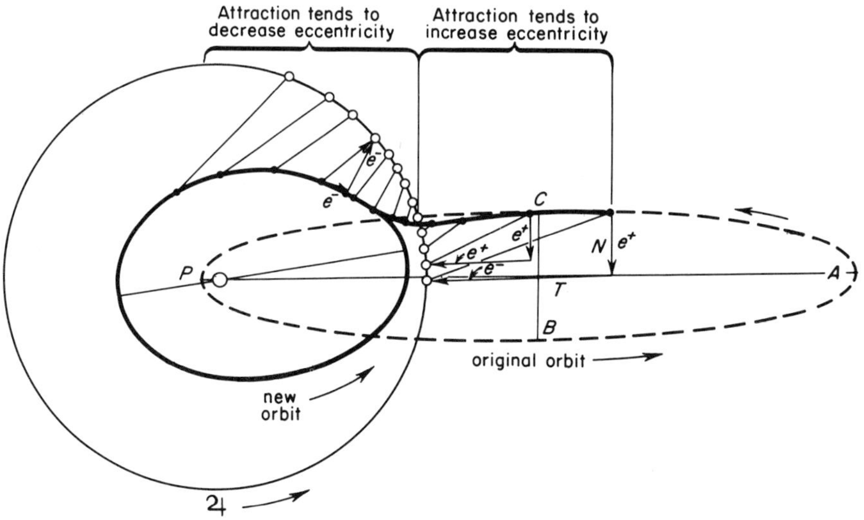

FIG. 2-41. Schematic Presentation of the Capture of a Comet by Jupiter.

this proximity increases the gravitational force significantly, so that as the comet proceeds on past the planet, the planet tends to pull the comet along with it, and the comet orbit eccentricity decreases. Using Tisserand's orbit criterion,

$$\frac{1}{a} - 2\sqrt{a(1 - e^2)} \cos i = C \tag{2-27}$$

where a is the semi-major axis of the ellipse, e the eccentricity, and i the inclination, we may identify some "captured" comets with other "vanished" ones. That is, taking the values with respect to the Sun-planet barycenter, and denoting the original and new orbits by subscripts, we may say,

$$\frac{1}{a_1} + 2\sqrt{a(1 - e_1^2)} \cos i_1 = \frac{1}{a_2} + 2\sqrt{a(1 - e_2^2)} \cos i_2 \tag{2-28}$$

There are many more high eccentricity comets ($e>0.9$) than periodic ($0.4<e<0.9$). The great periods and near-parabolic orbits of such comets make it difficult to identify them for longer than a single visit. A comet of hyperbolic orbit would be of extra-solar origin, or at least would be preparing to leave the solar system.

By reorienting the comet-planet situation for capture, we may accelerate a comet into hyperbolic orbit ($e>1$). Figure 2–42 illustrates this where the comet has passed through aphelion, is accelerated by the planet's

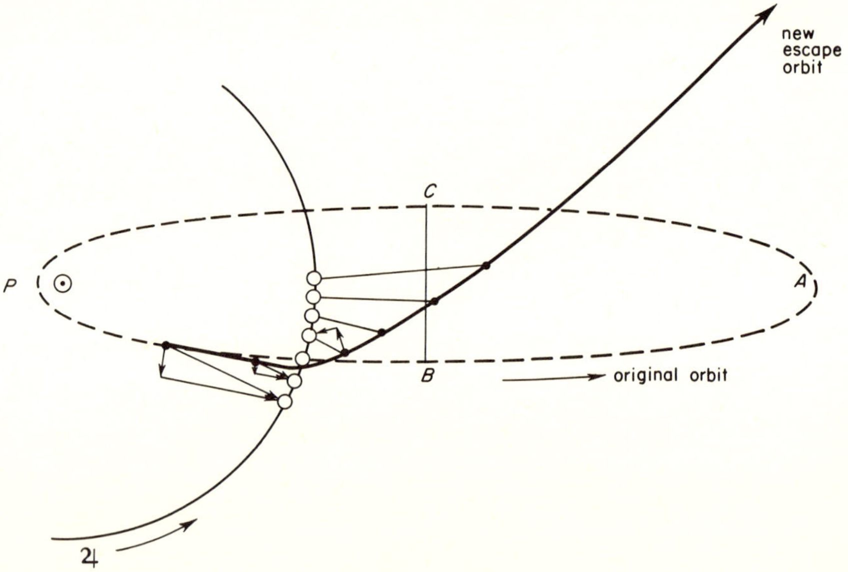

FIG. 2-42. Schematic Presentation of the Removal of a Comet from the Solar System by Jupiter.

gravitational pull into hyperbolic velocity and attains such kinetic energy that the planet cannot slow it down sufficiently on passing but merely deflects the comet's orbit as it departs the solar system.

Meteors and dust are generally part of the solar system. Table 2–18 presents orbital data on some better known meteor streams. Meteor streams may sometimes be associated with a particular comet having related occurrences, orbit inclinations, etc. Being solar system material, meteors will have velocities less than escape velocity (a considerable velocity on collision), and can be predicted somewhat as to occurrence distribution and size distribution from Earth atmosphere encounters (see Section 2.3.5).

2.4.3 Gravitational Field

The gravitational field of the solar system is dominated by the Sun's field which is the prime force holding the solar system together. However, local areas are present around all the significant bodies, in which the Sun's field does not predominate. The gravitational parameters of the Sun, planets and satellites have been presented in Tables 2–13 and 2–15. Figure 2–43 portrays the Sun's gravitational field in Earth surface gravity units versus distance from the Sun. This is basically a $1/r^2$ relation; that is, the field strength diminishes with distance from the Sun as the square of the distance increases. The function may be taken directly in terms of the gravitational parameter as:

$$g_\odot = \frac{K_\odot}{i^2}$$

The Sun is spherical in shape, with no apparent oblateness, so that higher order terms are very minor relative to the primary gravitational potential field strength, which may thus be taken

$$U_\odot = \frac{K_\odot}{r} = \frac{GM_\odot}{r} \tag{2–30}$$

In such a simple central force field, we may easily calculate the circular velocity necessary at any radius to balance the centripetal force with the gravitational force as,

$$V_c = \sqrt{|U|}\,\sqrt{|gr|} = \sqrt{\frac{K}{r}} \tag{2–31}$$

The parabolic velocity is related to the circular velocity by $\sqrt{2}$; this velocity being that necessary to remove a body from the influence of a field. Thus, the parabolic velocity is,

$$V_p = V_c\sqrt{2} = \sqrt{\frac{2K}{r}} \tag{2–32}$$

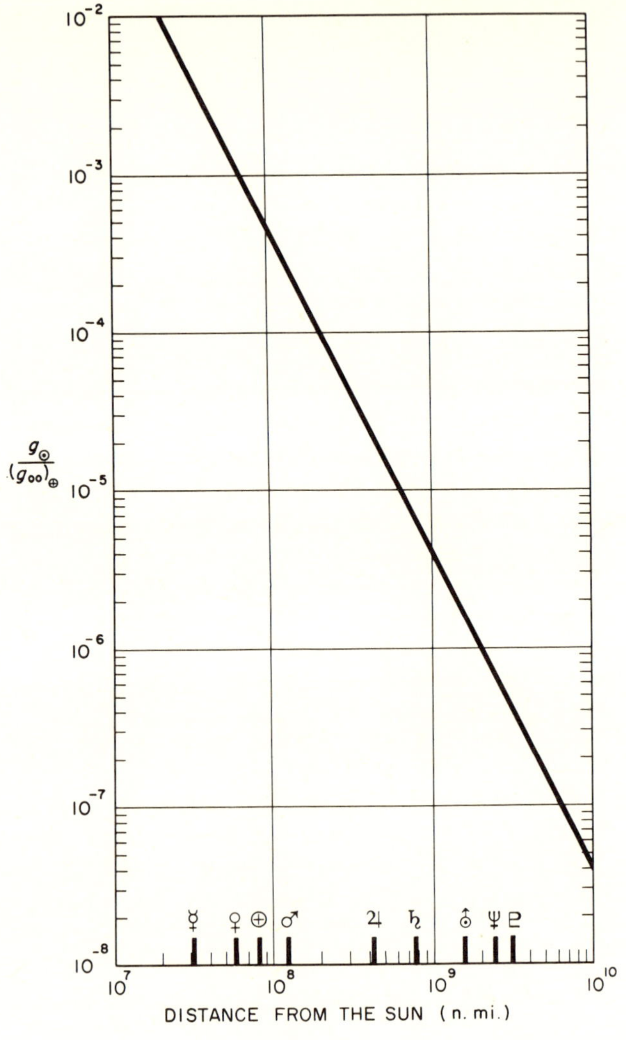

FIG. 2-43. The Gravitational Field of the Sun.

Figure 2–44 is a plot of the circular and parabolic velocity requirements of the solar gravitational field. Bodies in elliptic orbits will have velocities less than the circular requirement at aphelion, and between circular and parabolic at perihelion. Hyperbolic orbits require velocities greater than parabolic. We may think of the solar field as being a gigantic funnel, and classify these velocities as those necessary (1) to maintain one's radial distance from the funnel axis (circular velocity), or (2) to leave the funnel entirely (parabolic velocity). Similar central force fields occur around the individual planets, satellites and smaller bodies (neglecting oblateness and anomalous factors). Figures 2–45 and 2–46 show the circular and parabolic velocity requirements for the Earth, Moon, Venus, and Mars as a function of radial distance from the planet or satellite center in terms of the planet or satellite surface radius. Figure 2–46 portrays, in addition, the summation of Earth and Moon parabolic velocity requirements. This may be thought of as superimposing a small funnel-shaped dent on the Earth's funnel in terms of velocities, with the small funnel hole in rotation about the large hole. In a similar manner, the individual planetary gravitational fields or velocity requirements may be superimposed on the Sun's field, which yields, in parabolic velocity terms, Figure 2–47. This figure indicates that the large Sun funnel has several smaller funnels rotating around on it at velocities determined by their radial distances; and thus it indicates the aiming complexity of simple boost, free-flight, minimum energy trajectories (MET), and the variations in transit trajectories as the funnel or planetary constellations change.

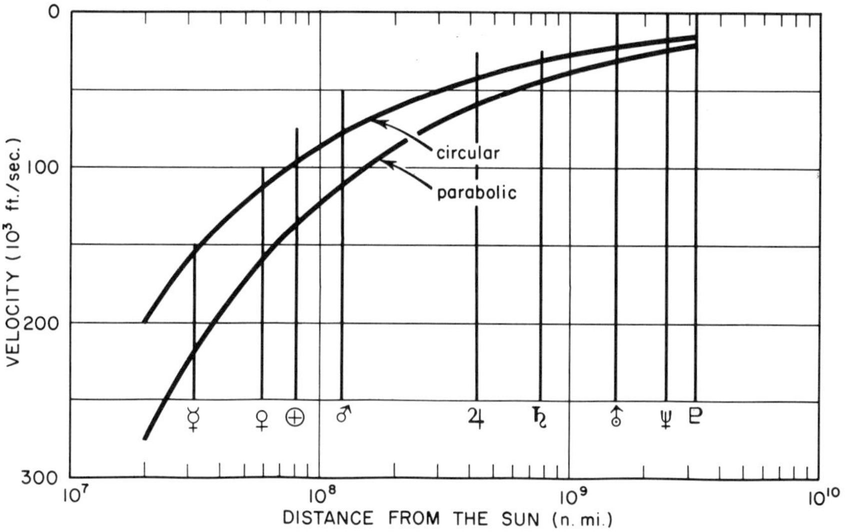

FIG. 2-44. Circular and Parabolic Velocity in the Solar Field.

2.4.4 Magnetic Fields

The solar magnetic field is known to be quite complex. The field strength is the largest in the solar system, and has been estimated at nearly 10 gauss (Ref. (*1*)) to 5 gauss (Ref. (*42*)). The magnetic field pattern is that of a simple dipole with poles coincident with the rotational poles. Superimposed on this simple field are very strong fields in the vicinity of sunspots, and large regions of low magnetic activity. The eruptive nature of the

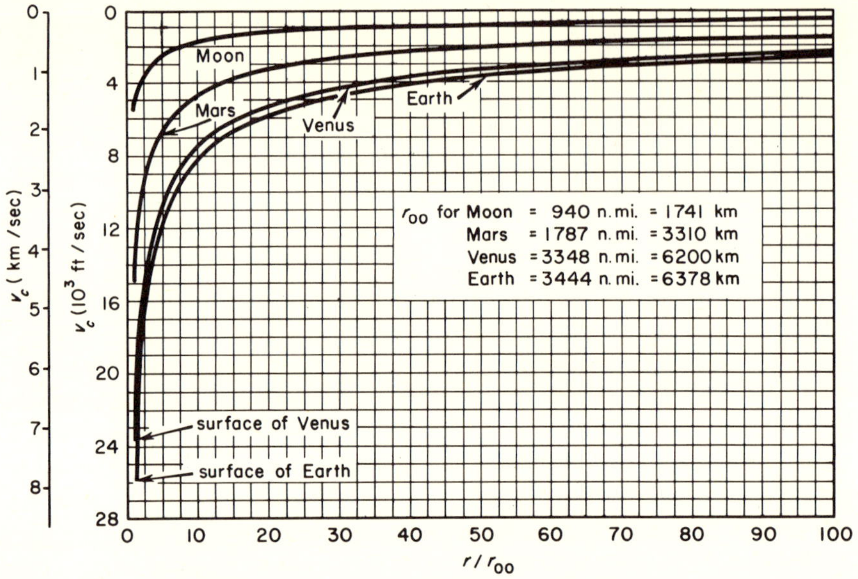

FIG. 2-45. Circular Velocity Requirements for Earth, Moon, Venus and Mars.

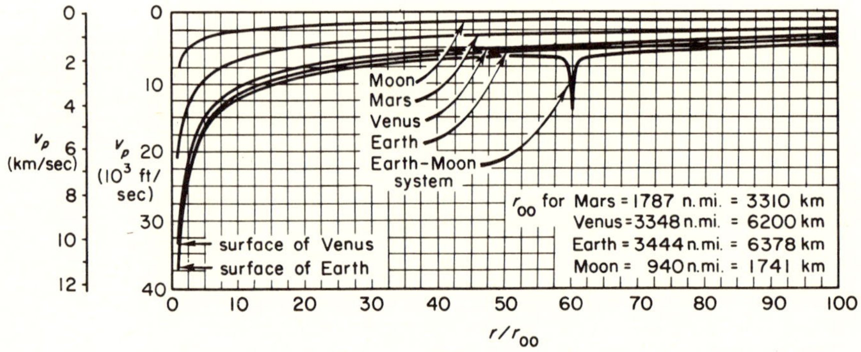

FIG. 2-46. Parabolic Velocity Requirements for the Earth, Moon, Venus, Mars and Earth-Moon System.

Sun's surface seems to have major effects on the magnetic field as well, producing magnetic storms throughout the solar system.

Very little is known of planetary magnetic fields; however, speculation on the existence of these and corresponding belts of trapped radiation around Mars and Venus has been corroborated somewhat by radio emissions from Jupiter which suggests a strong charged particle belt associated with that planet (Ref. (*7*) of Chapter 1).

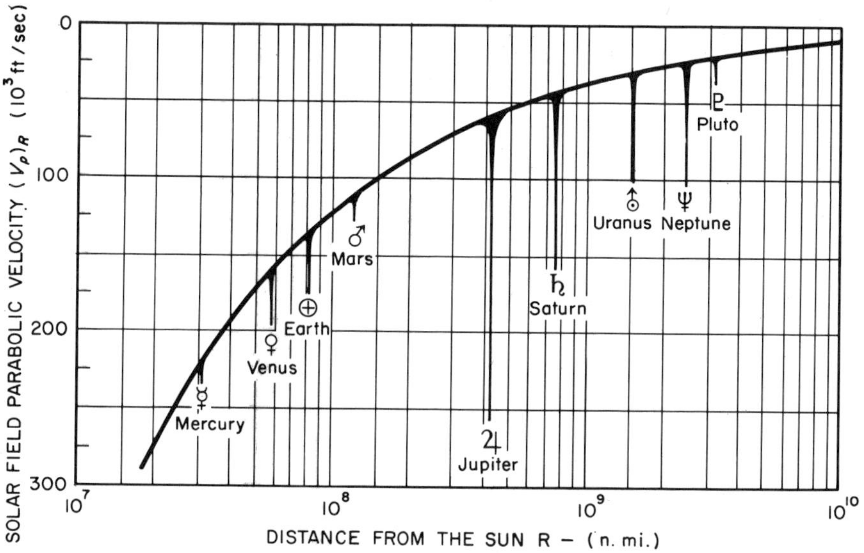

Fig. 2-47. Solar System Parabolic Velocity Requirements.

2.4.5 Radiation

The dominant radiating body in the solar system is the Sun. Radiations occur in many types: thermal, visual, radio, X-ray, ultraviolet, and particle. The Sun has a "linear" diameter of 109 Earth diameters or 1.39×10^6 km (870,000 mi.). The Sun's apparent "angular" diameter, as viewed from Earth, is about 1/2 degree. More precise values are:

Perihelion	32′36″
Aphelion	31′32″
Mean Distance	31′59″

The mean or average distance from the Earth is $149.5 \cdot 10^6$ km or 8.32 light minutes. Figure 2–48 shows the apparent diameter of the Sun as a function of distance.

Three recognizable solar "gaseous" strata are defined as the photosphere, chromosphere, and corona. The photosphere is a spherical shell of about 400 km thickness. It has a granular texture and exhibits a violent vertical and horizontal turbulence. The photosphere has an outer edge pressure of about 0.01 atmosphere (atm) and a temperature of about 5750°K. About 80 per cent of the observable solar radiation is produced in the photosphere; the remaining radiation coming from both below (10 per cent) and above (10 per cent). Radiation of wavelength (λ) greater than 910Å generally comes from the photosphere. The chromosphere is a layer gas at about 10^{-7} atm pressure which emits $\lambda\lambda$ (wavelength band) 100–910 radiation.

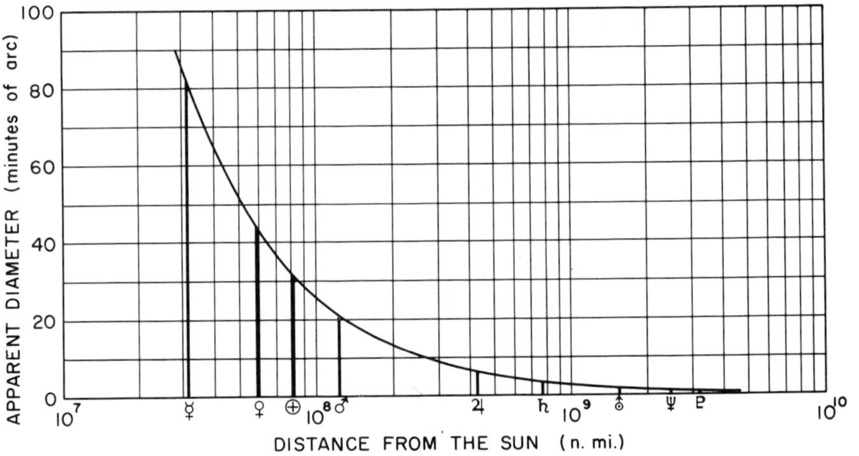

FIG. 2-48. The Apparent Diameter of the Sun.

The corona is an even thinner envelope extending out to about 3 Sun radii (altitude of 2 Sun radii from photosphere). Emissions of λλ20–100 come from the corona. Table 2–19 is a survey of the Sun's outer layers which presents the best current knowledge (Ref. (*43*)) of their physical characteristics including solar layer definition, altitude, number of particles per cubic centimeter, temperature, and whether hydrogen is ionized or neutral.

TABLE 2–19. SURVEY OF THE SUN'S OUTER LAYERS*

Solar Layer	*h* (Km)	*r* (R_0)	$\log \frac{N_p}{cm3}$	*T* (°K)	*Hydrogen is:*
Outer corona	1,400,000	3	5.3	1,500,000	Ionized
Medium corona	700,000	2	6.2	1,500,000	Ionized
Inner corona	140,000	1.2	7.6	1,500,000	Ionized
	21,000	1.03	8.3	1,000,000	Ionized
Upper chromosphere	14,000	1.02	8.7	300,000	Ionized
	7,000	1.01	9.7	25,000	Ionized
Lower chromosphere	4,000	1.006	10.0	6,100	Neutral
	1,000	1.0014	10.7	5,300	Neutral
	500	1.0007	11.0	4,900	Neutral
Photosphere	0	1.0000	12.1	4,500	Neutral
	−100	0.9999	12.3	4,500	Neutral
	−210	0.9997	12.7	5,000	Neutral
	−335	0.9995	13.8	6,400	Neutral
	−360	0.9995	14.3	7,000	Neutral
	−375	0.9995	14.7	7,300	Neutral

* Reprinted from *The Sun* by Gerard P. Kuiper by permission of The University of Chicago Press. Copyright 1953 by the University of Chicago.

Sunspots and other prominences appear in the Sun's outer layers from time-to-time. Sunspots are dark regions of 450 mi. to ten thousands of miles diameter, occurring generally in the 10 to 30 degree latitudes North and South of the Sun's equator. A nominal mean value for a large sunspot radius is 0.02 the Sun's radius, or 17,400 miles. Prominences are vast flame-like eruptions which are observable at the periphery of the solar disc. Table 2–20 classifies the various types of solar prominences.

TABLE 2–20 CLASSIFICATION OF SOLAR PROMINENCES
(according to Pettit)

Type	Designation	Description
I	Active	Most frequent type seen all over Sun. Filaments originate in photosphere; follow curved path into a point at base of chromosphere ("center of attraction").
II	Eruptive	Occurrence more frequent in sunspot regions. Often evolves from Type I; may rise several 100,000 mi. before being pulled back into a center of attraction; may separate and rise to great altitudes and fade while climbing; highest ascent observed: 1.22 solar radii.
III	Sunspot	Appear in vicinity of spots; may be of Type I, II or V.
IV	Tornado	Rare, resembling cyclones 3000–12,000 mi. in diameter, 15,000–60,000 mi. high; may evolve into Type I or II or disintegrate.
V	Quiescent	Longest lifetime of all prominences; minimum of activity; about 300 mi. thick, 25,000 mi. high, 120,000 mi. long.
VI	Coronal	Originate in corona; are long, curved strips of luminous gas. Descent at high speed (60–120 mi./sec) toward the Sun, possibly drawn in by sunspots.

The temperature at the Sun's center is about $20 \times 10^{6°}$ K at which the ionized matter behaves like a gas in spite of a pressure equivalent to millions of atmospheres. Radiation from the Sun's interior is in the X-ray and γ-ray region, $\lambda\lambda$10 to 0.1Å (1Å = 10^{-8} cm), which will not pass through to the surface without going through transformation to longer wavelengths. The maximum radiation at the photosphere is in the yellow portion of the visible spectrum at about λ5400Å.

Table 2–21 presents some additional data pertaining to solar radiation. The temperature of the Sun's outer layers is determined from the Stephan-Boltzman law,

$$E = \sigma T^4 \tag{2–33}$$

where E is the radiation intensity (energy per unit time per unit area) where σ is the Stephan-Boltzman constant [$5.669 \cdot 10^{-5}$ erg cm^{-2} (°K)$^{-4}$, or $17.3 \cdot 10^{-10}$ Btu hr^{-1} ft^{-2} (°R)$^{-4}$, or $1.374 \cdot 10^{-12}$ cal cm^{-2} sec^{-1} (°K)$^{-4}$], and T is the temperature.

Table 2–21 Data Pertaining to Solar Radiation

Stellar magnitude:	
photographic	−26.41
visual	−26.68
Total solar radiation	$3.86 \cdot 10^{33}$ erg/sec
Mean brightness of Sun's disc in space at 1 A.U. distance	$2.08 \cdot 10^{5}$ stilb = $6.53 \cdot 10^{5}$ lambert
Brightness of center of disc in space at 1 A.U. distance	$2.62 \cdot 10^{5}$ stilb = $8.23 \cdot 10^{5}$ lambert
Light flux in space at 1 A.U. distance	14.14 phot = 141,400 lux
Candlepower	$3.17 \cdot 10^{27}$ p
Solar constant in space at 1 A.U. distance	1.94 cal/cm^2 min

Using the solar constant, the radiation energy intercepted normal to the Sun's direction per unit time per unit area, the Sun's surface temperature may be calculated to be 5750°K. The use of Planck's equation,

$$E_\lambda = \frac{C_1 \lambda^{-5}}{e^{C_2/\lambda T} - 1} \tag{2–34}$$

yields other values (Ref. (*35*)), and solar strata and Earth's atmospheric absorption must be considered.

The solar radiation energy, S, may be computed as a function of the solar constant, $S_\oplus$, the astronautical unit, $R_\oplus$, and distance from the Sun, R,

$$\frac{S}{S_\oplus} = \left(\frac{R_\oplus}{R}\right)^2 = \frac{9.05 \times 10^{25}}{4\pi R^2} \ \frac{\text{cal/sec}}{\text{unit area}} \tag{2–35}$$

A plot of this relation is shown in Fig. 2–49. The pressure field of this radiation may be calculated using Einstein's relation between energy and mass,

$$E = mc^2 \tag{2–36}$$

The solar constant, S is thus

$$S = \dot{m}c^2 \tag{2–37}$$

where $\dot{m}$ is the photon mass intercepted per unit time per unit area. The momentum per unit area is,

$$M = \dot{m}tc = \frac{St}{c} \cos \alpha \tag{2–38}$$

where α is the angle of incidence (angle between normal and radiation vectors).

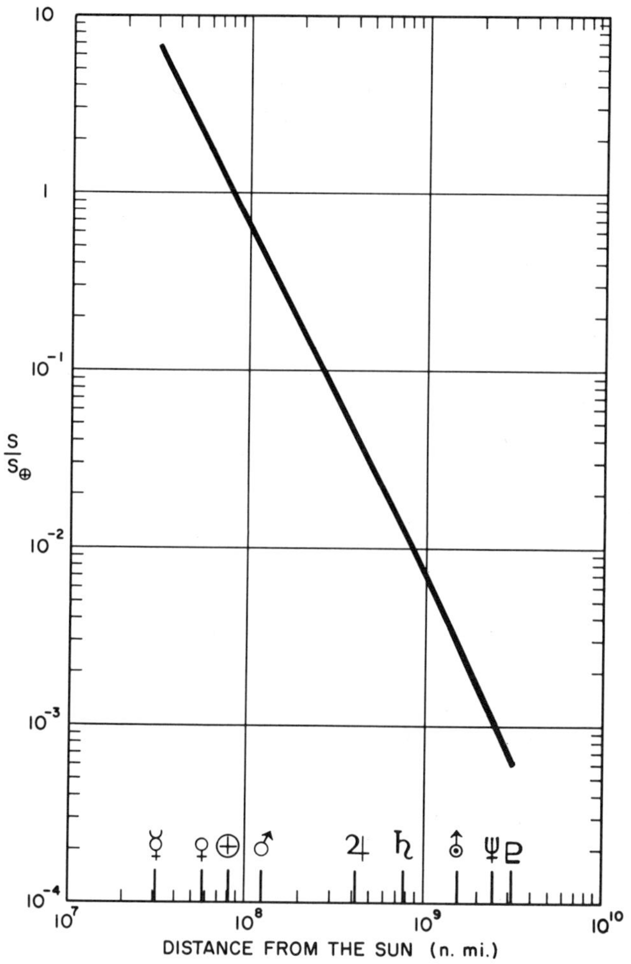

FIG. 2-49. Radiation Energy Field of the Sun.

The force per unit area is,

$$F = \frac{dM}{dt} = \dot{M} \cos \alpha = \frac{S}{C} \cos^2 \alpha \tag{2-39}$$

If the body has a reflectivity, σ, the force per unit area or radiation pressure is,

$$p_r = (1 + \sigma) \frac{S}{c} \cos^2 \alpha \tag{2-40}$$

Figure 2–50 is a plot of this function for a perfect reflector, $\sigma = 1$, and normal incidence, $\alpha = 0$. The ratio of outward radiation pressure to gravitational pull may be balanced for various shapes and specific densities.

For a spherical body, the ratio of radiation pressure to gravitational pull is $1/r$, where r is the sphere's radius. If the specific gravity is 1, the critical radius is 1.5 microns, indicating that particles of small size will literally be blown out of the solar system by radiation pressure.

A body in a radiation field generally absorbs radiated energy and converts it into heat thus increasing its own radiation until an equilibrium is reached between energy absorbed and radiated. This establishes a radiation equilibrium temperature. If we use the solar constant, S, the

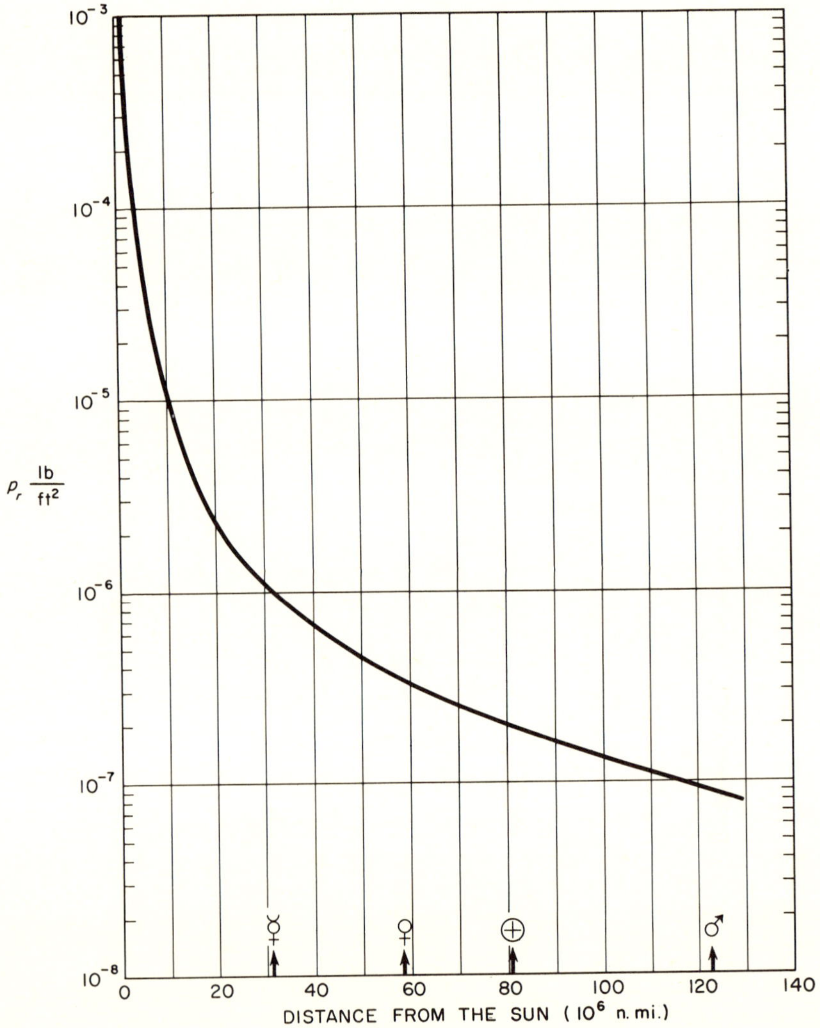

FIG. 2-50. Radiation Pressure Field of the Sun for Normal Incidence on a Perfect Reflector.

absorptivity of the material, α_s, and the effective absorbing area, A_α, we may calculate the energy absorbed,

$$E_\alpha = \alpha_s A_\alpha \, S \tag{2–41}$$

The Stephan-Boltzman law may be expanded to introduce emissivity, ϵ_T, at temperature, T, from emitting area A_ϵ to obtain energy emitted,

$$E_\epsilon = E_T A_\epsilon \, \sigma \, T^4 \tag{2–42}$$

These two equations yield the radiation equilibrium temperature,

$$T_{eq} = \left[\frac{\alpha_s}{\epsilon_T} \frac{A_s}{A_\epsilon} \frac{S}{\sigma} \right]^{1/4} = \left[z \frac{S}{\sigma} \right]^{1/4} \tag{2–43}$$

Figure 2–51 shows the temperature field of the sun plotted for various values of z which depends on the ratio of absorptivity to emissivity and the ratio of effective absorbing area to effective emitting area,

$$z = \frac{\alpha_s}{\epsilon_T} \frac{A_s}{A_\epsilon} \tag{2–44}$$

It should be noted that internal heat conductivity has not been introduced in the foregoing and that these equations are strictly valid only if the heat conductivity is infinite.

Particle radiation from the Sun may occur at a generally "low" level, interrupted by periods of intense radiation associated with solar flares. In the Chapman-Ferraro theory of magnetic storms (Ref. (*44*)), corpuscular streams of electrons and protons are emitted from active regions on the Sun, and proceed earthward to cause magnetic storms and aurorae. These particles travel to the Earth in about one day, indicating a velocity on the order of 10^8 cm/sec. The particle density per cm^3 decreases rapidly with radial distance from the Sun, being about 100 particles per cm^3 at one A.U. On entering the atmosphere, the Earth's magnetic field is supposed to guide the particles so as to cause aurorae and polar magnetic storms, while some particles are circulated in an equatorial ring current at a distance of a few Earth radii; the so-called Störmer current rings which bear a marked similarity to the Van Allen ionization belt (quite likely Van Allen measured what Störmer predicted). The particle occurrences are associated with solar prominences such as sunspots and flares. Swamm (Ref. (*44*)) has predicted that sun spots from giant stars (many times larger than our Sun) may be so large and have sufficiently strong peripheral electric fields, ($E = -R\Delta H/\Delta t$), as to be capable of accelerating particles into intergalactic space, thus producing the extremely high energy cosmic rays of extra-solar system origin. Solar particles of 1/3 the speed of light have been observed leaving the Sun in solar flare areas. Particle energies of 18 billion electron volts have been associated with solar storms. Present

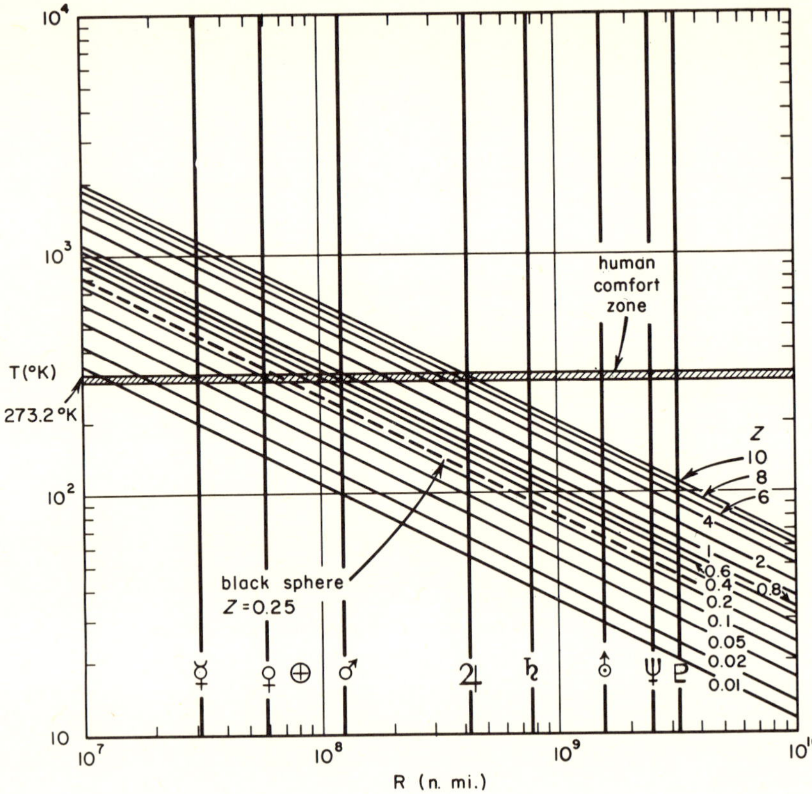

FIG. 2-51. Temperature Field of the Sun.

incomplete data indicates that in addition to protons and electrons, about 15 per cent of primary cosmic rays are α-particles, about one or two per cent are heavier nuclei (Li, Be and B are not found for some inexplicable reason) while less than one per cent are primary electrons, and less than 1/10 per cent are gamma rays.

The Sun emits radio noise as well as infrared, visual, and ultraviolet as indicated by the temperatures relation of Plank's equation,

$$E_\lambda = \frac{C_1\lambda^{-5}}{e^{C_2/\lambda T} - 1} \tag{2-45}$$

where E_λ is the energy flux in ergs/cm^2

$C_1 = 3.71 \cdot 10^{-5}$

$C_2 = 1.435$

λ is in cm

T is in °K

A plot of this relation for several temperatures is shown in Fig. 2–52. As may be observed from this figure, for a given temperature the energy radiated is a maximum at a particular wavelength. This is known as the Wien displacement law,

$$\lambda_{max} = \frac{C}{T} \tag{2–46}$$

where $C = 0.289$,

λ is in cm, and

T is in °$K_\odot$

Solar flares are known to increase the radio noise level as much as 1000 times above normal. The mechanism for this is unknown, but the environmental effect is known, and must therefore be circumvented in communication and solar tracking system design.

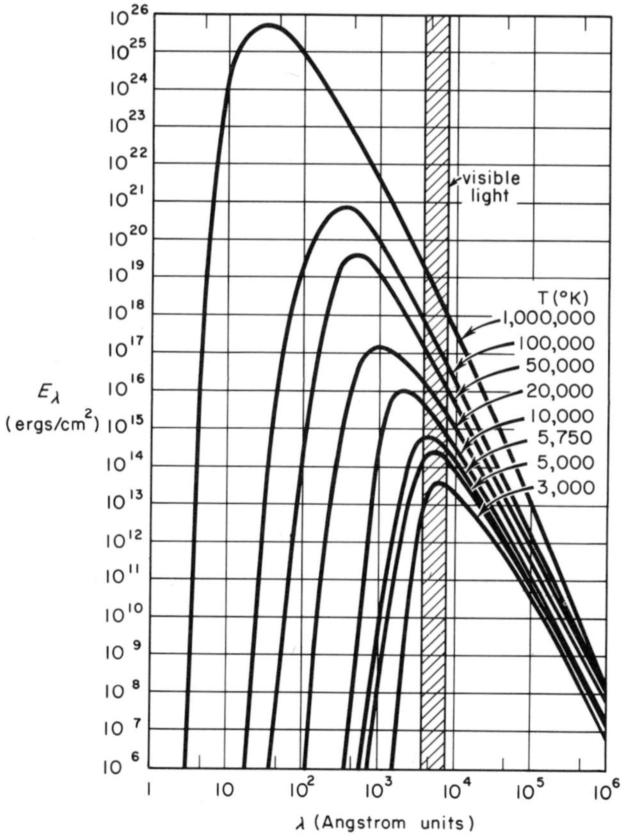

FIG. 2-52. Radiation Energy Curves of a Black Body.

2.4.6 Meteoric Activity and Dust

Knowledge of meteoric activity and dust within the solar system is pretty well confined to that observed in cislunar space, and previously discussed in Sections 2.3.5 and 2.2.7. The source of meteor streams within the solar system has been postulated as related to a mythical planet between Mars and Jupiter which blew up, or failed to form properly (possibly under Jupiter's strong influence) and thus resulted in asteroids and meteors. The high eccentricity of meteor orbits may indicate that some are related to comets, and have been deflected into highly eccentric orbits by near misses of large planets. In any event, the density of meteors and dust material beyond cislunar space must await further development of probes, measurement, and communication experiments.

2.4.7 Planet Characteristics

The surface and atmospheric characterics of the bodies orbiting the Sun are of particular interest to space explorations and operations. In addition to the Earth, Moon, and Sun, previously discussed, we should consider the particular characteristics of the remaining eight planets and a number of their satellites.

Mercury is the smallest planet and is nearest to the Sun. It is thought that its rotational period is equal to its orbital period, as with the Earth's moon, so that one side faces the Sun continually, the other side being illuminated by planetary and star light only. This situation would indicate a sunlit surface temperature of about 330°C (625°F), and a night side surface temperature of about 20°–30° above absolute zero. It is still disputable whether Mercury has any atmosphere at all. If so, it should be very rare (perhaps 1 mm Hg), and composed of heavy noble gases—krypton, argon and xenon.

Venus resembles the Earth in many ways, e.g., size, mass, and atmosphere. Some authors have postulated identical rotational and orbital periods. However, more recent evidence, such as radio signal modulation and thermal equality on day and night sides, has indicated a rotational period of about 22 hours 17 min (± 10 min). The Venutian atmosphere is covered by a dense cloud layer so that visual observations of the surface have not been feasible. Spectroscopic investigations of the upper atmosphere have indicated a high CO_2 concentration, and no O_2 or H_2O. However, this does not mean oxygen and water are not present at surface altitudes. Nitrogen is quite likely the major atmospheric constituent as on Earth. The question is whether there is a terrestrial-type atmosphere which is shielded by the dense cloud layer from the intense solar radiation, or a dry, dust-cloud atmosphere. It is not to be expected that the Venutian atmosphere will be identical to the Earth's, but it may be sufficiently similar to be hospitable or even habitable to terrestrial organisms. Polarization measurements of the Venutian atmosphere and similar measurements with water droplets have indicated some definite correlation as is shown in Fig. 2–53. To establish firmly the true nature of Venus—its atmosphere, surface structure (ocean or desert), cloud composition, etc.—requires much closer experimentation than has been available heretofore.

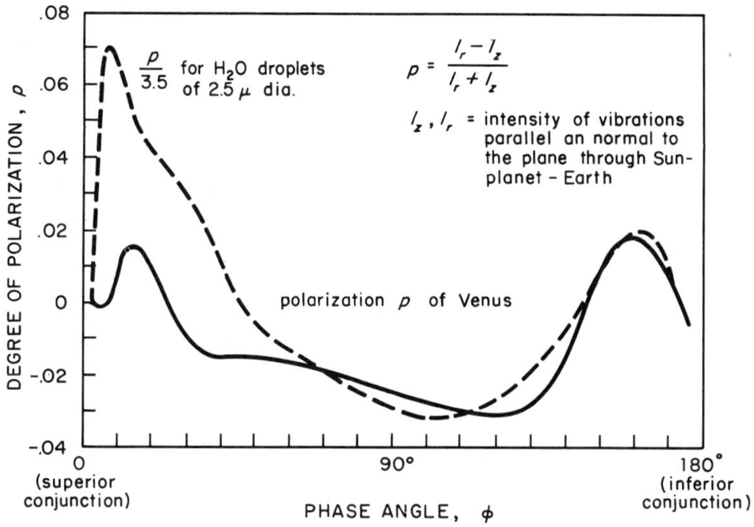

FIG. 2-53. Polarization Measurements of Venutian Atmosphere and Water Droplets.

Mars has exhibited some peculiar characteristics which indicate that it may be the most hospitable of all the extra-terrestrial bodies. Figure 2–54 is a "map" of Martian topography. The so-called mares are no longer believed to be seas. However, the seasonal variation of the polar ice caps indicates some water vapor in the thin atmosphere. The polar ice caps are quite likely merely a frost layer. The light colored areas are probably deserts. The nature of the dark areas is unknown, as is the nature of the canali which connect these mares on great circle routes. Mars has a low albedo, absorbing more solar radiation than the Earth. The Martian temperature averages about 50°C lower than Earth; mean values are postulated of −70°C in the polar regions and −20°C at the equator with higher temperatures in local areas. Noon temperatures at the equator go above freezing to about +25°C in the deserts, and +30 to +40°C in the mares. The Martian atmosphere is much thinner than the Earth's. The main constituent is probably nitrogen, with CO_2, O_2, H_2O, and A being present. Atmospheric composition has been postulated in Table 2–14.

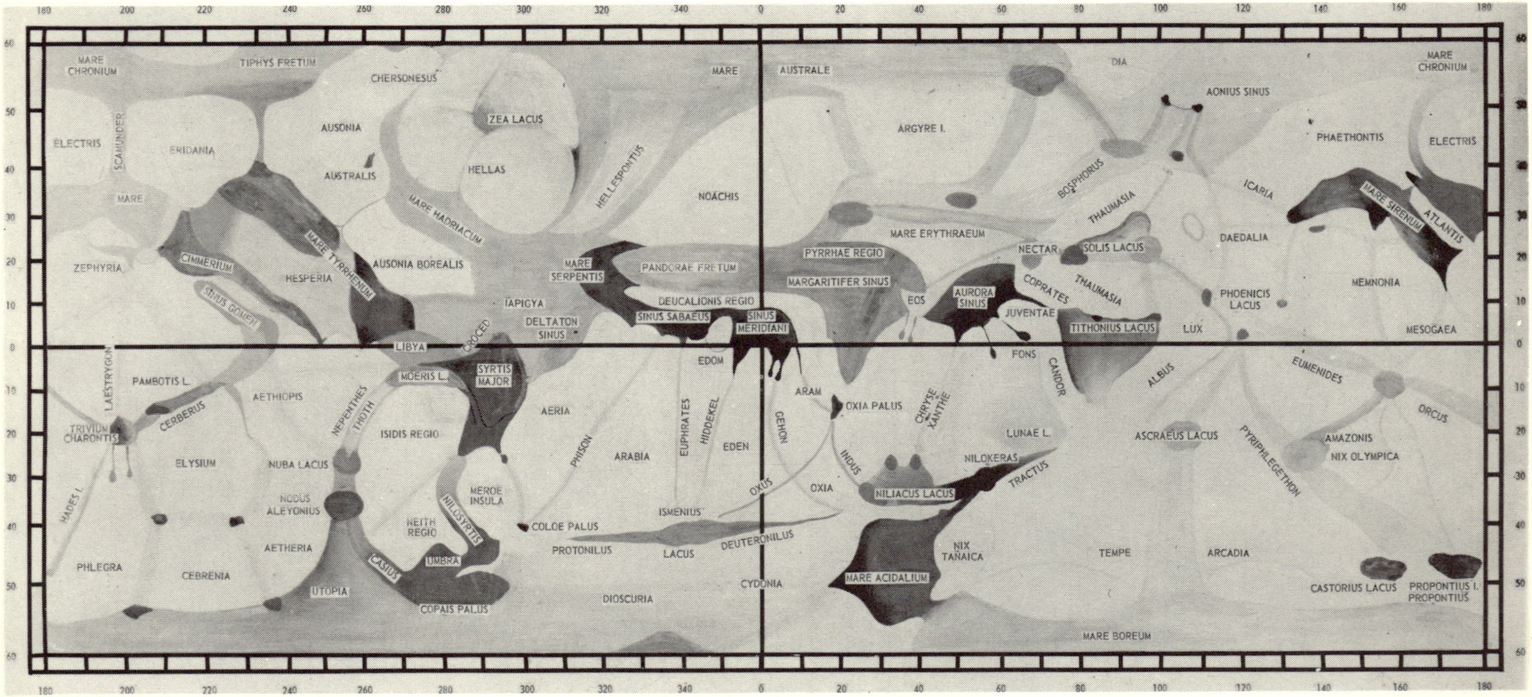

FIG. 2-54. Map of Martian Topography.

Figure 2–55 presents altitude profiles of temperature and pressure for Mars, with Earth profiles for comparison and a Venus pressure profile. It has been estimated that all the frost in the polar caps is no more than the water contained in Lake Erie, and that Mars is a very dry, cold place by Earth standards. The dark areas cover about 20 per cent of the planet's surface. These areas exhibit color changes and seasonal size or definition changes. It is not known whether this is geological, biological, or illusionary in nature. Much hypothesizing has been done on this phenomenon, but first-hand inspection appears necessary to establish clearly the true nature of these areas and their markings. The atmosphere shows some thin cloud layers, blue and violet hazes, and lower white or yellow clouds. It is speculated that the high altitude blue haze and white clouds may be ice crystals. The nature of the violet haze is unknown. Life in some form may be found on Mars. Of all the planets, it certainly is the most promising, although Venus is a distinct possibility.

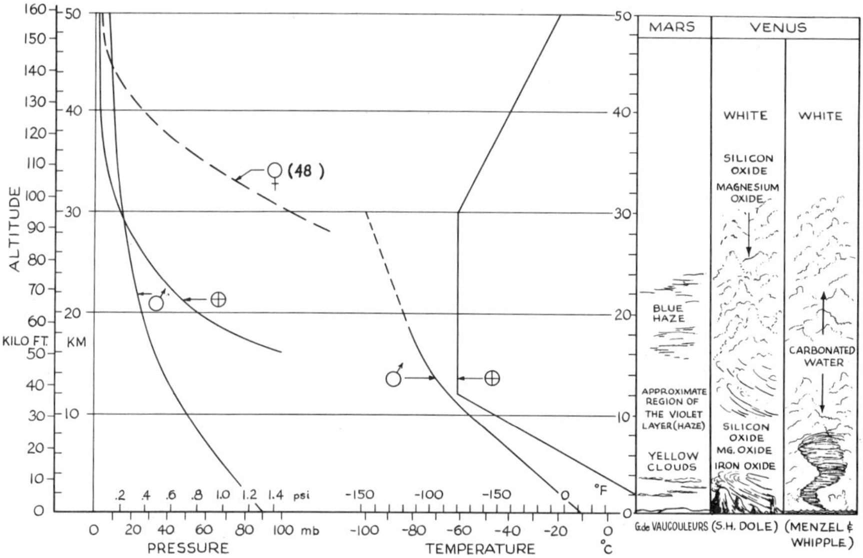

FIG. 2-55. The Atmosphere of Mars and Venus.

The two small satellites of Mars, Phobos and Deimos, are located at 2.77 and 6.95 Mars radii from planet-center. The synodic or apparent period of these satellites in crossing a single planetary meridian is about −11 hours and +131.5 hours, respectively, Phobos appears to rise in the West and set in the East when observed from Mars. This latter phenomenon reflects the relation of surface period, t, orbit period, T, and apparent period, x,

$$1/x = \frac{1}{t} - \frac{1}{T} \tag{2–47}$$

Both satellites are in direct orbits as stated in Table 2–15. The small diameters of these satellites make them attractive as ready-made satellite observatories and store-houses for surface explorations. They both exhibit an orbital precession caused by the planet oblateness, the lower satellite being the more affected. Phobos has an orbital precession period of about 2.25 years while Deimos orbital precession period is about 55 years. The oblateness of Mars is about 1/192 (which is greater than Earth's) and can be expected to perturb space vehicle orbits. The Martian equator is inclined 25 degrees from the ecliptic so that orbital precession will be less effective on an ecliptic plane orbit; however, the oblateness will cause more eccentricity in such an orbit (as compared with an equatorial orbit). The two Martian moons are within 2 degrees of equatorial orbits, or about 26 degrees inclination from the Martian ecliptic plane which indicates considerable energy must be expended to effect a rendezvous with one (easily justified if a moon should possess fuel ores).

Going beyond Mars and the asteroids to the outer solar system we come to the large planets of the solar system, Jupiter, Saturn, Uranus, and Neptune. These planets, and far off Pluto, are extremely distant from the Sun and hence quite cold by Earth standards. Figure 2–56 portrays a planetary temperature curve and individual planet variations throughout the solar system. This chart indicates that active terrestrial-type life will generally be feasible under ambient thermal conditions in the Venus-Earth-Mars realms. Also included in Fig. 2–56 are the impulse velocity requirements for minimum energy transfer from Earth, and minimum energy round trip mission period operating from Earth. Note that minimum energy round trips to the outer solar system will require from seven to over 90 years (without stopping at the orbit aphelion). This suggests that solar and chemical energy will be useful only in the inner solar system, and that nuclear energy will quite likely be the (first) power source for outer solar system exploration. This conclusion, of course, will not hold if natural fuel ores (chemical or nuclear) are discovered on the planets and their satellites. Some of the first exploratory trips to the Moon, Mars, the Martian moons, the asteroids, and the Jovian moons should therefore include a search for nuclear (or chemical) ores with a view towards establishing refineries, fuel storage, and refueling stations. Until proven wrong by actual exploration, the author optimistically predicts that suitable ores will be found and refueling will become a routine operation in space. That is, nuclear and chemical fuels of some type may be found, and engines may be converted to employ different oxidizing agents as well as fuels at the refueling stations. The cost of transfer energy is thus set aside as soluble by propulsion experts, and it is assumed that minimum transfer energy will cease to be a paramount constraint, and we may rationally discuss

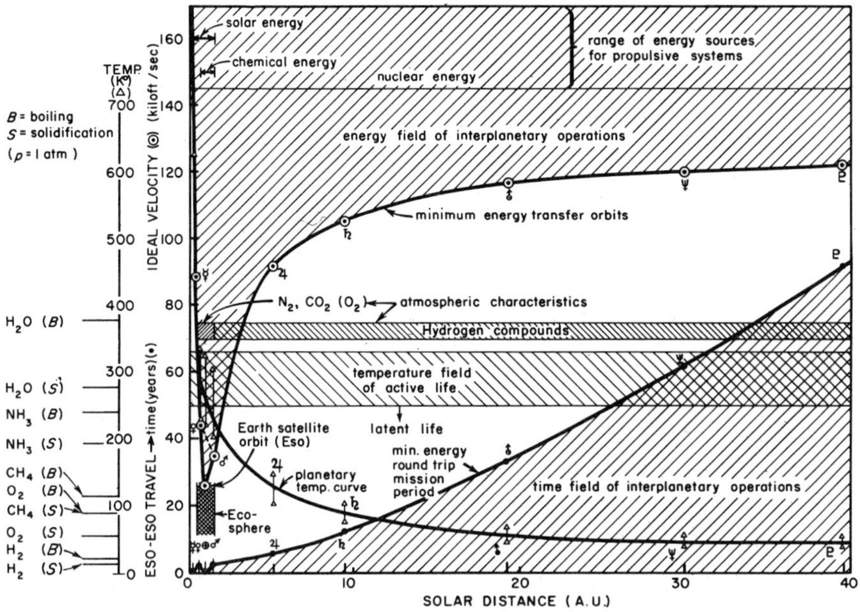

FIG. 2-56. Comparison of Conditions Throughout Solar System as Related to Space Flight Feasibility by Terrestrials.

exploration and transfer orbits, including stops, throughout the solar system in round trips of less than five or ten years.

The large planets, Jupiter, Saturn, Uranus and Neptune, all have low densities as indicated in Table 2–13. Thus, the gravitational parameters indicated in Table 2–13 are much lower than their volumes (relative to Earth) might indicate. Including the centrifugal relief from gravity available from the planetary rotations, it appears (from Table 2–14) that humans might be able to walk on all planets except Jupiter. Walking on Saturn and Neptune might feel as though one was carrying a heavy back-pack (and perhaps a heavy helmet).

The dominating physical characteristic of these planets will be the extreme cold (by Earth standards) which will result in unusual physical states of the standard chemical compounds.

Table 2–14 presents estimates of the atmospheric composition on the various planets. Jupiter has a very dense cloud layer, as do the other large planets. These "atmospheres" appear to be arranged in belts parallel to the planetary equators, and vary over weekly or monthly periods indicating turbulent motion in the lower atmospheres. Jupiter possesses a great red spot which is evident in Fig. 2–57. This spot is about 7,000 by 30,000 miles. There is a wide variety of color in the Jovian clouds. Radiometric measurements have indicated cloud temperatures of 135°–150°K. The atmospheres of Jupiter and Saturn are believed to be predominantly ammonia, methane, hydrogen, and helium. Uranus and Neptune should have similar atmospheres, but with ammonia frozen out by the low temperatures. The low densities of the four large planets indicate that the visible discs are really vast atmospheres surrounding much smaller solid cores.

The planet Pluto is more nearly Earth-size, but is so far away and so cold that very little can be determined directly about it. Its atmosphere, if any, must be predominantly hydrogen and helium.

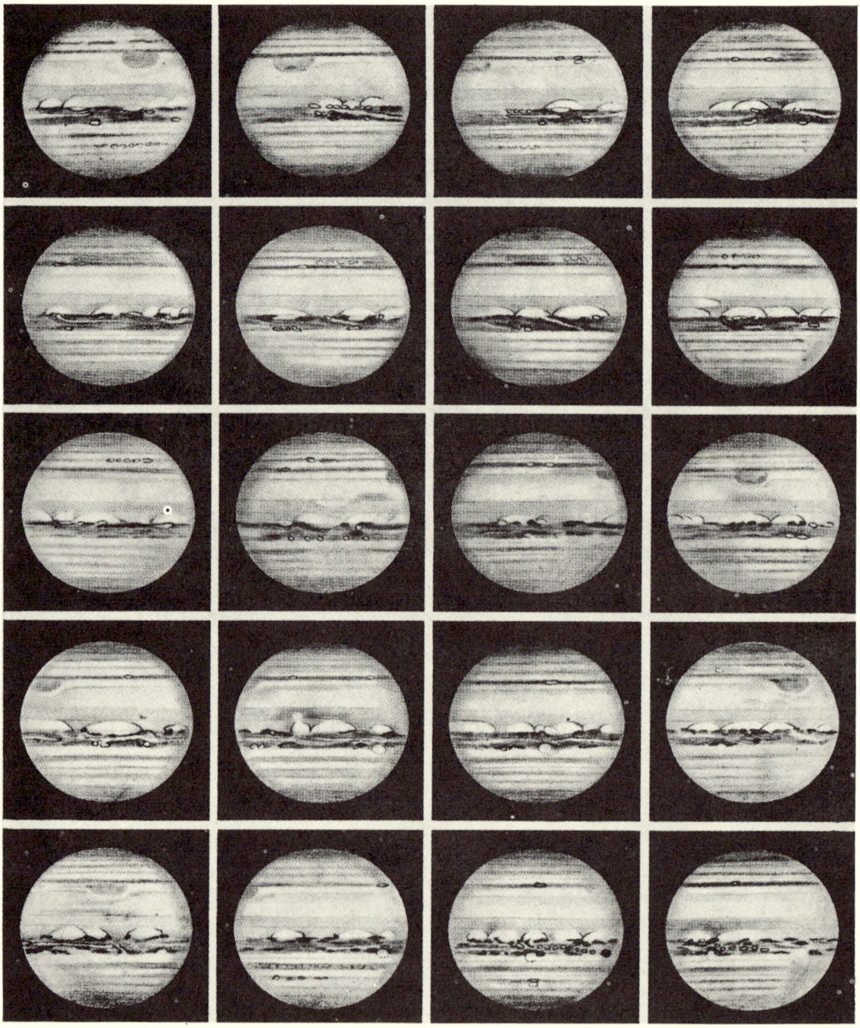

FIG. 2-57. The Appearance of Jupiter in 1938.

The Jovian moons form a miniature solar system about Jupiter as indicated in Fig. 2–38. The moons orbiting Saturn form a second such system, and to some extent, the moons of Uranus form a third. Four of the Jovian and one Saturnian moon have atmospheres, probably of very low boiling point gases such as their parent planets. These are indicated in Table 2–15. These satellites (as all satellites) may form prime observation and refueling stations for inter-planetary flight or even extra-solar flight. Of particular interest are Saturn's unique rings, designated A, B and C in decreasing radii order (A being the outermost). These rings show no trace of atmosphere, and are believed to be composed many individual globules less dense than water, either ice or rocks covered with frost. Ring C has an inner dimension of 1.18 Saturn radii, and an outer radius of 1.5. Ring B lies between 1.53 and 1.95 Saturn radii, and ring A between 2.04 and 2.3. There is a small dividing line occasionally apparent within the outer ring.

The asteroids present a unique array of solids lying generally between Mars and Jupiter. They are almost surely solid bodies, and have so little mass as to preclude holding any significant atmosphere. Figures 2–39 and 2–40 have portrayed the orbits of some of the larger asteroids. Table 2–16 has tabulated orbital data on many. Table 2–22 shows the low gravitational parameters of the five principal asteroids assuming both Earth-density and 0.6 Earth-density.

TABLE 2–22 GRAVITATIONAL DATA OF ASTEROIDS

	K ft³/sec²	v_c (ft/sec)	v_p (ft/sec)	$g/g_\oplus$
Ceres ($\oplus$-density)	$3.04 \cdot 10^{13}$	1550	2190	0.147
(0.6 $\oplus$-density)	$1.76 \cdot 10^{13}$	1180	1670	$8.6 \cdot 10^{-2}$
Pallas ($\oplus$-density)	$0.82 \cdot 10^{13}$	1020	1440	0.1
(0.6 $\oplus$-density)	$0.47 \cdot 10^{13}$	780	1100	$5.7 \cdot 10^{-2}$
Juno ($\oplus$-density)	$0.047 \cdot 10^{13}$	390	550	$3.1 \cdot 10^{-2}$
(0.6 $\oplus$-density)	$0.027 \cdot 10^{13}$	300	420	$2.1 \cdot 10^{-2}$
Vesta ($\oplus$-density)	$0.38 \cdot 10^{13}$	780	1100	$7.3 \cdot 10^{-2}$
(0.6 $\oplus$-density)	$0.22 \cdot 10^{13}$	590	840	$4.2 \cdot 10^{-2}$
Eros ($\oplus$-density)	$9.4 \cdot 10^{8}$	49	68	$4.6 \cdot 10^{-3}$
(0.6 $\oplus$-density)	$5.6 \cdot 10^{8}$	37	52	$2.8 \cdot 10^{-3}$

Comets are unique bodies in the solar spectrum. A comet has three parts; nucleus, coma and tail. The nucleus is small and round, and is of greater brightness than the other parts, but less bright than the planets. The coma is a thin veil of gases, around the nucleus and opening towards the tail. The tail is an extension of the coma, becoming bright as the comet approaches the Sun. The coma and tail vanish at extreme distance from the Sun indicating that solar radiation liberates and illuminates gases from the solid nucleus. Some comets may not exhibit all three parts but only one or any two. Coma dimensions may be of 10,000 to 100,000 miles in diameter. Tails may be tens to hundreds of millions of miles in length and always point away from the Sun. A typical nucleus will have a diameter of about 1 km. Halley's comet was extremely large; i.e., 20 km in diameter. The gases of the coma and tail cannot be retained by gravitational forces, and therefore must be liberated permanently. Nevertheless, the mass of comets is of the order of $6{\cdot}10^{16}$ to $6{\cdot}10^{12}$ tons; being sometimes larger than the smaller asteroids. Table 2–17 presents some data on periodic comets, and identifies their associated planets.

2.4.8 Planetary Radiation Belts

Although the presence of magnetic fields has not been established for the various planets, it seems reasonable to assume that some, if not all, have magnetic fields of various strengths. Within these magnetic fields, particularly for the inner planets, we may expect solar particle radiation to be trapped, forming radiation belts of the type predicted by Störmer and measured by Van Allen around the Earth, as depicted in Fig. 2–3. Singer (Ref. (*45*)) has predicted such belts for Venus and Mars in particular by assuming magnetic dipole moments in proportion to the planetary volumes (an arbitrary assumption since the core and perhaps rotation will affect the planetary magnetic field). This has led to predictions of Inner Belt intensity and Outer Belt intensity maximums as indicated in Tables 2–23 and 2–24. Similar belts may exist around Mercury, Jupiter, and the other planets; however, the conditions are so different from terrestrial in terms of magnetic field nature and strength and solar radiation intensity, as to make predictions inappropriate at this time.

TABLE 2–23 INTENSITY OF PROTONS IN INNER RADIATION BELT VERSUS ALTITUDE

Relative Intensity	*r/R*
0.0	1.0
48.0	1.1
76.5	1.2
89.3	1.4
76.0	1.6
31.0	1.8
0.0	2.0

TABLE 2–24 POSITION OF OUTER RADIATION BELT MAXIMUM

	Earth	*Mars*	*Venus*
Distance from Sun (A.U.)	1.00	1.52	0.72
Relative Density of Solar Stream	1.00	0.43	1.93
Max. of outer belt (planetary radii)	3.00	3.45	2.7

2.5 GALACTIC AND EXTRA-GALACTIC CONDITIONS

The Sun is one of a local cluster of stars in a much larger galaxy which, in turn, is one of many galaxies which exist throughout the universe. Although inter-galactic and extra-galactic travel are predicted by current science fiction writers, travel to even the nearest star (Alpha Centauri, 4.2 light years distant) seems very remote for today's technology. Major break-throughs in transportation methods, environmental control, and life duration and sustenance will be needed before such travel becomes feasible. However, astronomical observations have provided some estimates of the nature of this universe which are pertinent to our interest in space electronic systems. The statistics quoted in this section derive generally from Ref. (*46*).

2.5.1 Terrestrial Observations

The Earth has been our only observatory base through the centuries; hence our observations have been handicapped by the transmissability spectrum of the atmospheric blanket. This spectrum is depicted in broad form by Fig. 2–13. As evident from this figure, we have thus far been able to observe and chart the universe only through particular "windows" of the wavelength spectrum; even then we are subject to various types of weather restrictions. Observations from high altitude aircraft and balloons have eliminated many of the weather limitations, but have not been of such precision as to warrant being classed as first order quality astronomical data; particularly with regard to telescope size, resolving and collecting powers, lens or mirror quality, stabilization, and accuracy of position knowledge. An orbiting astronomical observatory is under development currently which should improve on this situation somewhat in that the atmospheric effects, absorption, scattering, refraction, attenuation, etc. will be eliminated. Also, much more precise control and knowledge of the vehicle's attitude and position will be available and fairly good quality telescopes will be employed (although not of the same size and quality as are available at fixed Earth-bound installations.)

The various stars of the universe are classified by stellar spectrum (Sp) in a stellar life scheme as:

```
                            R—N        nebular series R
     Q                     /
P—W—O—B—A—F—G—K—M          main series
                     \
                      S        nebular series S
  Early types        Late types
```

About 99 per cent of the stars lie in the B through M series, the remaining types being comparatively rare. There are subdivisions of these major divisions (e.g., B0, B1, B2) in use by astronomers. The general characteristics of these classes of stars are indicated in Table 2–25. Stars are also classified by luminosity (e.g., G5 IV or gG0). Table 2–26 indicates the luminosity designations. Stars are also classified by absolute visual magnitude, color, density, mass, radius, temperature, etc.

TABLE 2–25 CHARACTERISTICS OF STAR CLASSES

S_p	*Characteristics*
Q	Noval
P	Planetary nebulae
W	Wolf-Rayet stars; hot, with broad emission
O	Hot stars with He II absorption
B	HeI absorption; H developing later
A	Very strong H, decreasing later; CaII increasing
F	CaII stronger; H weaker; metals developing
G	CaII strong; Fe and other metals strong; H weaker
K	Strong metallic lines; bands developing
M	Very red; TiO band developing strongly
R, N	CN, CO (?C_2) bands instead of TiO
S	ZrO bands

TABLE 2–26 LUMINOSITY CLASSES OF STARS

Yerkes Luminosity Class	*Type*
Ia or sg Ib or c	Super giant
II	Bright giant
III or g	Giant
IV	Sub giant
V or d	Dwarfs
sd	Sub-dwarfs
w	White dwarfs

Still another classification of stars is by population type as indicated in Table 2–27. From these tables we may certainly infer that astronomers are working with as varied and diffused a population of items as may be encountered anywhere. Indeed a particularly interesting chart is the density distribution of stars as a function of galactic latitude. These data are presented in Table 2–28 where m is the relative stellar magnitude defined with reference to particular stars in the North Polar Sequence. Particular stars are used as photographic, photoelectric, and color standards.

TABLE 2–27 POPULATION TYPES

	Population I	*Population II*
Stars	Mean sequence Classical cepheid variables Supergiants G to M giants Rapidly rotating stars	Globular cluster sequences Cluster cepheids Sub-luminous O and B stars Sub-dwarfs RV Tan variables Noval
Other Objects	Diffuse nebulae Interstellar grains & absorption	Planetary nebulae
Occurrence	Arms of late type spirals Resolvable irregular nebulae Open clusters The Solar Neighborhood New Stellar systems	Early type elliptical nebulae Centers of late type nebulae Irresolvable irregular nebulae Globular clusters Nearby high-velocity stars Old stellar systems

TABLE 2–28 NUMBER OF STARS AT VARIOUS GALACTIC LATITUDES

N_m = number of stars per square degree brighter than magnitude m

A_m = number of stars per square degree within the brightness range $m + \frac{1}{2}$ to $m - \frac{1}{2}$

Variation of N_m with galactic latitude, b

$\log N_m$

m	Galactic latitude											Mean 0° to 90°	Mean 0° to 90°
	0°	±5°	±10°	±20°	±30°	±40°	±50°	±60°	±70°	±80°	±90°		
	photographic magnitudes												visual
0·0													$\bar{5}$·7
1·0													$\bar{4}$·4
2·0		$\bar{3}$·18			$\bar{4}$·8				$\bar{4}$·7			$\bar{4}$·96	$\bar{4}$·99
3·0		$\bar{3}$·68			$\bar{3}$·27				$\bar{3}$·0			$\bar{3}$·40	$\bar{3}$·55
4·0	$\bar{2}$·25	$\bar{2}$·17	$\bar{2}$·12	$\bar{3}$·99	$\bar{3}$·85	$\bar{3}$·75	$\bar{3}$·70	$\bar{3}$·66	$\bar{3}$·64	$\bar{3}$·62	$\bar{3}$·60	$\bar{3}$·89	$\bar{2}$·11
5·0	$\bar{2}$·72	$\bar{2}$·64	$\bar{2}$·57	$\bar{2}$·44	$\bar{2}$·32	$\bar{2}$·24	$\bar{2}$·20	$\bar{2}$·17	$\bar{2}$·14	$\bar{2}$·12	$\bar{2}$·11	$\bar{2}$·37	$\bar{2}$·60
6·0	$\bar{1}$·18	$\bar{1}$·10	$\bar{1}$·03	$\bar{2}$·90	$\bar{2}$·78	$\bar{2}$·71	$\bar{2}$·66	$\bar{2}$·63	$\bar{2}$·62	$\bar{2}$·60	$\bar{2}$·58	$\bar{2}$·86	$\bar{1}$·07
7·0	$\bar{1}$·61	$\bar{1}$·54	$\bar{1}$·47	$\bar{1}$·34	$\bar{1}$·23	$\bar{1}$·16	$\bar{1}$·11	$\bar{1}$·08	$\bar{1}$·06	$\bar{1}$·04	$\bar{1}$·03	$\bar{1}$·31	$\bar{1}$·54
8·0	0·05	$\bar{1}$·99	$\bar{1}$·91	$\bar{1}$·78	$\bar{1}$·68	$\bar{1}$·60	$\bar{1}$·55	$\bar{1}$·52	$\bar{1}$·49	$\bar{1}$·47	$\bar{1}$·46	$\bar{1}$·75	0·00
9·0	0·52	0·43	0·35	0·22	0·12	0·04	$\bar{1}$·99	$\bar{1}$·94	$\bar{1}$·91	$\bar{1}$·89	$\bar{1}$·88	0·19	0·45
10·0	0·97	0·88	0·80	0·66	0·54	0·46	0·40	0·35	0·31	0·29	0·27	0·62	0·92
11·0	1·43	1·33	1·23	1·08	0·96	0·87	0·80	0·75	0·70	0·64	0·66	1·05	1·34
12·0	1·88	1·77	1·65	1·50	1·37	1·26	1·19	1·12	1·08	1·04	1·03	1·46	1·77
13·0	2·30	2·19	2·07	1·90	1·76	1·64	1·54	1·47	1·42	1·40	1·39	1·87	2·17
14·0	2·72	2·61	2·48	2·28	2·12	1·98	1·88	1·79	1·75	1·71	1·71	2·26	2·56
15·0	3·12	3·00	2·88	2·65	2·46	2·31	2·20	2·10	2·03	2·00	1·97	2·62	2·95
16·0	3·40	3·41	3·24	3·00	2·77	2·61	2·48	2·38	2·28	2·25	2·24	2·98	3·30
17·0	3·83	3·78	3·60	3·33	3·07	2·89	2·75	2·64	2·55	2·52	2·48	3·33	3·64
18·0	4·20	4·10	3·93	3·63	3·35	3·14	2·99	2·87	2·78	2·75	2·72	3·64	3·96
19·0	4·5	4·4	4·3	3·9	3·6	3·4	3·2	3·1	3·0	2·9	2·9	3·90	4·20
20·0	4·7	4·7	4·6	4·2	3·8	3·6	3·4	3·3	3·2	3·1	3·1	4·17	4·45
21·0	5·0	4·9	4·8	4·5	4·0	3·7	3·6	3·4	3·3	3·3	3·2	4·4	

Distribution in absolute magnitude ranges $M \pm \frac{1}{2}$ for stars counted to a given apparent magnitude ($m \simeq 6$). The table gives percentage distribution

	M											
	−6	−5	−4	−3	−2	−1	0	1	2	3	4	5
Photographic. All stars	1	1	3	7	10	14	18	21	15	6	3	1
Visual. Class O	3	15	31	37	12	2	0	0	0	0	0	0
B	3	8	14	22	23	22	7	1	0	0	0	0
A	4	13	14	8	2	2	16	25	13	2	0	0
F	1	8	11	7	9	5	2	16	18	18	5	0
G	1	5	9	11	8	1	11	29	7	9	6	2
K	0	1	4	11	10	13	30	18	5	2	1	2
M	3	8	7	3	9	24	29	13	3	1	0	0

Of particular interest also is the solar motion in galactic coordinates:

Towards galactic center, longitude 328°
latitude 0°, $X = 10.1$ Km/s

In galactic plane towards longitude 58°
latitude 0°, $Y = 15.0$ Km/s

Towards galactic pole, latitude 90°, $Z = 7.0$ Km/s

The Sun is a member of a local system within the galaxy. This local system is sometimes called the Gould Belt and has the following characteristics:

N Pole in galactic coordinates: longitude 170°
latitude +74°

Center of system longitude 240°
latitude −3°

Diameter of system $\simeq$ 1000 pc (one parsec
= 206,265 A.U.
= 3.263 light years)

Sun's distance from center: $\simeq$ 100 pc

Sun's distance from local plane $\simeq$ 12 pc N of plane

Velocity of system rotation = $0.04R$ Km/s/pc
where R is radial distance from center

Mass of the system = 1×10^8 Mass of the Sun

System elongation:

Diameter (given above) is of the elongation in the direction of Cygnus (62° long) and Carina (242° long)

The Galaxy has a mass of 1.6×10^{11} the Sun's mass. The diameter is 25 Kpc (kiloparsecs). The diameter of the extended spherical system (which includes globular clusters) = 40 Kpc. The thickness of the galaxy at center = 5 Kpc. The overall density is 0.10 mass of Sun per cubic parsec $= 7 \times 10^{-24}$ g/cm³. The galactic pole in ecliptic plane coordinates is:

Epoch 1900	190.0 deg long
	+28.0 deg lat
Epoch 2000	191.23 deg long
	+27.45 deg lat

Longitude of ascending node in ecliptic = $(266.96 + 1.40T)$ deg

Longitude of ascending node on equator = $(280.00 + 1.23T)$ deg

Inclination to ecliptic = 60.55 deg

Inclination to equator = $(62.00 + 0.55T)$ deg,
where T is in centuries from 1900.0

The direction of the galactic center is:

galactic system: 327.9 ± 0.4 deg long
-1.3 ± 0.3 deg lat

ecliptic system (epoch 1900):
265 deg long
-29 deg lat

The Sun's distance from the galactic center = 8.2 ± 0.8 Kpc

The Sun's distance from the galactic plane = 14 ± 2 pc N of plane

The rotational velocity in the Sun's neighborhood = 0.57 seconds of arc per century

The period of rotation in the Sun's neighborhood (cosmic year) = 2.2×10^8 years

The central density of the galaxy = 0.5 Sun's mass per cu. pc.
$= 3.5 \times 10^{-23}$ g cm^{-3}

Escape velocity, from galactic center = 470 Km/s
from near the Sun = 310 Km/s
from galactic rim = 190 Km/s

The potential energy of the galaxy = 1.2×10^{59} ergs

Extra-galactic nebulae of galaxies are classified according to the scheme:

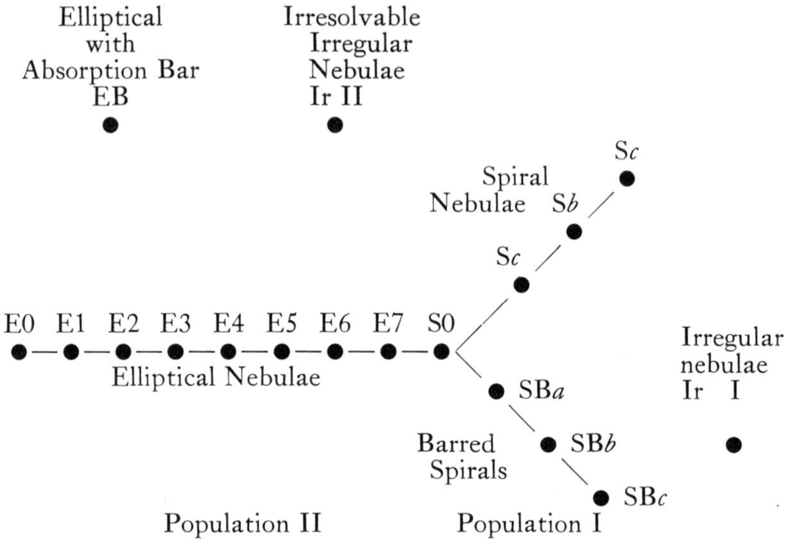

Elliptical nebulae, En, are classified by their ellipticity $= (a-b)/a = n/10$ where a and b are the greater and smaller diameters. The S and SB branch sequence 0, a, b, c represents increasing openness of the spiral arms.

The mean mass of extra galactic nebulae $= 5 \times 10^{10}$ sun's mass $= 10 \times 10^{43}$ g.

Stars per nebulae $= 10^9$ to 10^{11}

Speed of recession with distance = 290 Km/s/Mpc

$$\begin{aligned}(\text{Mpc} &= \text{mega parsec}\\ &= 3.263 \times 10^6 \text{ light years})\end{aligned}$$

Space density of nebulae (neb) $= 3$ neb/Mpc^3
$= 1.0 \times 10^{-73}$ neb/cm^3

Average diameter of clusters of galaxies = 1.6 Mpc

Average number of galaxies per cluster = 200

Average number of cluster centers per cubic parsec $= \log^{-1} - 19$

Pole of the local supergalaxy (cluster) in galactic coordinates = 15° long
+6° lat.

Center of the local supergalaxy in the direction of the Virgo cluster

$$\begin{pmatrix} 255° \text{ long} \\ +75° \text{ lat} \end{pmatrix}$$

Galaxies in the observable universe $= 10^{10}$ galaxies

Mass of the observable universe $> 10^{54}$ g

Radius of the observable universe

$$R = c/V_r = 3.2 \times 10^{27} \text{ cm} = 1040 \text{ Mpc}$$

Volume of the observable universe

$$4\pi R^3/3 = 1.4 \times 10^{83} \text{ cm} = 5 \times 10^9 \text{ Mpc}^3$$

Speed of recession of nebulae

$$V_r = 0.94 \times 10^{-17} \text{ cm/s/cm}$$
$$= 290 \text{ Km/s/Mpc}$$

Life of universe for constant recession speed $= 1.06 \times 10^{17}$ sec $= 3.4 \times 10^9$ years

Density of universe $= 10^{-29}$ g/cm^3
(assumes nebulae mass predominates over internebulae gas)

Density calculated from

$3\ V_r^2/8\pi G$ (where G is gravitational constant)
$= 1.5 \times 10^{-28}$ g/cm^3

Rate of volume increase of a portion of the universe

$$dV/V = 3V_r = 2.8 \times 10^{-17} \text{ sec}^{-1}$$

Radiation density of the universe $= 4 \times 10^{-13}$ erg/cm^3

From the foregoing statistics we may well be overwhelmed by the insignificance of a small race of people on a minor planet of a small sun in an average galaxy within a tremendous universe. No wonder the task of charting the heavens is literally astronomical! However, attempts have been made and will continue to be made, as discussed in the following sections.

2.5.2 Extra-Terrestrial Observations

The foregoing data has represented a small fraction of the masses of data available from terrestrial observations. It is to be expected that these data will be multiplied many times when extra-terrestrial observations become feasible, particularly from satellite and lunar observatories. The absence of the atmospheric blanket will make a major difference, of course, changing luminosity and spectral charts, and perhaps even stellar population density figures. Radiometric astronomy has already provided some indications of further stellar variations, and may be extended (along with visual astronomy) by use on satellites including the moon. The great infra-red, microwave, and ultra-violet spectrums offer even more potential when used extra-terrestrially.

2.5.3 Stellar and Constellation References

As indicated previously (see Fig. 2–36) astronomical references are measured on the celestial sphere with respect to either the equatorial, ecliptic or the galactic plane and with the zero meridian in the direction of Aries. Generally, references are taken at visual wavelengths, although radiometrical references are being established. The absorption, attenuation, and scattering of the atmosphere have been serious limitations. However, the brightest stars and galaxies observable through the atmosphere may generally be expected also to be the brightest observable in space and it appears that spatial references based on these far-away, essentially fixed, points would be useful for extra-terrestrial navigation references. This has been confirmed by the first orbital flight of Lt. Col. John Glenn. Table 2–29 tabulates the 100 nearest stars, and Table 2–30 the 100 brightest stars. These stars are tabulated according to name, equatorial coordinates at epoch 1900 with longitude (α) in hours and minutes from the Greenwich meridian and latitude (δ) in degrees and minutes, apparent visual magnitude (m_v), color index (c), absolute magnitude (M_v = apparent magnitude standardized to 10 parsecs), spectral classification (Sp) with luminosity class for the brightest stars, annual proper motion (μ) in seconds per year, the direction of the annual proper motion (θ) in degrees measured from North towards East, parallax (π) in seconds ($= 1/d$ with d in parsecs), and the sight line velocity (V_r) away from the Sun. These stars should be useful as good navigational references extra-terrestrially providing suitable angle and magnitude discrimination is available in the star tracking equipment.

Table 2–29 The Nearest Stars (following 3 pages)

V and $B-V$ = visual magnitude and colour, M_v = absolute magnitude, μ = proper motion, θ = direction of proper motion (from N towards E), π = parallax, v_r = radial velocity (+ = moving away from sun), $\mathfrak{M}$ = mass, R = radius. In the Sp column, w = white dwarf, sd = subdwarf. Since nearly all the near stars are on the main sequence the luminosity classification is not given.

The notes give the fairly recent separation of doubles or some orbital elements when known. P = period, a = semi-major axis of secondary relative to primary. Flare, indicates that the star has exhibited sudden variations of brightness (flare star). db and tr = double and triple star with invisible components. sp = spectroscopic.

TABLE 2–29 THE NEAREST STARS

Star		1900 α	1900 δ	V	B−V	M_v	Sp	μ	θ	π	v_r	$\mathscr{M}$	$\mathscr{R}$	Notes
		h m	° ′					″/y	°	0″·001	km/s	$\mathscr{M}_\odot$	$\mathscr{R}_\odot$	
Grm 34 = +43° 44	A	0 13	+43 27	8·08	1·55	10·30	M2	2·89	82	278	+ 14			} A = sp db
(= CC19)	B	”	”	11·05	1·78	13·27	M5	”	”	”	+ 21			} AB = 38″
β Hyi		0 20	−77 49	2·79	0·61	3·71	G2	2·25	82	153	+ 23			
η Cas	A	0 43	+57 17	3·44	0·57	4·73	G0	1·22	115	181	+ 9	0·85	0·84	} a = 12″·0
”	B	”	”	7·25		8·54	M0	”	”	”	+ 13	0·52	0·07	} P = 490 y
v. Maanen = Wolf 28		0 44	+ 4 55	12·34	0·55	14·19	wF5	2·949	155·4	235	+260 ?			
L726–8 [13]	A	1 34	−18 28	12·41	1·9	15·28	M6e	3·36	80	375	+ 29	0·044		} a = 5″, P = 200y
(= UV Cet)	B	”	”	12·95	1·9	15·87		”	”	”		0·035		} B = Flare st.
τ Cet		1 39	−16 28	3·50	0·72	5·70	sdG8	1·934	296·5	275	− 16			
82 Eri = e Eri		3 16	−43 27	4·23	0·71	5·19	G5	3·14	76	156	+ 87			
ε Eri		3 28	− 9 48	3·74	0·87	6·14	K2	0·990	271·7	303	+ 15			
o² Eri = 40 Eri	A	4 11	− 7 49	4·44	0·81	5·96	K1	4·08	213	201	− 42	0·11		A db ? 0″·01, 3 y
”	B	”	”	9·64		11·16	wA	”	”	”	− 42	0·44	0·018	} BC a = 6″·9
”	C	”	”	11·05		12·57	M5e	”	”	”	− 45	0·21	0·43	} P = 248 y
Kapteyn = −45° 1841		5 08	−44 59	8·9		10·9	sdM0	8·73	131	251	+242			Nearest subdwarf
HD 36395 = −3° 1123		5 26	− 3 42	7·97	1·48	9·03	M1	2·24	160	163	+ 11			
Ross 47		5 36	+12 29	11·58		12·65	sdM4	2·54	127	164	+103			
−21° 1377		6 06	−21 49	8·18	1·51	9·33	M1	0·72	188	170	+ 2			
Ross 614	A	6 24	− 2 44	11·25		13·23	M5	1·00	131	249	+ 24	0·14		} a = 1″·0
(= CC390)	B	”	”	14·8		16·8		”	”	”		0·08		} P = 16·5 y
Sirius = αCMa	A	6 41	−16 35	− 1·46	0·01	1·41	A1	1·32	204	375	− 8	2·31	1·8	} a = 7″·6
”	B	”	”	8·67		11·54	wA5	”	”	”		0·98	0·022	} P = 49·9 y
Wolf 294		6 48	+33 24	10·15		11·29	M4	0·86	240	169	+ 36			
Ross 986		7 03	+38 43	11·68		12·86	M5e	1·12	208	172	+ 39			
Luyten = +5° 1 668		7 22	+ 5 32	9·92		12·02	M4	3·75	171	266	+ 26			db ?
Procyon = αCMi	A	7 34	+ 5 29	0·35	0·40	2·64	F5	1·25	214	287	− 3	1·75	1·7	} a = 4″·5
”	B	”	”	10·8		13·1	wF	”	”	”		0·64	0·01	} P = 40·6 y
L745–46	A	7 36	−17 10	13·06		14·14	wF	1·26	117	164				} 21″
”	B	”	”	17·6		18·7	M	”	”	”				
L97–12		7 53	−67 30	14·9		16·1	wF-G	2·04	135	170				

TABLE 2–29 THE NEAREST STARS (Cont.)

Star		1900 α	1900 δ	V	$B-V$	M_v	Sp	μ	θ	π	v_r	$\mathscr{M}$	$\mathscr{R}$	Notes
		h m	° ′					″/y	°	0″·001	km/s	$\mathscr{M}_\odot$	$\mathscr{R}_\odot$	
Ross 619		8 06	+9 11	12·88		13·78	M5	5·40	167	151	−35			
LFT571 = L674–15		8 08	−21 15	13·8		14·9	M	0·73	175	166				
+53° 1320	A	9 08	+53 07	7·68	1·44	8·74	K8	1·68	248	163	+10			19″
+53° 1321	B	,,	,,	7·77		8·82	M0	,,	,,	162	+9			P = 1 000 y
Grm 1 618 = +50° 1725		10 05	+49 57	6·60	1·37	8·33	M0	1·464	249·8	222	−27			
AD Leo = +20° 2465		10 14	+20 22	9·41	1·55	11·04	M4	0·429	258	212	+10			Flare st. invis. comp.
Wolf 359		10 52	+7 36	13·66		16·80	M6e	4·71	235	425	+13			
Lal 21185 = +36° 2147		10 58	+36 38	7·47	1·51	10·42	M2	4·78	187	398	−87	0·35		Invis. comp. $\mathscr{M}$ = 0·03
+44° 2051	A	11 01	+44 02	8·76	1·54	9·94	M2	4·53	295	172	+64			28″
(= WX UMa)	B	,,	,,	14·8		16·0	M6	,,	,,	,,				B = flare star
I(UC37) = CC658		11 40	−64 17	12·5		14·0	wA	2·68	97	203				
AC 79° 3888		11 41	+79 14	10·92		12·41	sdM4	0·87	57	198	−119			
Ross 128		11 43	+1 23	11·13		13·50	M5	1·38	153	298	−13			
L68–28	A	12 23	−70 56	15·7		16·6	K–M	1·17	339	152				15″
L68–29	B	,,	,,	17·7		18·6	K–M	,,	,,	,,				
Wolf 424	A	12 28	+9 34	12·63		14·44	M6	1·87	276	230	−5			0″·5
,,	B	,,	,,	12·7		14·5	M7	,,	,,	,,				
+15° 2620		13 41	+15 26	8·49	1·44	10·01	M3	2·30	129	201	+15			
Proxima Cen		14 23	−62 15	10·7		15·1	M5e	3·85	282	763		0·1		Flare st. nearest st.
−11° 3759		14 29	−12 06	11·38		12·37	M4	0·69	334	158				
α Cen	A	14 33	−60 25	0·00	0·69	4·38	G2	3·69	281	752	−22	1·09	1·23	a = 17″·7
,,	B	,,	,,	1·38		5·76	K4	,,	,,	,,		0·88	0·87	P = 80·1 y
−20° 4125	A	14 52	−20 58	5·82	1·12	7·00	K5	1·99	149	172	+20			20″
−20° 4123	B	,,	,,	8·10		9·28	M2	,,	,,	,,	+25			
−40° 9712		15 26	−40 54	10·1		11·1	M4	1·55	229	166				
−12° 4523 = CC995		16 25	−12 25	10·07	1·60	12·01	M4	1·18	182	244	−13			
−8° 4352	A	16 50	−8 09	9·72	1·59	10·63	M3e	1·19	222	152	+19	0·38		AB 0″·20, 1·7 y
,,	B	,,	,,	9·8		10·7	M4	,,	,,	,,		0·34		
(= Wolf 629)	C	,,	,,	11·76		12·78	sdM4	,,	,,	160	+25			BC 72″
+45° 2505 [5]	A	17 09	+45 50	9·95		10·97	M3	1·58	172	160	−18	0·31		0″·7
(= Fu46)	B	,,	,,	10·31		11·33	M4	,,	,,	,,		0·25		P = 13·1
36 Oph	A	17 09	−26 27	5·07	0·85	6·31	K1	1·24	203	177	−1			AB 5″
(= −26° 12026)	B	,,	,,	5·11		6·35	K1	,,	,,	,,	0			BC 732″
(= −26° 12036)	C	17 10	−26 24	6·34	1·14	7·52	K5	,,	,,	172	0			
−46° 11540		17 21	−46 47	9·4	1·5	11·0	M4	1·04	147	213				

TABLE 2–29 THE NEAREST STARS (Cont.)

Star		1900 α	1900 δ	V	$B-V$	M_v	Sp	μ	θ	π	v_r	$\mathscr{M}$	$\mathscr{R}$	Notes
		h m	° ′					″/y	°	0″·001	km/s	$\mathscr{M}_\odot$	$\mathscr{R}_\odot$	
−44° 11909		17 30	−44 14	11·1		12·7	M5	1·16	217	209				
+68° 946		17 37	+68 26	9·13	1·52	10·67	M4	1·331	194·1	203	− 17			
I (UC48)		17 38	−57 14	12·9		14·0	M	1·72	219	167				
Barnard = +4° 3561		17 53	+ 4 25	9·53	1·75	13·21	M5	10·27	356	545	−108			? db. 0″·06, 1·1 y
70 Oph	A	18 01	+ 2 31	4·22	0·87	5·64	K0	1·13	167	192	− 7	0·89		4″·5 db
(= +2° 3482)	B	”	”	5·94		7·36	K5	”	”	”	− 10	0·68		87·8 y db
+59° 1915	A	18 42	+59 29	8·90	1·54	11·16	M4	2·28	324	283	+ 1			17″ (A, B)
(= Σ2398)	B	”	”	9·69	1·58	11·95	M5	”	”	”	+ 14			
Ross 154		18 44	−23 54	10·6		13·3	M5	0·72	103	350	− 4			Flare star
+4° 4048	A	19 12	+ 5 02	9·13	1·49	10·29	M3	1·46	203	171	+ 34			74″ (A, B)
”	B	”	”	18·0		19·2	M5	”	”	”				
L347–14		19 13	−45 42	13·7		14·8	M7	2·93	168	169				
σ Dra		19 33	+69 29	4·69	0·80	5·92	K0	1·84	162	176	+ 27			
Altair = αAql		19 46	+ 8 36	0·75	0·25	2·23	A7	0·66	54	198	− 26			
δ Pav		19 59	−66 26	3·56	0·75	4·71	G7	1·64	134	170	− 22			
−36° 13940	A	20 05	−36 21	5·33	0·85	6·51	K4	1·63	164	172	−130			7″ (A, B)
(= HR7703)	B	”	”	11·5		12·7	M5	”	”	”				
−45° 13677		20 07	−45 28	8·0	1·44	9·0	M0	0·78	100	159	− 30			
61 Cyg	A	21 02	+38 15	5·20	1·21	7·53	K5	5·20	52	292	− 64	0·59		a = 24″·6, 720 y,
”	B	”	”	6·03	1·4	8·36	K7	”	”	”	− 63	0·50		? invis. trip.
−39° 14192		21 11	−39 15	6·69	1·42	8·73	M0	3·47	251	256	+ 20			
−49° 13515		21 27	−49 26	8·9		10·6	M3	0·81	185	219	+ 18			
ε Ind		21 56	−57 12	4·73	1·05	7·01	K5	4·69	123	286	− 40			
Krüger 60 = DO Cep	A	22 24	+57 12	9·83	1·63	11·82	M3	0·86	246	250	− 24	0·27	0·51	a = 2″·4, 45 y
(= + 56° 2783)	B	”	”	11·37		13·36	M4e	”	”	”	− 28	0·16		B = Flare st.
L789-6		22 33	−15 52	12·58		14·93	M6e	3·260	45·2	296	− 60			
−21° 6267	A	22 33	−21 08	9·3		11·0	M2	0·46	99	219	− 8			23″ (A, B)
”	B	”	”	11·0		12·7	Me	”	”	”				
+43° 4305		22 42	+43 49	10·05	1·39	11·53	M5e	0·83	238	198	− 1			Flare star
−15° 6290 = Ross 780		22 48	−14 47	10·16	1·60	11·73	M5	1·11	123	206	+ 9			
−36° 15693		22 59	−36 26	7·39	1·50	9·57	M2	6·90	79	273	+ 10			
+56° 2966		23 08	+56 37	5·58	1·01	6·49	K3	2·09	82	152	− 18			
Ross 248		23 37	+43 39	12·25	1·8	14·75	M6e	1·82	176	316	− 81			
+1° 4774		23 44	+ 1 52	8·99	1·49	10·05	M2	1·38	134	163	− 64			
−37° 15492		23 59	−37 51	8·59	1·48	10·29	M3	6·08	113	219	+ 24			

2.5.4 Interstellar Space

Interstellar space contains irregular shaped clouds of gas and grains. All that can be quoted at this time are average and approximate figures as follows:

Number of clouds = 5×10^{-5} per pc^3
Radius of clouds = 8 pc
Distance between clouds = 40 pc

The proportion of space near the galactic plane occupied by clouds is 7 per cent. Radiation from hot stars is capable of ionizing hydrogen; 9 per cent of space near the galactic plane lies sufficiently near to such hot stars. The product of these or 0.6 per cent of space near the galactic plane is occupied by ionized clouds.

Mean cloud mass = 400 × Sun's mass
Gas density in clouds = 10 atoms per cm^3
Gas density between clouds $\leq$ 0.1 atoms per cm^3

The relative abundance of atoms in interstellar gas is very similar to that in stars.

Table 2–30 The Brightest Stars (following 3 pages)

The list contains 100 visually brightest stars. For multiple stars the data refer to the combined system, and the multiplicity is given in the notes where db = double, tr = triple, qu = quadruple, sp = spectroscopic, vis = visual, var = variable, ecl = eclipsing. Approximate separations of visual doubles and periods of spectroscopic doubles and variables are also given in the notes.

The star magnitudes V and colours $B - V$ are given on the B, V system, and the spectral classifications Sp are on the MKK system (sometimes smoothed by averaging). μ = proper motion per year, and θ = direction of μ (from N through E). The distance d is from the parallax π when $\pi > 0''.030$/y, and from the luminosity spectral class when $\pi < 0''.015$/y. Some averaging has been introduced. In the radial velocity column v_r, v = variable, + = distance increasing, i.e. red shift.

The absolute magnitudes M_v are derived from the V and d columns. In a few cases an interstellar absorption correction has been introduced from measured colour excess.

TABLE 2–30 THE BRIGHTEST STARS

Star		1900 α		1900 δ		V	$B-V$	M_v	Sp	μ	θ	d	v_r	Notes
		h	m	°	′					0″·001/y	°	pc	km/s	
Alpheratz	α And	0	03	+28	32	2·07	−0·07	−0·5	B9p III	211	139	31	−12 v	sp. var, 96·7 d, Mn star
Caph	β Cas	0	04	+58	36	2·26	+0·34	+1·5	F2 IV	557	109	14	+11·7	
Ankaa	α Phe	0	21	−42	51	2·37	+1·07	+0·2	K0 III	444	154	27	+75 v	sp. and vis. db, 3849 d
Schedar	α Cas	0	35	+55	59	2·20	+1·16	−1·3	K0 II–III	59	121	50	− 3·7	irreg. var.
Diphda	β Cet	0	39	−18	32	2·04	+1·01	+0·8	K0 III	234	80	18	+13·1	
Cih	γ Cas	0	51	+60	11	2·15	−0·2	−0·9	B0e IV	28	94	40	− 7 v	vis. db 2″, irreg. var.
Mirach	β And	1	04	+35	05	2·07	+1·62	+0·2	M0 III	213	122	24	+ 0·3	
Polaris	α UMi	1	23	+88	46	2·02	+0·6	−4·5	F8 Ib	45	95	200	−17 v	sp. db, 30 y, var. 4·0 d
Achernar	α Eri	1	34	−57	45	0·49	−0·17	−2·2	B5 IV	96	108	35	+19	
Almach	γ And	1	58	+41	51	2·16	+1·3	−2·3	K3 II–III	70	138	80	−11·7	tr. vis. and sp. db, 10″, 55 y
Hamal	α Ari	2	02	+22	59	2·00	+1·17	+0·3	K2 III	241	127	22	−14·2	
Mira	o Cet	2	14	− 3	26	2·0	+1·5	−1·0	M6e III	234	181	40	+63	var. 332 d, vis. db (close)
Menkar	α Cet	2	57	+ 3	42	2·53	+1·16	−1·0	M2 III	75	187	50	−25·9	
Algol	β Per	3	02	+40	34	2·10	−0·05	−0·5	B8 V	8	125	31	+ 5 v	ecl., sp. tr., 2·87 d, 1·87 y
Mirfak	α Per	3	17	+49	30	1·80	+0·48	−4·1	F5 Ib	36	132	150	− 2·4	
Aldebaran	α Tau	4	30	+16	19	0·80	+1·55	−0·8	K5 III	203	160	21	+54·2	vis. db, 31″
Capella [9]	α Aur	5	09	+45	54	0·09	+0·81	−0·6	G8, G0	436	168	14	+30 v	sp. db, 104 d
Rigel	β Ori	5	10	− 8	19	0·11	−0·05	−7·0	B8 Ia	2	135	270	+22 v	qu., vis. db 9″, sp. db 22 d
Bellatrix	γ Ori	5	20	+ 6	16	1·63	−0·22	−4·1	B2 III	17	200	140	+18·2	
El Nath	β Tau	5	20	+28	32	1·65	−0·13	−2·9	B7 III	179	169	80	+ 8·0	
Mintaka	δ Ori	5	27	− 0	22	2·19	−0·21	−6·0	O9·5 II	3	146	450	+17 v	ecl., tr., vis. db 53″, sp. db 5·7 d
Arneb	α Lep	5	28	−17	54	2·58	+0·22	−4·8	F0 Ib	6	30	300	+24·7	
Alnilam	ε Ori	5	31	− 1	16	1·70	−0·18	−6·8	B0 Ia	1	180	500	+26·1	
Alnitak	ζ Ori	5	36	− 2	00	1·79	−0·21	−6·2	O9·5 Ib	7	131	400	+18·1	tr., vis. db 2″·5, sp. db 19 d
Saiph	κ Ori	5	43	− 9	42	2·06	−0·16	−7·1	B0·5 Ia	6	149	700	+20·8	
Betelgeuse	α Ori	5	50	+ 7	23	0·4	+1·85	−5·9	M2 I	30	74	180	+21 v	semi-reg. var. 5·8 y, db 40″
Menkalinan	β Aur	5	52	+44	56	1·89	+0·04	−0·2	A2 V	49	264	26	−18 v	ecl. sp. db 3·96 d
Mirzam	β CMa	6	18	−17	54	1·96	−0·23	−4·5	B1 II–III	4	270	200	+34 v	sp. var. 0·25 and 42 d
Canopus	α Car	6	22	−52	38	−0·72	+0·16		F0 I–II	24	56		+20·4	
Alhena	γ Gem	6	32	+16	29	1·93	0·00	−0·5	A0 IV	66	136	30	−12 v	sp. db 2175 d

Star		1900 α		1900 δ		V	B−V	M_v	Sp	μ	θ	d	v_r	Notes
		h	m	°	′					0″·001/y	°	pc	km/s	
Sirius	α CMa	6	41	−16	35	−1·44	−0·01	+1·41	A1 v	1 324	204	2·7	− 8 v	vis. db 49·9 y, 9″
Adhara	ε CMa	6	55	−28	50	1·48	−0·17	−5·0	B2 II	4	135	200	+27·4	vis. db 8″
Wezen	δ CMa	7	04	−26	14	1·85	+0·63	−7·0	F8 Ia	5	292	600	+34·3	
Aludra	η CMa	7	20	−29	06	2·42	−0·07	−7·1	B5 Ia	8	294	800	+40·2	
Castor	α Gem	7	28	+32	06	1·56	+0·05	+0·8	A1, A5	202	237	14	+ 3 v	vis. tr., each sp. db
Procyon	α CMi	7	34	+ 5	29	0·36	+0·41	+2·7	F5 IV–V	1 247	214	3·5	− 3·2	vis. db 4″ 41 y
Pollux	β Gem	7	39	+28	16	1·15	+1·01	+1·0	K0 III	624	264	10·7	+ 3·4	nearest giant
Naos	ζ Pup	8	00	−39	43	2·23	−0·27	−7·3	O5	33	282	800	−24	
	γ Vel	8	06	−47	03	1·85	−0·25	−4·2	WC7	8	290	160	+35	vis. db 41″
Avior	ε Car	8	20	−59	11	1·94	+1·2	−3·1	K0, B	30	296	100	+11·5	db, composite sp.
	δ Vel	8	42	−54	21	1·93	+0·04	+0·1	A0 V	89	170	23	+ 2·2	vis. tr. 3″, 69″
Suhail	λ Vel	9	04	−43	02	2·23	+1·7	−4·3	K5 Ib	25	281	200	+18·4	
Miaplacidus	β Car	9	12	−69	18	1·68	−0·01	−0·4	A0 III	186	301	26	− 5	
Scutulum	ι Car	9	14	−58	51	2·24	+0·18	−4·2	F0 Ib	21	266	180	+13·3	
	κ Vel	9	19	−54	35	2·45	−0·16	−3·0	B2 IV	12	275	120	+22 v	sp. db 117 d
Alphard	α Hya	9	23	− 8	14	2·05	+1·43	−0·7	K4 III	35	333	35	− 4·3	
Regulus	α Leo	10	03	+12	27	1·34	−0·11	−0·8	B7 V	248	269	26	+ 3·2	vis. tr. 177″, 3″
Algeiba	γ Leo	10	14	+20	21	2·02	+1·2	−0·5	K0p III	349	119	32	−36·7	vis. db 400 y, high vel.
Merak	β UMa	10	56	+56	55	2·36	−0·02	+0·6	A1 V.	88	71	23	−12·0	
Dubhe	α UMa	10	58	+62	17	1·81	+1·06	−0·6	G9 III	138	240	30	− 8·9	close vis. db
Zosma	δ Leo	11	9	+21	04	2·55	+0·12	+0·8	A4 V	204	135	23	−21	
Denebola	β Leo	11	44	+15	08	2·13	+0·08	+1·6	A3 V	509	256	13	− 1	
Phecda	γ UMa	11	49	+54	15	2·43	0·00	−0·1	A0 V	94	88	32	−12	
Gienah	γ Crv	12	11	−16	59	2·58	−0·09	−2·4	B8 III	161	275	100	− 4·2	
Acrux	α Cru	12	21	−62	33	0·83	−0·26	−3·7	B1, B3	44	228	80	− 6 v	vis. db 5″, each sp. db
Gacrux	γ Cru	12	26	−56	33	1·68	+1·58	−2·5	M3 II	272	176	70	+21·3	optical pair
Muhlifain	γ Cen	12	36	−48	25	2·16	−0·01	−1·7	A0 IV	198	266	60	− 7·5	vis. db 85 y
Mimosa	β Cru	12	42	−59	09	1·29	−0·25	−4·3	B0 III	51	240	130	+20·0	
Alioth	ε UMa	12	50	+56	30	1·78	−0·02	−0·2	A0p	115	95	25	−10	sp. db 4·15 y, Cr-Eu star
Mizar	ζ UMa	13	20	+55	27	2·12	+0·03	0·0	A2 V	129	103	26	− 9 v	tr., vis db 14″, sp. db 20 d
Spica	α Vir	13	20	−10	38	0·97	−0·23	−3·1	B1 V	53	229	65	+ 1 v	ecl. sp. db 4·0 d
	ε Cen	13	34	−52	57	2·34	−0·23	−3·6	B1 IV	35	229	150	+ 5·6	
Alcaid	η UMa	13	44	+49	49	1·86	−0·19	−2·3	B3 V	120	260	70	−10·6	
Hadar	β Cen	13	57	−59	53	0·63	−0·24	−5·0	B1 II	36	220	130	−12 v	vis. db 1″·2
Menkent	θ Cen	14	01	−35	53	2·07	+1·02	+0·9	K0 III–IV	738	225	17	+ 1·3	

Star		1900 α		1900 δ		V	B−V	M_v	Sp	μ	θ	d	v_r	Notes [1, 3, 7]
		h	m	°	′					0″·001/y	°	pc	km/s	
Arcturus	α Boo	14	11	+19	42	−0·05	+1·24	−0·2	K1 III	2 285	209	11	− 5·2	
	η Cen	14	29	−41	43	2·39	−0·21	−3·0	B2 V	48	224	120	− 0·2	db, composite sp.
Rigil Kent	α Cen	14	33	−60	25	−0·27	+0·71	+4·2	G2, K1	3 678	281	1·3	−22 v	vis. tr. 80·1 y, 2°·2
	α Lup	14	35	−46	58	2·5	−0·22	−2·5	B1 V	33	220	100	+ 7·3	
Izar	ε Boo	14	41	+27	30	2·39	+0·93	−0·6	K1, A	49	280	40	−16·5	vis. db 3″, composite sp.
Kochab	β UMi	14	51	+74	34	2·04	+1·49	−0·6	K4 IIII	31	280	33	+16·9	
Alphecca	α CrB	15	30	+27	03	2·22	−0·02	+0·5	A0 III	156	130	22	+ 2 v	ecl. sp. db 17·36 d
Dzuba	δ Sco	15	54	−22	20	2·32	−0·14	−4·0	B0 V	35	200	180	−15	
Acrab	β Sco	16	00	−19	32	2·52	−0·09	−4·0	B0·5 V	27	200	200	− 7 v	vis. tr., 14″, 1″
Antares	α Sco	16	23	−26	13	0·94	+1·83	−4·7	M1, B	30	192	130	− 3·2	vis. db 3″
	ζ Oph	16	32	−10	22	2·56	0·00	−3·4	O9·5 V	22	35	160	−19	
Atria	α TrA	16	38	−68	51	1·93	+1·43	−0·4	K2 III	43	148	29	− 3·6	
	ε Sco	16	44	−34	07	2·29	+1·15	+0·6	K2 III–IV	664	248	22	− 2·5	
Sabik	η Oph	17	05	−15	36	2·44	+0·05	+0·8	A2·5 V	96	23	21	− 1·0	close db
Shaula	λ Sco	17	27	−37	02	1·60	−0·23	−3·2	B1 V	33	180	90	0	sp. db 5·6 d
	θ Sco	17	30	−42	56	1·86	+0·38	−4·0	F0 Ib	12	110	150	+ 1·4	
Ras-Alhague	α Oph	17	30	+12	38	2·07	+0·15	+0·9	A5 III	262	154	17	+13	
	κ Sco	17	36	−38	59	2·39	−0·21	−3·3	B2 IV	30	203	140	−10	
Eltanin	γ Dra	17	54	+51	30	2·21	+1·54	−0·8	K5 III	23	197	40	−27·7	
Kaus Australis	ε Sgr	18	18	−34	26	1·81	−0·02	−1·7	B9 IV	137	198	50	−11	
Vega	α Lyr	18	34	+38	41	0·03	0·00	+0·5	A0 V	346	36	8·1	−13·7	
Nunki	σ Sgr	18	49	−26	25	2·09	−0·20	−2·4	B2 V	62	173	80	−11	
	ζ Sgr	18	56	−30	01	2·57	+0·09	−0·4	A2 IV	20	270	40	+22	vis. db 21 y
Altair	α Aql	19	46	+ 8	36	0·77	+0·22	+2·3	A7 IV–V	658	54	4·9	−26·2	
Peacock	α Pav	20	17	−57	03	1·94	−0·20	−2·9	B3 IV	87	177	90	+ 2 v	sp. db 11·8 d
Sadir	γ Cyg	20	19	+39	56	2·22	+0·66	−4·8	F8 Ib	1		250	− 7·5	
Deneb	α Cyg	20	38	+44	55	1·25	+0·08	−7·2	A2 Ia	3		500	− 4·6	
Gienah	ε Cyg	20	42	+33	36	2·46	+1·03	+0·6	K0 III	483	47	24	−10 v	
Alderamin	α Cep	21	16	+62	10	2·43	+0·23	+1·5	A7 IV–V	159	72	15	− 9	
Enif	ε Peg	21	39	+ 9	25	2·38	+1·56	−4·6	K2 Ib	26	92	250	+ 4·9	vis. db 82″
Al Na'ir	α Gru	22	02	−47	27	1·75	−0·14	−0·2	B5 V	197	141	25 ?	+11·8	
	β Gru	22	37	−47	24	2·16	+1·62	−2·6	M3 III	133	99	90	+ 1·6	
Fomalhaut	α PsA	22	52	−30	09	1·16	+0·09	+1·9	A3 V	367	117	7·0	+ 6·5	
Scheat	β Peg	22	59	+27	32	2·50	+1·7	−1·4	M2 II–III	234	55	60	+ 8·7	
Markab	α Peg	23	00	+14	40	2·49	−0·04	0·0	B9·5 III	73	125	32	− 3·5	

Radiation density in interstellar space is 12×10^{-13} erg per cm^3. The equivalent black body temperature is 3.5° K. The spectral distribution of radiation density is:

Wavelength in microns (μ)	0.05	0.1	0.2	0.3	0.4	0.5	0.6	0.8	1.0
Radiation in 10^{-14} erg $cm^{-3}\mu^{-1}$	3	11	26	34	45	62	82	121	162

The interstellar magnetic field is approximately 10^{-5} gauss.

Cosmic radio emission presents an interesting picture. The surface intensity of an extended area is related to equivalent temperature and frequency

$I_r = 3.0715 \times 10^{-4}\, T\, \mu^2$

I is in watt in^{-2} $sterad^{-1}$ $(c/s)^{-1}$

and includes the two polarization components

T is in °K

μ is in c/s

The equivalent radio temperature on the galactic equator at 100 mc/s appears as:

Longitude in degrees	0	30	60	90	120	150	180	210	240	270	300	330
Temperature in K	4000	1700	1300		1300	1000	740	730	1100	2000	4200	6000

(NOTE: Galactic center is at 328° Longitude.)

Frequency equivalent temperature and radio intensity are presented in Table 2–31.

TABLE 2–31. FREQUENCY EQUIVALENT TEMPERATURE AND RADIO INTENSITY AT GALACTIC CENTER AND AT POLES

Frequency	*Wavelength*	*Galactic Center*		*Cold Sky Near Celestial Poles*	
		log T	log T_U	log T	log I_U
mc/s	cm	T in °K	I in watt in^{-2} $sterad^{-1}$ $(c/s)^{-1}$		
10	3000	5.4	−20.1	5	−20.5
20	1500	5.3	−19.5	4.6	−20.2
50	600	4.6	−19.5	3.6	−20.5
100	300	3.8	−19.7	2.7	−20.8
200	150	3.1	−19.8	2.0	−20.9
500	60	2.1	−20.0	1	−21
1000	30	1.4	−20.1	0	
2000	15	0.8	−20.2		
5000	6	0.0	−20.2		

Table 2–32 presents data on the brighter discrete radio stars.

Source Name, etc.	I.A.U.	1950 α		δ		*f* at 100 Mc/s	*f* at 1 000 Mc/s	*x* at 100 Mc/s	Size	log Dist.	Identifications and notes
		h	m	°	′	10^{-26} watt $m^{-2}(c/s)^{-1}$			′	in pc	
Cas B	00N6A	0	23	+63	52	250	56	−0·7	5·5	2·8?	Tycho's SN I, 1572
And A	00N4A	0	40	+40	50	190		−0·5	140	5·68	Andr. neb., M31
Per A	03N4A	3	16	+41	21	130	22	−0·8	2·4	7·85	Col. gal., NGC 1275
For A	03S3A	3	20	−37	21	240			large		Pec. gal., NGC 1316?
Per	04N3A	4	34	+29	36	280	70	−0·85	P		
HB9	05N4A	4	58	+46	26	140	150	+0·1	60+h		Gal. neb., SN II
Pic A	05S4A	5	18	−45	52	500			large		
Tau A	05N2A	5	32	+21	59	1 700	1 100	−0·26	5	3·02	Crab neb., SN I, 1054
Ori	05S0A	5	33	− 5	24	50	500		10	2·7	Orion neb., M42
Gem	06N2A	6	15	+22	38	500	200	−0·4	30+h	3·1	IC443, SN II
Mon		6	29	+ 4	59	500	320	−0·17	70	3·15	Rosette neb.
Pup A	08S4A	8	21	−42	58	560	180	−0·8	40+h	2·7?	Gal. neb., SN II
Hya A	09S1A	9	16	−11	53	400	65	−0·9	P	8·32	Col. gal.
CVn	12N4A	12	15	+47					3		NGC 4258
Vir A	12N1A	12	28	+12	40	1 700	290	−0·73	4	7·04	Pec. gal., M87
Cen A	13S4A	13	22	−42	46	1 750	2 000	−0·66	5+h	6·7	Pec. gal., NGC 5128
Cen B	13S6A	13	30	−60		700					
Boo	14N5A	14	10	+52	26	100	30	−0·6	P		NGC 5457??
Tr A	16S6A	16	10	−60	45	800					
2C 1402	16N4A	16	27	+39	38	100	7	−1·4	1		Four gal. NGC 6166
Her A	16N0A	16	49	+ 5	07	580	74	−0·8	2·5	8·53	Pec. gal.?
2C 1473		17	18	− 0	52	400	84	−1·2	4		
2C 1485		17	28	−21	16	80			P	2·8?	Kepler's SN I, 1604
Sgr A	17S2A	17	43	−28	50	3 800	2 000	−0·6	70	3·91	Galactic centre
Sgr		18	01	−24	21		140	*T*			Gal. neb. M8
Oph	18S0A	18	13	− 7		200					
Sgr	18S1A	18	18	−16	18		700	*T*	10	3·2	Gal. neb. Ω = M17
Cyg A	19N4A	19	58	+40	35	11 800	2 150	−0·68	1·2	8·34	Col. gal.
Cyg X	20N4A	20	30	+41			900	*T*	100	3·1	IC1 318 (γ Cyg)
2C 1725		20	42	+50	20	510	180	−0·5	100		SN II
Cyg loop		20	48	+30	30	430	200	−0·35	150	2·88	Cyg. loops SN II
America		20	52	+44	00		400	*T*	150	3·0	America neb.
Cas A	23N5A	23	21	+58	32	19 000	3 100	−0·80	4	3·35	Gal. neb. SN II

TABLE 2–32 SELECTED DISCRETE RADIO SOURCES

Discrete radio sources are labelled by various names and catalogue numbers and also by an I.A.U. designation. Some sources have a central nucleus and an extended halo (+ h in the size column). There is a tendency for the halo to be included in the high frequency *f* measurements but not in the low, and this can account for some discrepancies between *f* and *x* values. The identifications of the sources include several supernovae SN of types I and II, colliding (or double) galaxies, peculiar galaxies, and galactic nebulosities. Thermal sources are labelled *T* in the *x* column. *P* = point source.

Cosmic rays outside the earth's magnetic field have a flux of 0.6 primary particles $cm^{-2}s^{-1}$ and an energy of 5 Bev $cm^{-2}\ s^{-1}$ (= 0.007 erg $cm^{-2}s^{-1}$). In interstellar space we find 0.8×10^{-10} particles $cm^{-2}s^{-1}$ with an energy of 1.0×10^{-12} erg $cm^{-2}s^{-1}$. The mean energy of primary cosmic ray particles is 7 Bev = 0.011 erg. The relative abundance of cosmic ray nucleus is distributed as follows:

Energy in Bev	0.2	1	2	3	5	10	20	50	100	1000
Intensity in/10^{-5} particles cm^{-2} s^{-1} sterad^{-1} Bev^{-1}	4000	5000	4000	2500	1200	400	110	12	2	0.005

Extragalactic space exhibits an internebular absorption of 0.005 magnitude/Mpc. The density of grain material is about 10^{-31} g/cm. The gas density is about 10^{-29} g/cm. The internebular hydrogen density necessary to account for the absorption is one hydrogen atom per cm^3. Data on extragalactic space are, of course, quite questionable at this time. Further detailed data are available in Ref. (*46*) on extra-terrestrial environment.

References to Chapter 2

1. K. A. Ehricke, *Space Flight*, Volume I, D. Van Nostrand Company, Inc., Princeton, N.J., 1960.
2. W. A. Heiskanen and F. A. Vening Meinesz, *The Earth and Its Gravity Field*, McGraw Hill Book Co., New York, pp. 51–53.
3. *Space Facts*, a handbook published by the General Electric Company, Philadelphia, Pa., 1960.
4. Fritz Zwicky, "Is Newton's law of gravitation really universal?", *Astronautics*, Vol. 4, No. 1, January 1959, p. 18.
5. Herman F. Michielsen, "The Odd Harmonics of the Earth's Gravitational Field," pre-print 61–25, Seventh Annual Meeting, American Astronautical Society, January 1961.
6. Larry Booda, "Transit IV-A Proves Equator is Elliptical," *Aviation Week and Space Technology*, October 9, 1961, p. 33.
7. W. M. Kaula, "Estimation of Longitudinal Variations in the Earth's Gravitational Field from Satellite Observations," pre-print 61–26, Seventh Annual Meeting, American Astronautical Society, January 1961.
8. C. W. Allen, *Astrophysical Quantities*, University of London Athlone Press, 1955, p. 128.
9. William Knorr, "A Study of Flight Instrumentation for Vehicles Operating in the Fringe of, or Outside of, the Earth's Atmosphere," Part III, Definition of Operating Environment, pp. 22–23, WADC Technical Report 59–567, Part III, ASTIA No. 236175, Oct. 1959.
10. R. Saunders, "Geomagnetic and Geoelectric Activity as a Function of Magnetic Latitude," *ARS Journal*, October 1961, pp. 1438–1439.
11. S. Perlman, W. Russell and F. Dickson, "Concerning Optimum Frequencies for Space Vehicle Communications," IRE-PG-MIL Transactions of April–July 1960, pp. 184–192.
12. Allen, *op. cit.*, pp. 116–119.
13. Allen, *op. cit.*, pp. 114–115.
14. M. J. Aucreamanne, "The Ionosphere Beacon Satellite, S-45," NASA Technical Note D-695 of January 1961.

15. "S-48 to Sound High Ionosphere," *Missiles and Rockets*, June 26, 1961, pp. 12–13.
16. W. W. Berning, "Earth Satellite Observations of the Ionosphere," *Proceedings of the IRE*, February 1959, pp. 280–288.
17. D. F. Martyn, "The Normal F Region of the Ionosphere," *Proceedings of the IRE*, February 1959, pp. 147–155.
18. Sir Edward V. Appleton, "The Normal E Region of the Ionosphere," *Proceedings of the IRE*, February 1959, pp. 155–159.
19. J. J. Gibbons & A. H. Waynick, "The Normal D Region of the Ionosphere," *Proceedings of the IRE*, February 1959, pp. 160–161.
20. J. O. Thomas, "The Distribution of Electrons in the Ionosphere," *Proceedings of the IRE*, February 1959, pp. 162–175.
21. K. I. Maeda & T. Sato, "The F Region During Magnetic Storms," *Proceedings of the IRE*, February 1959, pp. 232–239.
22. C. O. Hines, "Motions in the Ionosphere," *Proceedings of the IRE*, February 1959, pp. 232–239.
23. E. N. Parker, "Auroral Phenoma," *Proceedings of the IRE*, February 1959, pp. 239–244.
24. B. Nichols, "Auroral Ionization and Magnetic Disturbances," *Proceedings of the IRE*, February 1959, pp. 245–254.
25. S. Chapman, "Sun Storms and the Earth: The Aurora Polaris and the Space Around the Earth," *American Scientist*, Vol. 49, No. 3, Sept. 1961, pp. 249–284.
26. W. Knorr, *op. cit.*, p. 25.
27. S. I. Akasofu & S. Chapman, "New Theory of the Aurora Polaris," *ARS Journal*, Vol. 31, No. 6, June 1961, pp. 775–783.
28. J. Lee, H. Pratt, R. Cluttenden, K. Foster, "Natural Interference Effects in Space Communications," RADC-TN-59-373, Nov. 1959, pp. 73–82.
29. R. A. Helliwell & M. G. Morgan, "Atmospheric Whistler's," *Proceedings of the IRE*, February 1959, pp. 200–208.
30. Roger M. Gallet, "The Very Low-Frequency Emissions Generated in the Earth's Exosphere," *Proceedings of the IRE*, February 1959, pp. 211–231.
31. "Space Dust Brought Back by Rocket," *Missiles and Rockets*, June 19, 1961, p. 31.
32. J. A. O'Keefe, "Tektites and the Cyrillid Shower," *Sky and Telescope*, January 1961.
33. J. A. O'Keefe, "Tektites as Natural Earth Satellites," *Science*, Feb. 24, 1961, pp. 562–566.
34. T. N. Nazorova, "Investigation of Meteoric Particles by the Third Soviet Satellite," translation printed in *ARS Journal*, Sept. 1961, pp. 1341–1344.
35. W. Knorr, *op. cit.*, p. 25.
36. F. L. Whipple, "The Meteoric Risk to Space Vehicles," 8th International Astronautical Congress, Barcelona, Oct. 1957.
37. F. G. Watson, *Between the Planets*, The Blakiston Co., Philadephia, 1941.
38. N. P. Barabashov, A. A. Mikhaylov, Yu. N. Lipkiy, "Atlas Obratnoy Storony Luny (Atlas of the Far Side of the Moon)," Academy of Sciences, U.S.S.R., March 1960, Translated by Technical Information Center, MCLTD, Wright-Patterson AFB, Ohio.

39. J. P. Ferrara and M. Chomet, "Preliminary Investigation of Lunar Surface Propagation," Sixth Symposium on Ballistic Missile and Aerospace Technology, republished in *Topics in Space* by Airborne Instruments Laboratory.

40. E. H. Vestine, "Utilization of a Moon-Rocket System for Measurement of the Lunar Magnetic Field," Rand Report No. RM-1933 of July 9, 1957.

41. "New Gap: U.S., USSR Astronomical Units," *Missiles and Rockets*, May 29, 1961, p. 23.

42. Allen, *op. cit.*, p. 130.

43. G. P. Kuiper, *The Sun*, University of Chicago Press, 1953, p. 211.

44. E. H. Vestine, "Physics of Solar-Terrestrial Space: Lunar Flight," Rand Report No. P-1344 of 24 Feb. 1958, pp. 6–10.

45. S. F. Singer, "Some Considerations of Expected Radiation Belts of Planets Mars and Venus," *Advances in the Astronautical Sciences*, Vol. 6, American Astronautical Society, New York, 1960, pp. 781–793.

46. Allen, *op. cit.*, Chapters 10–14.

CHAPTER 3

CELESTIAL MECHANICS

3.1 INTRODUCTION

Celestial Mechanics, as employed by astronautical engineers, generally deals with the determination of the motion of celestial bodies in a gravitational field. Ehricke (Ref. (*1*)) has subdivided the subject matter into:

1. Central force field, undisturbed orbits (conics), and their elements.
2. Coordinate transformations; orbit determination and ephemeris computation.
3. Perturbation theory.

In this chapter, an attempt is made to define the more essential elements and problems of interest to the Electronic Engineer. The approach used is to attack the subject from an instrumentation point of view. That is, we emphasize coordinate systems that may be instrumented directly in determining a vehicle's motion in a central force field, with a view to achieving accuracy and/or simplicity in this instrumentation.

Initially, we consider the nature of general trajectories in a gravitational field, describing the nature of both free-falling or coasting vehicles (as in orbit) and powered-flight trajectories. Considerations of relativity effects when operating at extremely high velocities, are discussed to provide some insight into problems for future research as a prelude to interstellar travel. Relativity effects will certainly be of immediate interest in communication problems (as with probes) long before transportation missions approach any significant fraction of the velocity of light.

3.2 COORDINATE SYSTEMS

In selecting coordinate systems we may say firstly that external reference systems, such as earth-located vehicle tracking nets, are obviously tied to an Earth (and someday Lunar, Martian, etc.) coordinate system such as latitude and longitude, Map-zones, polar coordinates based on a tracking radar location, etc. Interest in this group of coordinate systems is naturally great in our first probe and satellite experiments; this interest will continue in terms of terminal operations in the future. However, the purpose of this chapter is to describe those coordinate systems in which the space vehicle may sensibly operate by itself; that is, by self-contained instrumentation which is sensitive to both internal and external references. Tracking of a space-vehicle (even from another vehicle) and command guidance links are held to be communication problems. Conversely, the coordinate system employed in vehicle operation must be self-oriented in some way if control of the vehicle is to be maintained, and vehicle-borne sensors which are capable of determining the vehicle's progress in some celestial-coordinate system are deemed to present celestial mechanics problems.

The various celestial coordinate references available to a space vehicle include:

1. Inertial
2. Stellar and Constellation References
3. Clocks and Time
4. Local Body References
5. Sun, planet and satellite tracking

Further scrutiny will show that the favored reference systems will generally be those most closely related to determining a vehicle's motion in a gravitational field without requiring precise knowledge or computation from external tracking station data in the determination of a vehicle's location, path, and motion.

3.2.1 Inertial References

Inertial references are best explained by Newton's Laws of Motion:

(1) Every body continues in its state of rest or of uniform motion in a straight line except insofar as it may be compelled to change that state by the action of some outside force.
(2) Change of motion is proportional to force applied and takes place in the direction of application of the force.
(3) To every action there is always an equal and opposite reaction.

Basic inertial sensing devices are accelerometers, pendulums, and gyroscopes. These are a few of the many variations of these devices:

(1) Linear accelerometers
(2) Angular accelerometers
(3) Integrating accelerometers
(4) Bubble levels (a form of pendulum)
(5) Pendulous accelerometers
(6) Stable platforms (a form of synthetic pendulum)
(8) Rate gyroscopes
(9) Integrating rate gyroscopes

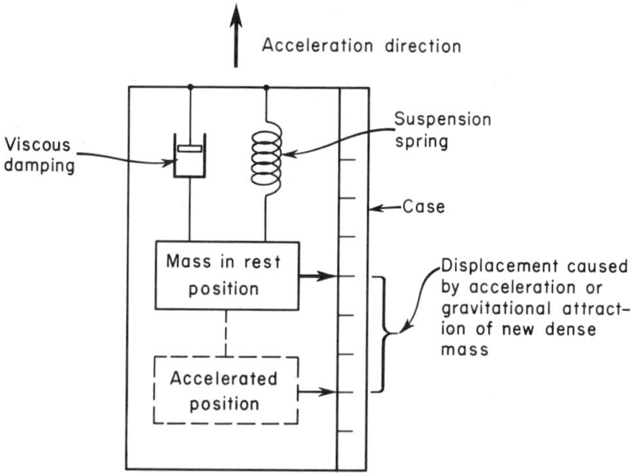

FIG. 3-1. Linear Accelerometer.

In general, we may say of accelerometers that they are a direct manifestation of Newton's First Law. As depicted in Fig. 3–1, an accelerometer is a device in which a mass body is suspended such that it will be displaced relative to a reference scale in proportion to a newly applied acceleration or gravitational force. If we apply an acceleration of $\ddot{x}$ to the case of an accelerometer, the case-mounted scale will move from the initial or rest position as $x = \iint \ddot{x}(dt)^2$, whereas the accelerometer sensing mass will be displaced relative to the scale. If we designate the case position as x_i; the mass, m; the mass position relative to the case as x_o; the spring constant as k; and the viscous damping constant as c, we may write the differential equation of motion for the mass as

$$m\,[\ddot{x}_i + \ddot{x}_o] + c\dot{x}_o + kx_o = 0 \qquad (3\text{–}1)$$

This may be rearranged in Laplace form, to,

$$s^2\, x_o(s) + \frac{c}{m}\, sx_o(s) + \frac{k}{m}\, x_o\,(s) = -s^2\, x_i\,(s) \qquad (3\text{–}2)$$

The input/output relation may be taken as,

$$\frac{x_o}{x_i}\,(s) = \frac{-s^2}{s^2 + \dfrac{c}{m}\, s + \dfrac{k}{m}} \qquad (3\text{–}3)$$

which may be placed in conventional form,

$$\frac{x_o}{x_i}\,(s) = \frac{-s^2}{s^2 + 2\zeta\omega_n s + \omega_n{}^2} \qquad (3\text{–}4)$$

where the natural or resonant frequency,

$$\omega_n = \sqrt{\frac{k}{m}} \tag{3–5}$$

and the damping ratio,

$$\zeta = \frac{c}{2\sqrt{mk}} \tag{3–6}$$

The response characteristics of Eq. (3–4) are well known, and may be found in any number of references (e.g. pp. 340–349 of *Guidance* by Locke, D. Van Nostrand Co., Inc., 1955). The essential characteristics are those of natural or resonant frequency and damping ratio.

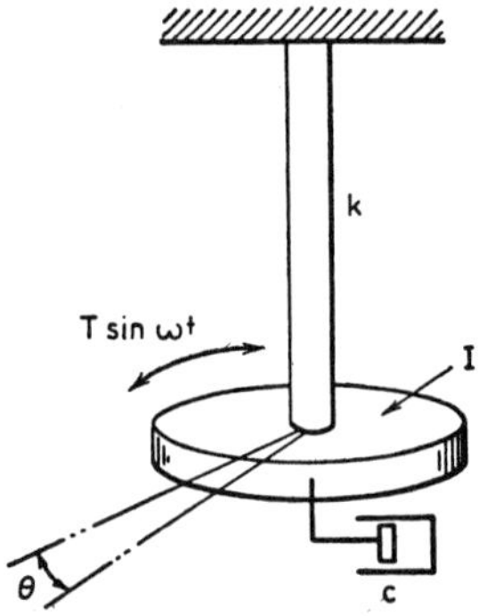

Fig. 3-2. Angular Accelerometer.

Angular accelerometers may be treated similarly by considering the device as a rotating mass of inertia, I, as illustrated in Fig. 3–2. In this case, the differential equation for angular motion is,

$$\ddot{\theta}_0 + \frac{c}{I}\dot{\theta}_0 + \frac{k\,\theta_0}{I} = -\ddot{\theta}_i \tag{3–7}$$

This equation will become identical with equation (3–4) if we merely let

$$\omega_n = \sqrt{k/I} \tag{3–8}$$

and

$$\zeta = \frac{c}{2\sqrt{kI}} \tag{3–9}$$

From the foregoing, particularly Eq. (3–4), we see that if we can bracket the time or frequency response of interest with our accelerometer response characteristic (denominator of Eq. (3–4)), we may obtain a good 1:1 measure of acceleration within any vehicle by internal instrumentation. This measurement of acceleration may be made in any set of orthogonal axes desired, including both straight line and angular. Limitations on accelerometer use will be, null sensitivity (or minimum detectable reading), linearity, maximum sensitivity, and frequency response.

Integrating accelerometers are merely those in which the first integration is done internally (as opposed to being done in some external computer), thus providing a velocity function measurement.

A particular peculiarity of accelerometers is that they cannot distinguish between an acceleration force and a gravitational force. Thus attention must be paid to their orientation during alignment and calibration.

Pendulums as reference devices for determining the local vertical are used quite successfully under static conditions. However, the motion of the body carrying the pendulum introduces an acceleration force (as in an orbiting vehicle) which must be added vectorially to the gravitational force on which the pendulum is trying to operate in order to determine the local vertical. Obviously, maneuvering perturbations must be accounted for precisely, in using the output of a pendulum reference. The single pendulum which will not deflect upon vehicle acceleration is the Schuler pendulum which has its bob at rest at the Earth's center, and acts over the Earth's radius. While this type of pendulum is impractical to build, it is effectively synthesized, for many Earth-oriented applications, in the so-called stable platform. This is a level platform on which accelerometers are mounted, the platform being maintained level (which defines the local vertical) through torquing from an appropriate combination of accelerometer and gyroscope measurements. For platforms designed to operate somewhat near the surface of the Earth (at Earth's radius from the Earth mass-center), "Schuler tuning" of the accelerometer feed-back loops is employed; that is, the platform is made to resonate with the same frequency as a Schuler pendulum—approximately an 84 minute period; actually

$$\text{Schuler period} = 2\pi \sqrt{\text{Earth's radius/gravity at Earth's surface}}$$

For Schuler tuning, the platform-mounted accelerometers are coupled to the platform torquers such that small displacements of the platform will produce a restoring acceleration equal to gravity acceleration, g, times the small displacement, and with a period of 84-minutes. This is illustrated in Fig. 3–3, a transfer function diagram of an inertial platform

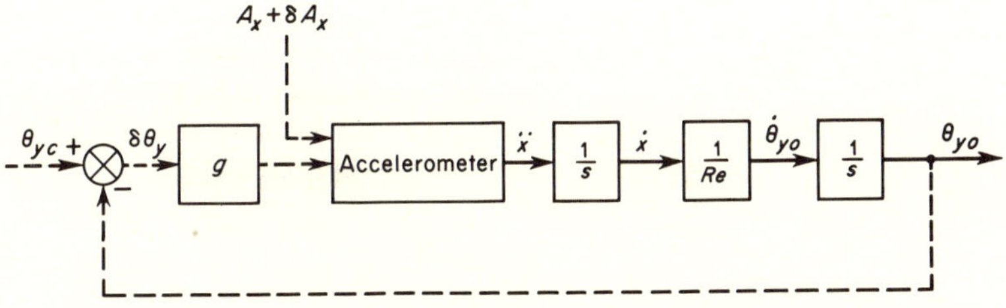

FIG. 3-3. Inertial Platform Stabilization Loop.

stabilization loop. In Fig. 3–3, we see illustrated the platform tilt angle, θ_{yo}, which may be commanded from some external source (e.g. navigation computer) to assume some new position, θ_{yc}. The true acceleration, $R_E\ddot{\theta}_y$, and its measurement error, $R_E\delta\ddot{\theta}_y$, are set equal to the accelerometer reading, A_X, its measurement error, δA_X, and the platform tilt error, $g\delta\theta_y$

$$R_E\ddot{\theta}_y + R_E\delta\ddot{\theta}_y = A_X + \delta A_X - g\delta\theta_y \tag{3-10}$$

$A_X = R_E\ddot{\theta}_y$ since both are taken as the true acceleration, and the error terms must be equal, or,

$$\delta\ddot{\theta}_y + \frac{g}{R_E}\,\delta\theta_y = \frac{\delta A_x}{R_E} \tag{3-11}$$

If we set $\dfrac{g}{R_E} = \omega_0^2$

we may say in complex frequency, s, notation,

$$\delta\theta_y(s)[s^2 + \omega_0^2] = \frac{\delta A_x(s)}{R_E} \tag{3-12}$$

or,

$$\delta\theta_y(s) = \frac{\delta A_x(s)}{R_E}\cdot\frac{1}{s^2 + \omega_0^2} \tag{3-13}$$

This expression defines the platform loop characteristics as having a resonant frequency at $\omega_0 = \sqrt{g/R_E}$, the Schuler frequency, and an unit input response of $\dfrac{1}{R_E\omega_0}$ sin $\omega_0\, t$. These are the desired characteristics of a Schuler pendulum and thus the goals in designing stabilized platforms for inertial navigation.

It should be noted that this has been a very simple discussion of the Schuler-pendulum nature of stabilized platforms. The use of gyros, their errors, the integration of accelerometer data to "dead-reckon" one's position, the various contributing errors, their interaction and statistical history has been omitted, and left to a companion book of this present series, *Inertial Navigation*, by Richard H. Parvin. The principal purpose here has been to illustrate the availability of a technique to achieve a local vertical, equivalent to a Schuler pendulum in characteristics.

A gyroscope is a spinning wheel suspended in orthogonally related gimbals as illustrated in Fig. 3–4. The wheel is precisely balanced, and spins at a high rate to achieve a large angular momentum. The gimbals are also precisely balanced, and the bearings are made as frictionless as possible. The spinning wheel and the gimbals may take many shapes, including spherical as well as cylindrical and square rings, as illustrated. The suspension may be on metal, air or gas bearings (injected through tubes), flotation in a fluid of similar density to the metal, etc. The spinning wheel may be driven electrically, by gas jets or merely coast after some initial impulse as by a spring. Potentiometers, synchros or reaction devices are employed to monitor the motion of each gimbal about its axis. Thus, during flight, any small motion of the case about one of the sensitive axes may be measured, the spinning wheel providing the inertia to maintain a fixed spatial reference direction. The limits or stops of gyros in their freedom of movement may be very small (as in those mounted on stabilized platforms), or through 360 degrees in one axis (as in aircraft directional gyros used to measuring heading bearings). Gyros may have one or two degrees of freedom. Gyros may also have spring-type restrainers on their gimbals so that angular rates rather than positions are sensed. Sensing of motion may be through the precessional effect as illustrated in Fig. 3–5. Precession is the result of Newton's Law that: "The time rate of change of angular momentum on a body about any given axis is equal to the torque applied about the given axis." As shown in Fig. 3–5, precession is about an axis orthogonal to both the input axis and the

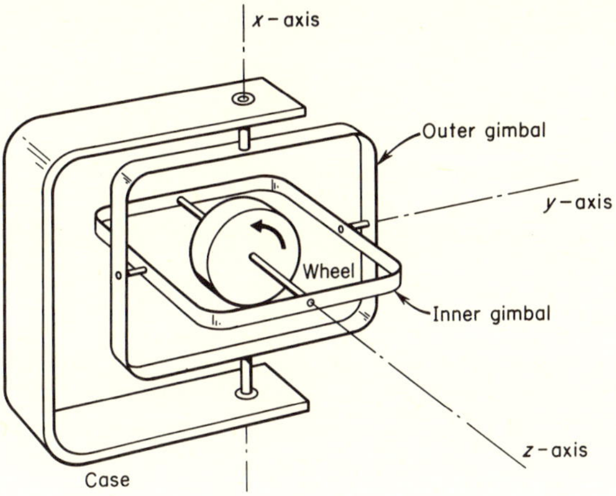

FIG. 3-4. Schematic of Two-Degree-of-Freedom Gyro.

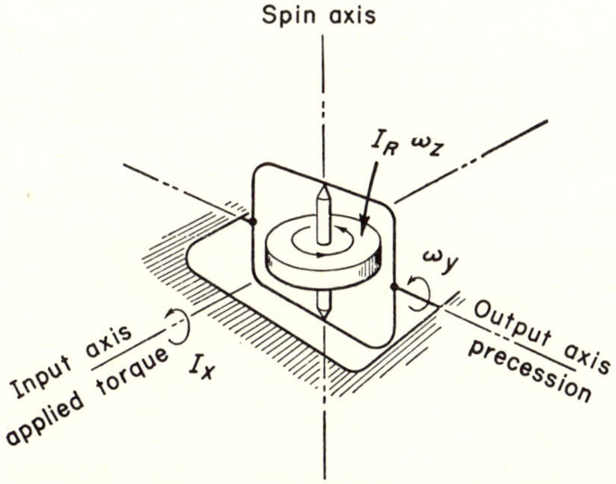

FIG. 3-5. Gyroscopic Precession in a Free Gyro.

spin axis, and in such a direction as to align the spin axis with the input axis so that the direction of rotation of the rotor is aligned with the direction of rotation of the applied torque. The following simple relation holds if we neglect error sources:*

$$T_x = I_R \omega_z \omega_y \tag{3-14}$$

where T_x = applied torque,
I_R = gyro rotor inertia about spin axis,
ω_z = rotor speed, and
ω_y = angular velocity of precession

A more complete expression which includes gimbal inertia is derivable from the vector relation (Newton's law),

$$T = \frac{dH}{dt} \tag{3-15}$$

in which T is the total torque applied, H is the total angular momentum,

$$H = iI_x\omega_x + jI_y\omega_y + kI_z\omega_z \tag{3-16}$$

or, taking the vector cross-products into account, and expressing in component form, we may say,

$$T_x = I_x\dot{\omega}_x + (I_z - I_y)\omega_y\omega_z \tag{3-17}$$

$$T_y = I_y\dot{\omega}_y + (I_x - I_z)\omega_x\omega_z \tag{3-18}$$

$$T_z = I_z\dot{\omega}_z + (I_y - I_x)\omega_x\omega_y \tag{3-19}$$

These three equations may be made more explicit by distinguishing between gimbal inertia, I_G, and rotor inertia, I_R, to provide

$$T_x = I_{Gx}\dot{\omega}_x + (I_{Gz} - I_{Gy})\omega_y\omega_z + I_R\omega_y\omega_z \tag{3-20}$$

$$T_y = I_{Gy}\dot{\omega}_y + (I_{Gx} - I_{Gz})\omega_x\omega_z - I_R\omega_x\omega_z \tag{3-21}$$

$$T_z = I_{Gz}\dot{\omega}_z + (I_{Gy} - I_{Gx})\omega_x\omega_y \tag{3-22}$$

The terms containing rotor inertia, I_R, are generally much larger than the other terms so that the approximate precessional equations may be quickly obtained,

$$T_x = I_R\omega_z\omega_y \tag{3-23}$$

$$T_y = -I_R\omega_z\omega_x \tag{3-24}$$

*Note: $I_R\omega_z$ is actually an angular momentum term.

If we assume no external torques, and ω_z to be zero for gimbal motion (i.e. ω_z has a rotor spin value, but the z-axis gimbal is stationary), we obtain

$$0 = I_{Gx}\dot{\omega}_x + I_R\omega_y\omega_z \tag{3–25}$$

$$0 = I_{Gy}\dot{\omega}_y - I_R\omega_x\omega_z \tag{3–26}$$

which describes gyro nutation, or conical motion of the spin axis. The solution of these equations shows that a point on the spin axis describes an ellipse.

A single-degree-of-freedom gyro contains a single gimbal which can rotate about the output axis relative to the case (the spin axis is perpendicular to the input and output axes). If we take the x-axis as the input, y-axis as the output, and z-axis as the spin axis, the input-output relation involves motion of the case relative to inertial space about the input axis. The rotation forces a precession of the spinning wheel which causes it to exert a torque (called gyroscopic torque) on the gimbal about the output axis. Balancing this torque are position, and first order and second order restraining torques; the positional (or spring) restraint being proportional to the gimbal angle relative to the case, the first order (or viscous) restraint being proportional to the gimbal angular velocity relative to the case, and the second order restraint being the inertial reaction torque of the gimbal. The predominant term of these three restraining torques will determine whether the single-degree-of-freedom gyro is called (and used as) a position or integrating gyro, a rate gyro, or an undamped gyro. The input-output relation (θ) requires a coordinate conversion between spatially-fixed and case axes,

$$\omega_x = \omega_a \cos\theta - \omega_c \sin\theta \tag{3–27}$$

$$\omega_y = \dot{\theta} + \omega_b \tag{3–28}$$

$$\omega_z = \omega_c \cos\theta + \omega_a \sin\theta \tag{3–29}$$

The output torque, T_y, is

$$T_y = -k\theta - d\dot{\theta} + T_G \tag{3–30}$$

Substituting these four equations in our previous equation for T_y (for a two-degree-of-freedom-gyro) we obtain,

$$-k\theta - d\dot{\theta} + T_G = I_{Gy}(\ddot{\theta} + \dot{\omega}_b) - I_R\omega_z(\omega_a \cos\theta - \omega_c \sin\theta) + (I_{Gy} - I_{Gx})\,(\omega_a\omega_c \cos 2\theta + (\omega_a^2 - \omega_c^2)\,\tfrac{1}{2} \sin 2\theta] \tag{3–31}$$

The $(I_{Gy} - I_{Gx})$ term is important only in some gyro drift analyses, and is generally neglected for single-degree-of-freedom gyros to yield for small output angles

$$I_{Gy}\ddot{\theta} + d\dot{\theta} + k\theta + I_R\omega_z\omega_c\theta = I_R\omega_z\omega_a - I_{Gy}\dot{\omega}_b + T_G \tag{3–32}$$

The terms, $I_R\omega_z\omega_c\theta$ and $I_{Gy}\dot{\omega}_b$ represent cross-coupling terms which can generally be considered as part of the gyro drift, and may be omitted to yield the Laplace transfer function for a rate gyro

$$\theta(s) = \frac{I_R\omega_z\omega_a + T_G}{I_{Gy}s^2 + ds + k} \tag{3–33}$$

Damping factors of 0.5, and undamped natural frequencies of 20 to 30 cps are typical of rate gyros used in missile stabilization systems. For an integrating or position gyro the spring constant, k, is negligible, and we obtain,

$$\theta(s) = \frac{I_R\omega_z\omega_a}{s(I_{Gy}s + d)} + \frac{T_G}{s(I_{Gy}s + d)} \tag{3–34}$$

or,

$$\theta(s) = \frac{I_R\omega_z\omega_a}{sd\left(\frac{I_{Gy}}{d}s + 1\right)} + \frac{T_G}{sd\left(\frac{I_{Gy}}{d}s + 1\right)} \tag{3–35}$$

where $\frac{I_{Gy}}{d}$ is generally a time constant on the order of 10^{-3} sec. Taking the rotor inertia as very large when compared with the gimbal inertia, and choosing a time constant, $\frac{I_{Gy}}{d}$, which will be unimportant in the application, we obtain,

$$\theta(f) = \frac{I_R\omega_z}{d}\int \omega_a dt \tag{3–36}$$

indicating we may have an output directly proportional to a small angle input measured from the reference position.

Undamped gyros usually have a gas or viscous output axis bearing in which the viscous and spring constant terms are small. This yields,

$$\theta(s) = \frac{I_R\omega_z\omega_a}{I_{Gy}\,s^2} \tag{3–37}$$

or

$$\theta(t) = \frac{I_R\omega_z}{I_{Gy}}\iint \omega_a(dt)^2 \tag{3–38}$$

which indicates an output signal proportional to the time integral of the input angle.

Gyros thus may take many forms, depending on the application, heading reference, vertical, platform stabilization, rate measurement, etc. The major error sources in gyros are mass unbalance, bearing friction, windage, pick-off sensitivity, and gimbal limits. If used properly (e.g., to measure within limits and to good sensitivities), the major error result in depending on gyroscopes is termed "drift." Gyro drift is usually taken as linear with time (e.g. 0.1 degrees per hour), which indicates that generally, for space guidance applications, where long transport times are required, gyroscopes will not be suitable except for their short-term accuracy relative to some time-insensitive reference system.

3.2.2 Stellar and Constellation References

For long-time transport missions, and astronomical observations, stellar or constellation references are quite likely to be employed. In general, the "fixed stars," although moving at tremendous speeds relative to our solar sun, do indeed remain relatively fixed in space, offer an excellent reference framework for interplanetary navigation and eventually must be used for inter-stellar navigation. A table of the brightest magnitude stars has been presented in Table 2–30. Once free of Earth (and planetary) atmospheric hindrances, it is to be expected that we shall be able to measure the positions of stellar references to rather precise angles. Von Pahlen Fedoroff in Ref. (*2*) reports that angular resolutions of 10^{-4} seconds of arc may be expected by improving equipment, and using known techniques. Lt. Col. John Glenn, in his first orbital manned space flight, used partially opaque windows to reduce the light level to the order of Earth-surface intensity, and reported that the stars were just as they appeared on a clear night in the desert except that they did not twinkle.

The fundamental system technique for using stellar references is illustrated in Fig. 3–6. The technique is to select 6 stars which are sufficiently

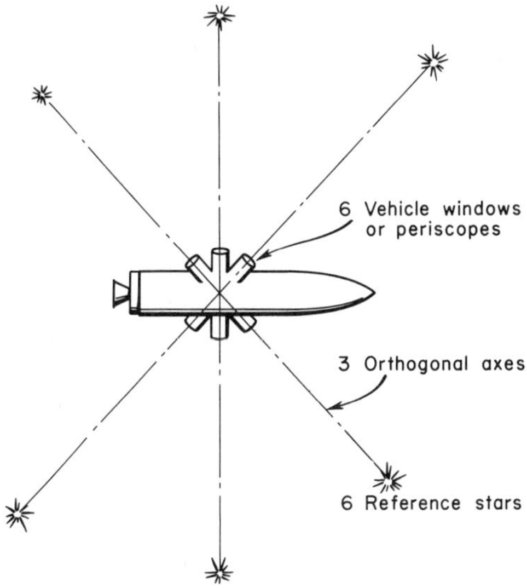

Fig. 3-6. Stellar Reference Coordinate System.

bright as to be easily distinguishable from the stellar background. The six stars must be so situated as to define three orthogonal axes relative to the solar system. They must be distant enough that their positions on the celestial sphere will not change significantly during the space vehicle journey. Thus, a permanent fixed reference coordinate system may be employed through a vehicle's journey, and may even be used to measure and stabilize or program vehicle attitude as well. It is expected that this technique will be employed to stabilize the NASA Orbiting Astronomical Observatory (OAO).

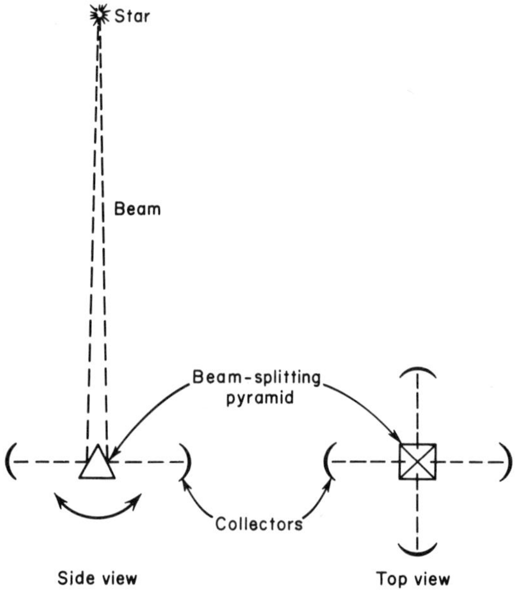

FIG. 3-7. Beam-Splitter Technique.

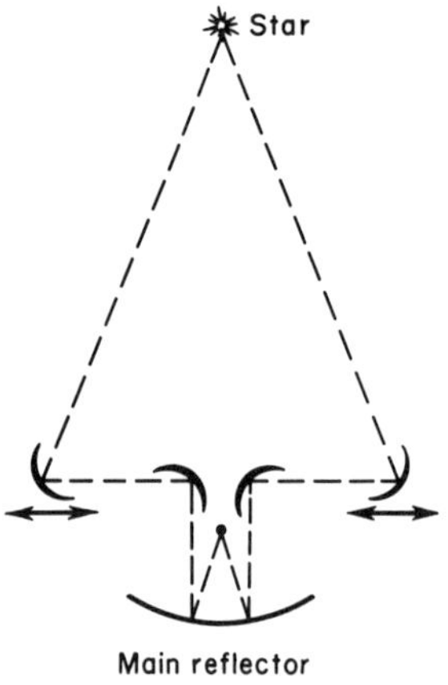

FIG. 3-8. Interference Technique.

Two general methods exist for measuring the angles relative to stars; geometrical and wave optics. Geometric techniques have shown their best capabilities in beam-splitter devices. Laboratory accuracies of 1 second of arc have been achieved with 0.1 sec anticipated with further development. Beamsplitting is the focusing of the stellar image on a prismatic pyramid such that the light intensity is split equally into four parts as illustrated in Fig. 3–7. If the prism is tilted slightly in any direction, the beam will be split unevenly, and a comparison of relative light intensity at the collectors is a direct measure of the angular deviation.

The wave or interferometric technique employs a large reflector, two inner mirrors in front of the reflector objective and tilted at 45 degrees, and two adjustable outer mirrors as illustrated in Fig. 3–8. In this technique (sometimes called "interference technique"), two small mirrors are mounted in front of the main reflector objective. Two other mirrors are moved outward until they no longer focus the star's image on the inner mirrors, hence the change in interference level. This technique has been used to measure the angular dimensions of stars by astronomical observatories to accuracies of 0.01 second of arc resolution, and may some day be employed to measure space vehicle attitudes to such accuracies.

The use of constellations (e.g., Ursa Major) as references appears to be useful only as a crude method of locating a particular star (e.g. Polaris). Once a 6-star reference system has been established and is being tracked, the remainder of the constellation would appear to be superfluous (as it is with mariners).

3.2.3 Clocks and Time References

The requirement for time references divides itself into long- and short-period references. For long-period references, one may select the Earth-year or the periods of the Jovian moons as has long been done by mariners, the main requirement being that some means be available for observing (or receiving) these references. For short-time references, as required for synchronization of reference datums (e.g., for space Loran-type systems), something similar to WWV radio time transmissions may be employed with appropriate corrections for path length transit time delays. Atomic clocks may be employed for frequency reference, as a time reference for precision integration in "dead-reckoning" navigation calculations and as a coding system reference. The general technique is to slave a quartz-crystal oscillator to a frequency determined by an atomic or molecular transition. Three techniques are currently employed; cesium-beam, rubidium-vapor, and ammonia-maser. All three techniques are reputed (Ref. (*3*)) to have exhibited stabilities of 1 part in 10^{10} with 1 part in 10^{11} predicted for the near future, and 1 part in 10^{12} within a few years.

It is expected that atomic clocks will be employed in satellite experiments to check the General and Special Theories of Relativity when $1{:}10^{12}$ is realized. The General Theory of Relativity predicts a frequency shift Δf of

$$\frac{\Delta f}{f} = \Delta U/c^2 \tag{3–39}$$

where f is the frequency employed and ΔU is the difference in gravity potential at the locations of the two standards. This shift should be on the order of a few parts in 10^{12} if one standard is in a balloon at the equator, and the other in a submarine at the North Pole (taking advantage of the Earth's ellipticity).

An orbiting satellite would exhibit a second-order Doppler frequency shift as predicted by the Special Theory of Relativity in addition to the gravitational shift. The Doppler shift has been demonstrated in the laboratory, and can be regarded as a known correction. The Doppler shift is,

$$\frac{\Delta f}{f} = \frac{U_E}{c^2}\left[\frac{1}{2} \times \frac{R_E}{a}\right] \tag{3–40}$$

where U_E is the gravitational potential at the Earth's surface, R_E is the Earth's radius, and a is the semi-major axis of the satellite orbit. The total shift of the satellite clock with respect to the Earth-located clock will be the summation of Eq. 3–39 and 3–40,

$$\frac{\Delta f}{f} = \frac{U_E}{c^2}\left[\frac{3}{2} \times \frac{R_E}{a} - 1\right] \tag{3–41}$$

Bender (Ref. (*3*)) predicts that with an 8000-mile circular orbit, a 10-day experiment should provide sufficient data from the two clocks to validate the effects of the General Theory as well as the Special Theory of Relativity. More knowledge of these effects must certainly be obtained before such clocks can be used with confidence for inter-planetary space navigation.

3.2.4 Local Body References

Local body references are defined as those references pertinent to vehicle operation in a locale dominated by the gravitational field of a single mass. These references include local vertical, magnetic, horizon-sensing, check-point and map comparison, altimeter, and doppler devices. In inter-planetary operation, one may determine the local vertical by a solar tracking device as discussed in the following section; whereas planet and satellite centered devices for determining the local vertical are deemed to be of particular importance in space navigation in the vicinity of such a body.

The use of a pendulum to determine the local vertical while on the surface of the Earth is traditional in surveying, but is limited by the knowledge of local gravitational anomalies. These may cause unknown deflections up to 5 minutes of arc, unless prior gravimetric mapping data have been obtained. The use of a pendulum under dynamic conditions, as in orbiting a body or when maneuvering in space, is clearly impractical when one considers that the pendulum must respond to the accelerations imposed as well as the gravitational force. The resultant error in a direct pendulum measurement of local vertical will be an angular deflection error as illustrated in Fig. 3–9. An error in local vertical determination by pendulous methods results when an acceleration, a, acting at an angle, α, with respect to the local vertical, is imposed as illustrated. The error angle is,

$$\rho = \tan^{-1}\frac{g + a\cos\alpha}{a\sin\alpha} \tag{3–42}$$

As explained in Section 3.2.1 there is one pendulum which will operate under induced accelerations without errors other than instrument limitations. This is the Schuler pendulum which places the bob at the Earth's center, which, although impractical to do directly, is what is attempted in the design of inertially stabilized platforms. For a synthetic Schuler pendulum it is necessary that the pendulum period be,

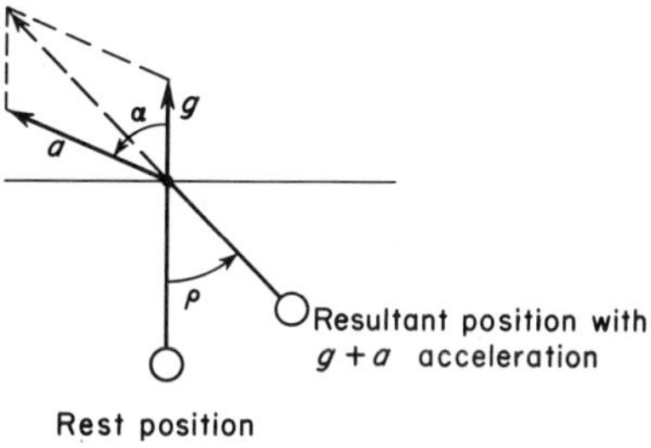

$$\tau_0 = 2\pi\sqrt{R_E/g_0} \qquad (3\text{–}43)$$

FIG. 3-9. Pendulum Error.

to operate successfully at the surface of the Earth. This period, which is nominally 84 minutes at the Earth's surface, is thus used to "Schuler-tune" inertial platforms for operation in the lower atmosphere surrounding the Earth. Such platforms are employed to establish a local vertical by means of accelerometer damping in company with gyro references. The navigation problem then becomes one of "dead reckoning" integration of accelerometer measurements taken in gyro-maintained, orthogonal axes. A transfer function block diagram of an inertial navigation as a position servo is illustrated in Fig. 3–10. This figure shows that,

$$\left[(\theta_{yc} - \theta_{yo})g + R_E s^2\theta_{yc} + \delta a_x\right]\left(\frac{1}{R_E}\right)\left(\frac{1}{s^2}\right) = \theta_{yo} \qquad (3\text{–}44)$$

with

$$\theta_{yo} = \theta_{yc} + \delta\theta_y \qquad (3\text{–}45)$$

and

$$\omega_0^2 = \frac{g}{R_E} \qquad (3\text{–}46)$$

we may obtain,

$$\delta\theta_y(s) = \frac{\delta a_x(s)}{R_E}\cdot\frac{1}{s^2 + \omega_0^2} \qquad (3\text{–}47)$$

which provides an error sensitivitv transfer function resonant at $s = \pm j\omega_0$, and a unit impulse response of $\frac{1}{R_E\omega_0}$ sin $\omega_0 t$. In dealing with

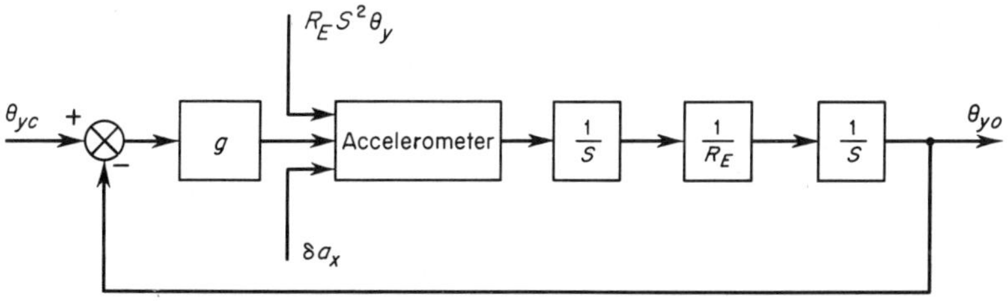

FIG. 3-10. Inertial Navigator as a Position Servo.

Schuler-tuned platforms, we must note that they are artificially tuned for operation at some specific distance from the dominant force-field mass center. If we examine the function of gravity at altitude, h, above the Earth's surface,

$$g_h = g_0 \left(\frac{R_E}{R_E + h} \right)^2 \tag{3-48}$$

we may see that the necessary natural frequency must be,

$$\omega_h = \frac{g_h}{R_E + h} \tag{3-49}$$

or

$$\omega_h = \omega_0 \left(\frac{R_E}{R_E + h} \right)^3 \tag{3-50}$$

From this relation, it is obvious that to have a useful stabilized platform for local vertical determination at any significant altitude above the Earth's surface, we must use a servo system in which the natural tuning frequency is adjusted as a function of altitude. Similarly, if we wish to employ stabilized platforms when operating near any planet or satellite, we must establish the planetary gravitational constant and radius precisely, align our platform (e.g. through optical or gyrocompassing techniques on the surface, or horizon sensors at altitude) and program the natural frequency into the stabilization system as a function of some precise altitude measurement so as to satisfy the relation;

$$\omega_h = \omega_0 \left(\frac{R_0}{R_0 + h} \right)^3 \tag{3-51}$$

or

$$\omega_h = \frac{g_0 R_0^2}{(R_0 + h)^3} \tag{3-52}$$

These equations apply directly to and will be useful for any externally supported vehicle. That is, any vehicle in which the outer frame is supported by surface contact aerodynamics or rockets, and in which powered flight is requisite to movement. Thus, the vertically-registering accelerometer can measure the local gravitational force relative to the vehicle frame, and the laterally (and vertically) registering accelerometers can measure the engine thrust effects. When in a free-flight orbit, the inertial platform cannot measure the actual orbit ephemeris as the accelerometer masses will not deflect relative to the cases or vehicle frame. Accelerometers can, of course, be used to measure powered deviations from the free flight orbit. With sufficient ephemeris knowledge, a computer should be able to determine the orbit change.

Magnetic sensors for navigational reference relative to local bodies are subject to two major flaws: (1) the knowledge of magnetic fields around the various bodies is not sufficiently precise in terms of general magnitude, anomalies, diurnal, seasonal and storm effects, and (2) the instruments themselves have not been developed to operate with the necessary accuracies. The first point is amply illustrated in Chapter 2 by the lack of fundamental data and the various (in)accuracies quoted on magnetic deflection charts. Magnetic guidance as an elementary mid-course guidance technique has been employed successfully in manned aircraft and missiles of the V-1 variety, generally for azimuth determination. The lack of precision in magnetic instruments and field charts forces reliance on some other technique for terminal guidance.

Current magnetometers sense magnetic field effects by use of phenomena such as the absorption process of a helium absorption cell at 1.08 micron wavelength (Ref. (*4*)), in which the relative number of helium atoms at each of three energy levels is a function of the external magnetic field. Such instruments provide useful data in charting magnetic fields, and may ultimately be employed as magnetic coordinate sensors if and when the staticity or periodic nature of local body magnetic fields can be accurately established. The use of permanent magnet bars as a "despin" device for Earth-satellites is a proven technique (Ref. (*5*)), having been employed successfully in the TRANSIT satellite program. This, however, is basically a means of stabilizing with respect to the ambient magnetic field, is subject to inertial effects as well as field fluctuations, and therefore is not considered as a direct coordinate reference-determining mechanism.

Horizon sensors offer a good solution to the local vertical sensing problem, being useful primarily in orbits at high altitudes around some local body. They are not subject to the altitude dependence of stabilized platforms, the inertial effects on pendulums, the non-functioning of accelerometers in free orbit and the lack of correlation between magnetic and gravitational coordinates. Horizon sensors are generally designed with infra-red sensing elements because the visible wavelength horizon is often obscured at night or by clouds, whereas the infra-red (or thermal) horizon is less sensitive to such conditions. As illustrated in Fig. 3–11, a single sensor may be rotated at various deflection angles, or, a prism may be scanned at various deflection angles using a single IR detector. If the scan axis is off-vertical, the detector will sense a modulation at the scan frequency with amplitude a function of the off-vertical angle, and phase angle directly related to the azimuth of the off-vertical error angle. As an alternate scheme, several sensors may look in various fixed directions, and their detector signals compared. Elevation programming of a pencil beam may be used, or one may employ a narrow (receiving or sensing) beam in azimuth with a wide beam (or field of view) in elevation as shown in Fig. 3–11. A second technique is illustrated in Fig. 3–12, in which two IR scanning sensors are employed; one for pitch and one for roll attitude determination. In this technique, two pencil beams are conically-scanned, the sensors sighting the planet's emissive surface during some fraction of the scan. The mid-point of this radiation sighting in each of the two axes defines the local vertical. This is sufficient to establish the local vertical to better than ±0.3 degrees (Ref. (*6*)). This technique may be developed into a right-left side-looking pair of scanning detectors and a fore and aft looking pair. Detector wavelength sensitivity is generally selected to bracket the Earth's self-emission spectrum between 5 and 20 microns, and to exclude the solar reflected spectrum below 5 microns; thus making the system insensitive to diurnal variations. A desensitizer is generally needed to protect the detector when looking directly at the Sun (sunrise and sunset effect).

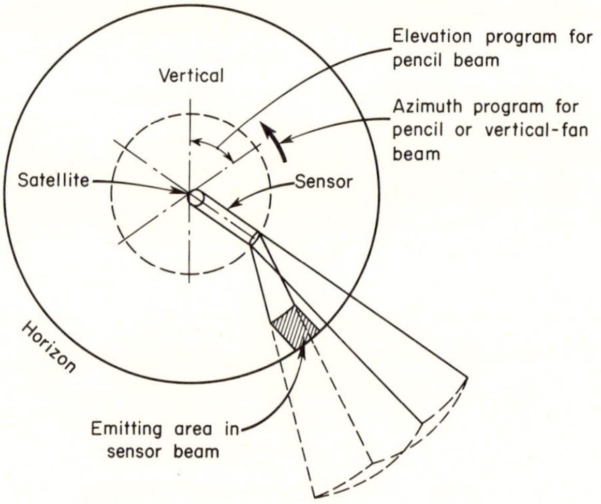

FIG. 3-11. Horizon Sensing in a Satellite.

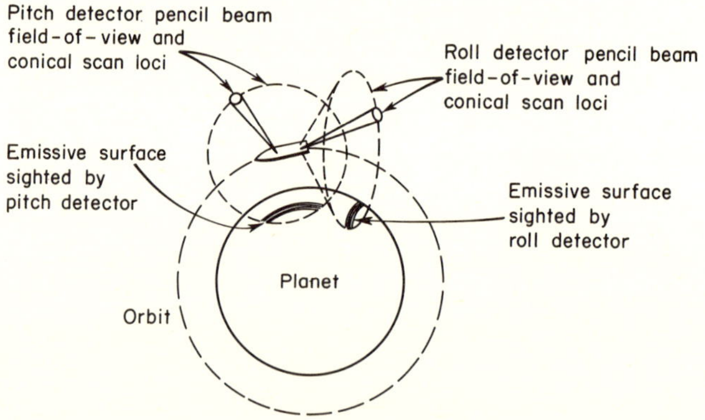

FIG. 3-12. Pitch and Roll Attitude Horizon Sensor Technique.

If we look at Fig. 3–11, we see that an intensity or amplitude modulation on the IR detector output may be expected in which the modulation frequency equals the scan frequency, the modulation amplitude is proportional to the pitch attitude error (from normal to the local vertical) and the phase of the modulation is equal to the roll attitude as illustrated in Fig. 3–13. This relation may be expressed as,

$$A = A_0 [1 + k\theta e^{i(\omega_\psi + \phi)}] \tag{3–53}$$

where A_0 and k are dimensionalizing constants, θ is pitch attitude, ω_ψ is the scanning frequency, and ϕ is the roll attitude. The system which employs pitch and roll conical scanning of small field-of-view (essentially spot) sensors, may be expressed for each axis, ϕ and θ, as measuring attitude by the interruption or presence of the Earth's return as a phase

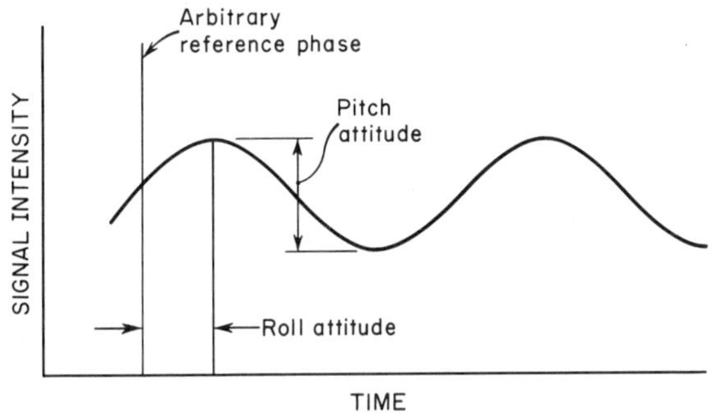

FIG. 3-13. Signal Intensity Versus Time for Azimuth Horizon Scanner

angle relative to the scan phase angle. The detected signal will be a square-shaped pulse or gate, occurring at the scanning frequency, with a width that is a function of altitude and a mid-point that is a direct measure of attitude, θ or ϕ. Radar "horizon" scanners offer an alternate to the infra-red technique. An active radar horizon scanner may be developed which uses monopulse resolution improvement in the elevation plane, and thus may provide angular accuracy competitive with infra-red devices. Such a system will not be subject to "crescent moon" effects on infra-red sensors but may track its own, range gate established artificial horizon. This offers the possibility of combining the orbital vertical determination function with a landing radar sensor thus achieving an overall weight saving by eliminating the infra-red device. The utility of horizon sensors in measuring altitude depends on how accurately altitude must be known. Recalling that the Earth, in perspective, is smoother than an orange (highest land-elevation is roughly 0.1 per cent of the radius), altitude may be measured by the horizon or disc size of the globe. This method is accuracy-limited by the angle measurement precision available; the disc radius, R, being related to altitude, h, by a tangent function from a high altitude ($h \gg R$). At low altitudes, ($h < R$), the relation must take into account global curvature effects.

Other altimeter devices for space operation employ radar echo techniques. These altimeters may be of the pulse, continuous-wave, or many hybrid forms. A limitation arises in altitude (or ranging) precision from the finite beam generated by the space-borne antenna which may be wide enough to obtain returns from several surface features rather than just the nadir point. Phase or echo-time measurement precision is a second limitation. In general, altimeter ranges are limited by the transmitted power available, antenna gains, antenna isolation in CW systems, and surface reflectivity coefficient. The square-law radar equation will apply.

$$h_{\max} = \sqrt{\frac{Pf^2kA}{4\,\lambda S_{\min}}} \tag{3–54}$$

where P is the transmitted power, k the efficiency of surface reflection, A the effective antenna area, f the antenna efficiency, $S_{\min}$ the minimum power required for successful detection, and λ the wavelength. Depending on the precision and instantaneousness of measurement required, much can be done to improve $S_{\min}$ by signal gating, integration and processing to extend the maximum range. Radar altimeters with precisely defined beams will almost surely be employed in descent control on atmosphereless bodies such as the Moon. The narrowness of the beams will necessitate the slaving of such altimeters to the local vertical (e.g., as may be established by an inertial memory of a horizon-sensor determination). Modulation techniques, such as ramp-modulated FM-CM and interrupted CW, are common hybrid altimeter forms. Leading-edge pulse tracking is another prevalent method. Such conventional altimeters are not precise enough to solve the unmanned vehicle rendezvous mating problem and more sophisticated techniques will be necessary.

Doppler measurements of vehicle velocity relative to the local body permit another navigation aid. Although "dead-reckoning" integration of doppler data is employed in some aircraft navigation systems, this method is not considered sufficiently dependable for exclusive use in space navigation; however, doppler data to update position fixes may prove quite useful. Basically, doppler data are obtained from the frequency shift experienced from the relative motion of the electromagnetic wave with respect to the surface as illustrated in Fig. 3–14. The fundamental relation in doppler measurements is,

$$\frac{f_d}{f} = \frac{2v}{c}\cos\gamma \tag{3–55}$$

where f_d = doppler frequency,
f = propagated frequency,
v = vehicle velocity relative to reflecting surface,
c = propagation velocity, and
γ = angle between velocity and propagation vectors.

Two-, three-, and four (Janus)-beam doppler systems are usually employed in doppler navigation systems as described in Ref. (*7*). The employment of doppler in landing and rendezvous operations is deemed quite likely in that these maneuvers are generally optimally controlled as a function of line-of-sight, range and range rate, all of which may be measured directly through a common multipurpose microwave equipment.

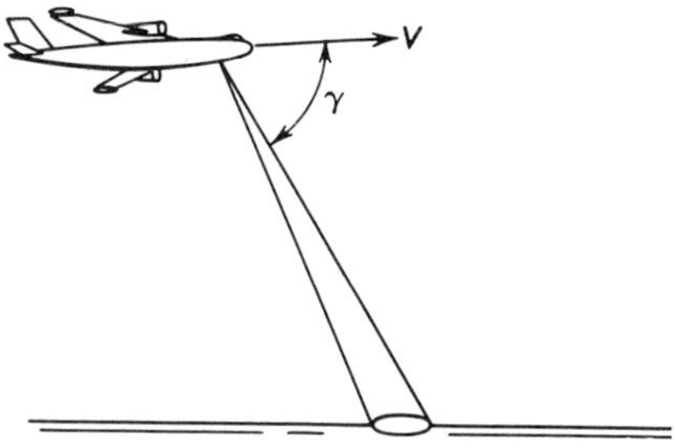

FIG. 3-14. Basic Doppler Beam Configuration.

In addition, orbital doppler measurements permit determination of pitch and yaw attitude of the vehicle relative to its own velocity vector. Thus a common radar measurement is feasible which can determine:

1. Pitch and yaw attitudes in orbit by doppler measurements
2. Local vertical, hence pitch and roll attitudes in orbit by radar, artificial horizon scanning techniques
3. Orbit parameters by direct altimetry measurements
4. Descent/ascent trajectories by direct altimetry and doppler velocity measurements
5. Landing closure by three-components of velocity and altimetry measurements
6. Rendezvous closure by line-of-sight range, rate, vector and vector rate measurements (discussed in Chapter 4).

Surface check-points and map-navigation as a navigation or guidance technique will undoubtedly be employed on the various planets and satellites as their significant features become better charted. These techniques will be particularly important when it becomes desirable to return to selected locations either for measurements and observations or to establish contact with surface bases. Beacons may be employed to establish significant artifacts as navigational references, as is done on Earth presently with Loran, Tacan, Vortac, Decca, etc. Descriptions of such systems are readily available in the literature. Homing beacon references may be used quite profitably as a direct aid in lunar-approach and landing guidance and lunar landings may be facilitated by employing a beacon-responder network to tri-laterate a vehicle's position during the final descent at a lunar port.

3.2.5 Sun, Planet, and Satellite Tracking

The tracking of a solar system body for coordinate orientation purposes appears quite attractive for interplanetary operation. Firstly, the local vertical anywhere in the solar system must be towards the solar disc centroid unless one is in the proximity of a planet or satellite. The local vertical is, of course, a primary reference in orbit determination. Sun trackers for space use may operate virtually anywhere in the electromagnetic spectrum, since the Sun is such a magnificent source of radiation. Generally, infrared, visual, or ultraviolet detectors will be employed. For planet and satellite tracking, point source tracking devices may generally be employed, such as beam-splitting prisms with detector elements sensitive in the emission spectrum of the body. Infrared cells, insensitive to the Sun's reflection may generally be employed (as in Earth horizon trackers) in tracking the inner planets when approaching from farther out. Infrared cells sensitive to the solar reflection spectrum may generally be employed in tracking bodies farther away from the Sun where phasing (e.g. lunar phase) effects will not be experienced. Upon near approach to any planet or satellite, horizon trackers may be employed. The tracking of bodies other than the Sun (which establishes the local vertical), is essential as an aid to "homing" on such bodies. Indeed, mid-course and terminal guidance considerations would seem to make planet and satellite tracking a "must" in achieving reliable manned transport to such bodies (although such tracking might be manual, as with telescopes, rather than automatized. The use of planet tracking to enable navigation other than to the destination planet appears useful only in controlling communication propagation direction, and in the unusual case where a near miss is planned to take advantage of the gravitational acceleration and higher velocity available from a near-miss. Lunar near-miss accelerations for extra-lunar flight may be the principal use of this technique. Solar (and other) disc trackers may use some conical scanning system which, in essence, is designed to detect the sharp intensity gradient at the disc periphery and center on a null position

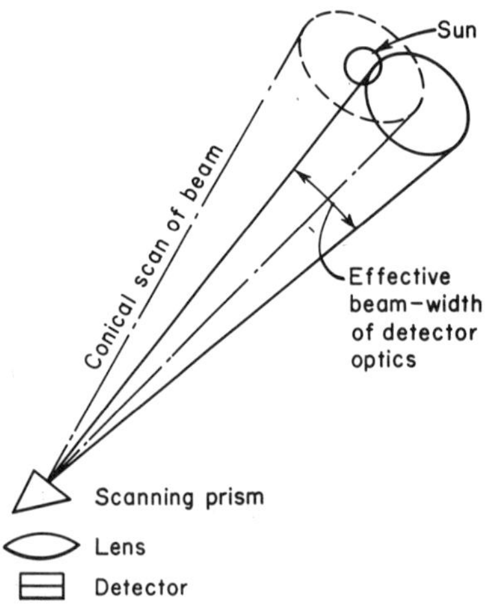

FIG. 3-15. Sun Tracker Principle.

such that the periphery is equidistant (in angle). This conically-scanning beam principle is illustrated in Fig. 3–15 where a rotating prism is used to scan an effective beam around the Sun's periphery. The intensity modulation experienced by the detector will be similar to that illustrated in Fig. 3–13 for the horizon scanner, with possible lapses in detection if the modulation exceeds 100 per cent (indicating the scanner is looking completely off the sun at some time during the scanning period). Accuracies of 0.2 degree have been indicated, with predictions of 0.001 degree in the future.

3.3 VEHICLE TRAJECTORIES

The previous section has dealt with vehicle-borne sensors which are capable of measuring some component of a vehicle's motion, position, or attitude as a function of time and relative to either the dominant gravitational field source, the vehicle inertial vector, or a spatial body reference. Understanding the use of these data in redirecting a vehicle's trajectory in space requires a background in the nature of orbital characteristics in a central force field and the effects of these characteristics on vehicle trajectories in space. Further, consideration must be given to the requirements of minimum energy versus minimum transit time in determining the actual nature of trajectories.

3.3.1 Central Force Field

The gravitational force, F, acting on a body of mass, m, by another body of mass, M, at a distance, r, is,

$$F = -\frac{GMm}{r^2} \tag{3–56}$$

where G is the universal constant of gravitation,

$$G = 6.673 \times 10^{-8} \frac{\text{cm}^3}{\text{sec}^2\, g_{\text{mass}}} \tag{3–57}$$

If we consider the predominant mass, M, to be concentrated at its center of gravity, and assume no variation of F with vector orientation, we define a central force field of concentric, equipotential spheres of radius, r. Thus, the gravitational potential gradient for a unit mass is,

$$\frac{\partial U}{\partial r} = F = \frac{-GM}{r^2} = -g \tag{3–57}$$

or integrating,

$$U = Fr = \frac{GM}{r} = -gr \tag{3–58}$$

The gravitational parameter of a body is,

$$K = -Ur = gr^2 = \text{a constant.} \tag{3–59}$$

Values of the gravitational parameter, for various solar system bodies, have been tabulated in Chapter 2. As indicated previously in that chapter, the nature of the Earth's gravitational field is not truly that of a point-mass concentration. Rather, higher order terms occur which are inverse functions of higher powers of the distance, r, from the mass center. However, the coefficients of these terms are of the order of 10^{-3}, and 10^{-6} relative to the primary field term (at least for the Earth) so that studies of orbits of short duration or subject to control corrections need consider only the predominating point-mass concentration form.

The gravitational potential, U, has the dimension of velocity-squared, and represents the field energy at a distance, r, from the field center. A body may stay on a single equipotential gravity sphere by having sufficient tangential (to the sphere) velocity to fly a circular path about the sphere as illustrated in Fig. 3–16. This velocity will be,

$$V_c = \omega r \tag{3–60}$$

By taking the body location as

$$s = re^{i\omega t} \tag{3–61}$$

we may determine,

$$v_c = ri\omega e^{i\omega t} \tag{3–62}$$

and,

$$a_c = -r\omega^2 e^{i\omega t} \tag{3–63}$$

Setting the required acceleration of a circular path, a_c, equal to the gravitational acceleration, g, associated with the equipotential sphere, U, we may maintain an equilibrium with a constant circular velocity, V_c.

The circular velocity required of potential U is,

$$|V_c| = |r\omega| = \sqrt{|gr|} = \sqrt{|U|} = \sqrt{\left|\frac{k}{r}\right|} \tag{3–64}$$

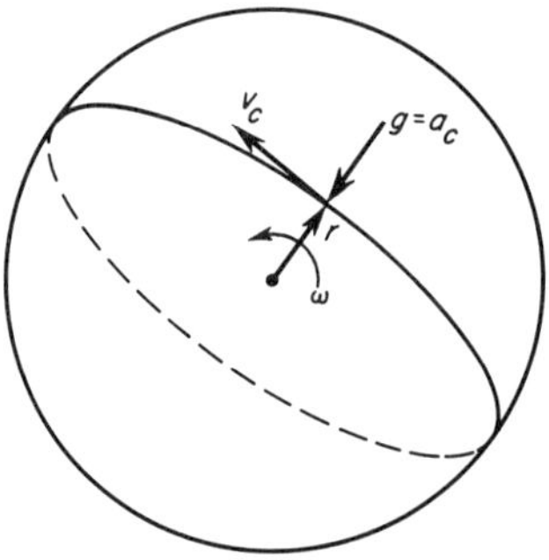

FIG. 3-16. Circular Velocity on an Equipotential Surface.

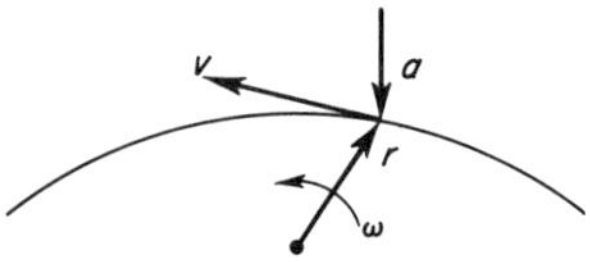

FIG. 3-17. Velocity in an Eccentric Orbit or Trajectory.

Thus, we have defined the circular velocity, V_c, required to describe a great circle path on any equipotential gravitational sphere of magnitude U. If the magnitude or vector of V_c is inappropriate to the gravitational potential, a non-circular path will be followed. If we take the equilibrium situation depicted in Fig. 3–17, we may say that the velocity, which is always tangential to flight path, is also,

$$V = \left(\frac{\partial r}{\partial t} + ir\omega\right) e^{i\omega t} \tag{3–65}$$

where r and ω are no longer constant, but are functions of time. By differentiation:

$$a = \frac{\partial v}{\partial t} = \left[\left(\frac{\partial^2 r}{\partial t^2} - r\omega^2\right) + i\left(2\frac{\partial r}{\partial t}\omega + r\frac{\partial \omega}{\partial t}\right)\right] e^{i\omega t} \tag{3–66}$$

This expression for the acceleration of the body may be separated into two expressions; one in-line with the central force field which must have a finite value and one normal to the central force field which must be zero. These considerations yield,

$$\frac{\partial^2 r}{\partial t^2} - r\omega^2 = g = \frac{-U}{r} \tag{3–67}$$

and,

$$2\frac{\partial r}{\partial t}\omega + r\frac{\partial \omega}{\partial t} = 0 \tag{3–68}$$

These two equations represent the equations of motion of a free body in a central force field. To obtain the path equation independent of time, this parameter must be eliminated from the equations of motion. Defining the angle, η, as,

$$\omega = \frac{\partial \eta}{\partial t} \tag{3–69}$$

we may use the relation,

$$\frac{\partial r}{\partial t} = -r^2\frac{\partial(r^{-1})}{\partial t} = -r^2\frac{\partial \eta}{\partial t}\frac{\partial(r^{-1})}{\partial \eta} \tag{3–70}$$

From Kepler's second law, the area swept by the radial vector per unit time is a constant, and we may say,

$$r^2\omega = r^2\frac{\partial \eta}{\partial t} = C \tag{3–71}$$

Thus,

$$\frac{\partial r}{\partial t} = -C\frac{\partial(r^{-1})}{\partial \eta} \tag{3–72}$$

and,

$$\frac{\partial^2 r}{\partial t^2} = -C\frac{\partial^2(r^{-1})}{\partial \eta^2}\frac{\partial \eta}{\partial t} = \frac{-C^2}{r^2}\frac{\partial^2(r^{-1})}{\partial \eta^2} \tag{3–73}$$

Also, from Eq. (3–71)

$$r\omega^2 = \frac{C^2}{r^3} \tag{3–74}$$

Substituting Eq. (3–73) and (3–74) in Eq. (3–67),

$$-\frac{C^2}{r^2}\frac{\partial^2(r^{-1})}{\partial\eta^2} - \frac{C^2}{r^3} = \frac{-U}{r} \tag{3–75}$$

or

$$\frac{\partial^2(r^{-1})}{\partial\eta^2} + \frac{1}{r} = \frac{Ur}{C^2} = \frac{K}{C^2} \tag{3–76}$$

Integration yields,

$$\frac{1}{r} - \frac{K}{C^2} = C_1 \cos(\eta - C_2) \tag{3–77}$$

or

$$r = \frac{1}{\frac{K}{C^2} + C_1 \cos(\eta - C_2)} \tag{3–78}$$

which may be converted to the polar equation for a conic,

$$r = \frac{p}{1 + \epsilon \cos \eta} \tag{3–79}$$

where $p = \frac{C^2}{K}$, (3–80)

$$\epsilon = \frac{C^2}{K} C_1 = pC_1 \tag{3–81}$$

and $C_2 = 0$ if integration starts at the point of closest approach to the focal point (e.g. perihelion or perigee). Thus, for a

circle:	$\epsilon = 0$	
ellipse:	$\epsilon < 1$	(3–82)
parabola:	$\epsilon = 1$	
hyperbola:	$\epsilon > 1$	

The orbit criterion may now be defined in terms of velocity or energy by comparing the velocity or kinetic energy at radius, r, with the parabolic (or escape) velocity defined by the gravitational parameter, K, operating at distance, r. Thus,

$$\frac{v^2}{2} - \frac{K}{r} = \frac{h}{2} \tag{3–83}$$

where

$$h > 0 \text{ for a hyperbola}$$
$$h = 0 \text{ for a parabola}$$
$$h < 0 \text{ for an ellipse} \qquad (3\text{–}84)$$
$$h = \frac{K}{r} \text{ for a circle}$$

or

$$v^2 = K/r \text{ for a circle}$$
$$\frac{K}{r} < v^2 < \frac{2K}{r} \text{ for an ellipse}$$
$$v^2 = 2\frac{K}{r} \text{ for a parabola} \qquad (3\text{–}85)$$
$$v^2 > 2\frac{K}{r} \text{ for a hyperbola}$$

This derivation of a free-body flight path is intended to show the nature of a central force field by its effects as well as by its definable characteristics. We may thus observe that free bodies in a single, spherical, central force field will follow planar conic trajectories or orbits. While not explicitly demonstrated herein, the center of the force field will always lie on one of the focii of the conic path. In lunar, solar, and planetary explorations

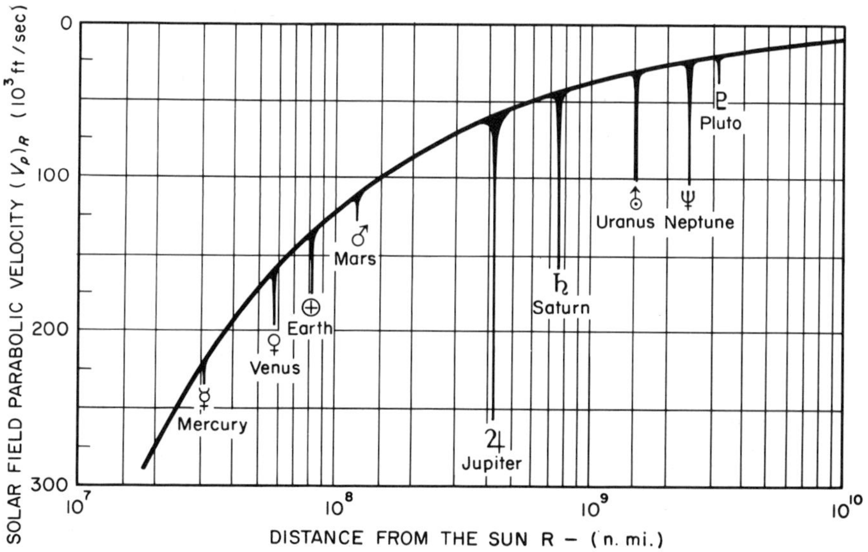

FIG. 3-18. Gravity Field of the Sun and Planets in Terms of Parabolic Velocity.

we must consider the effect of super-position of two or more central force fields as previously illustrated in Fig. 2–25, page 95. A second illustration of this super-position appears in Fig. 3–18 where the gravity field of the sun is depicted with the fields of the various planets superimposed. Note that in this figure, parabolic (or escape) velocity, v_p, is employed. This is related to circular velocity by,

$$v_p = v_c \sqrt{2} \tag{3–86}$$

The parabolic velocity required to escape from a particular location in a planetary gravitational field will thus merely place a body in heliocentric orbit with the initial circular (or elliptic) velocity of the source planet. Variations on this, such as employing a local orbit velocity vector to go in a particular direction, that is, to augment or negate part of the planetary heliocentric velocity, are possible.

3.3.2 Minimum Energy vs. Minimum Transit Time

The considerable expenditure of energy necessary to ascend into space has caused great interest in developing minimum energy trajectories (MET) in order to minimize the cost of performing a particular space flight experiment or operation, or, conversly, to obtain the maximum space flight performance or capability from an available rocket booster. Minimum energy trajectories will be of interest for some time to come as a limiting factor on the extent or scope of man's exploration of the solar system. However, as the cost of putting payloads into space is reduced and as peaceful and military space operations become more common-place, it is to be expected that minimum transit time trajectories will become more important and will govern many space operations.

All trajectories may be expressed by the vector integral relation

$$\vec{s}(t) = \vec{s_0} + \int_0^t \vec{v_0}\, dt + \int_0^t \int_0^t \vec{a}\ (t)\ (dt)^2 \tag{3–87}$$

Although this equation is quite general, certain axioms may be derived from it by inspection. First, one may expect to arrive at a particular destination in space in a minimal time if one selects an initial position in conjunction with or in close proximity to the destination, takes advantage of any available initial velocity, and imparts any available acceleration forces (roughly equivalent to impulse or energy available) at an early time. This statement is somewhat analogous to stating that a bullet fired from a gun will arrive at the earliest time if the muzzle is brought into close proximity to the target, if the muzzle velocity is as high as possible and if all the acceleration is accomplished while traversing the barrel. The rocket equivalent is to consider the trajectory as essentially ballistic in a central force field from the time of last stage burn-out; that is, burn-out velocity is equivalent to muzzle velocity in conventional fire control problems, and the space vehicle may be expected to follow a conic trajectory from then on. The control problems are, therefore, the common ones associated with fire control: range, target apparent motion, and the instrumentation and measurement of these quantities. The conservation or minimization of energy expenditure for a particular space flight dictates also that the maximum velocity be attained as early as possible in flight, but that this maximum velocity be only that necessary barely to reach the desired destination.

The circular velocity associated with any particular orbit has already been shown to be Eq. (3–64),

$$|v_c| = \sqrt{|U|} = \sqrt{\left|\frac{k}{r}\right|}$$

If we desire to transfer from one potential level to another, we must impart a velocity increment which has a minimum value,

$$\Delta v_{c\ \min} = |v_{c1}| - |v_{c0}| \tag{3–88}$$

This minimum transfer velocity must be augmented for any requirement to re-orient the orbit plane, reverse direction of the orbit (as direct to retrograde), or to attain some particular phasing of position in the new orbit. Thus,

$$\overrightarrow{\Delta v_c} = \overrightarrow{v_{c_1}} - \overrightarrow{v_{c_0}} \tag{3–89}$$

The transit time will be,

$$t_{1-0} = \frac{\overrightarrow{s_1} - \overrightarrow{s_0}}{\overrightarrow{\Delta v_{\text{ave}}}} = \frac{\overrightarrow{\Delta s}}{\overrightarrow{\Delta v_{\text{ave}}}} \tag{3–90}$$

Thus, t_{1-0} will approach a minimum as Δv_{ave}, the average velocity change from the original orbit velocity, approaches a maximum. We may say,

$$\Delta \vec{v} = \int_0^t \vec{a}\ (t)\ dt \tag{3–91}$$

Thus,

$$\overrightarrow{\Delta v_{\text{ave}}} = \frac{1}{t}\int_0^t \overrightarrow{\Delta v dt} = \frac{1}{t}\int_0^t\int_0^t \vec{a}(t)dt \tag{3–92}$$

Assuming no drag or friction effects, the earlier $\Delta \vec{v}$ reaches its maximum value (the impulse velocity imparted by a specific rocket to its payload), the greater $\overrightarrow{\Delta v_{\text{ave}}}$ will be. This indicates, therefore, that the earlier the total acceleration or propulsive force is imparted to a vehicle, the higher the average velocity increment will be, the shorter the transit time will be between two points or orbits and the minimum amount of energy will be expended.

Of particular interest is the Hohman transfer ellipse, the co-tangential transfer between co-planar circular orbits. This transfer is illustrated in Fig. 3–19. This form of transfer is easily seen as the minimum energy transfer because a lower energy transfer attempt could not intersect the target orbit and a higher energy transfer would overshoot the target orbit, that is, intersect it twice and carry the vehicle to a further out orbit as illustrated in Fig. 3–20. This figure illustrates a typical free flight orbit for a system attempting to reduce transfer time by the expenditure of more than the minimum energy requirement. From the Hohman transfer orbit, we may immediately observe an axiom; the Hohman transfer ellipse

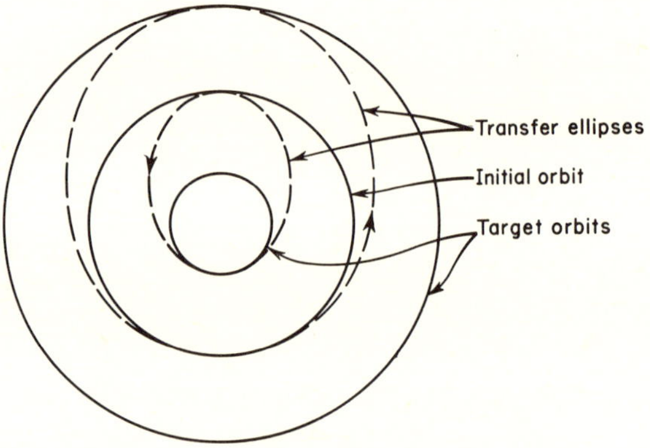

FIG. 3-19. Hohmann Transfer Ellipses.

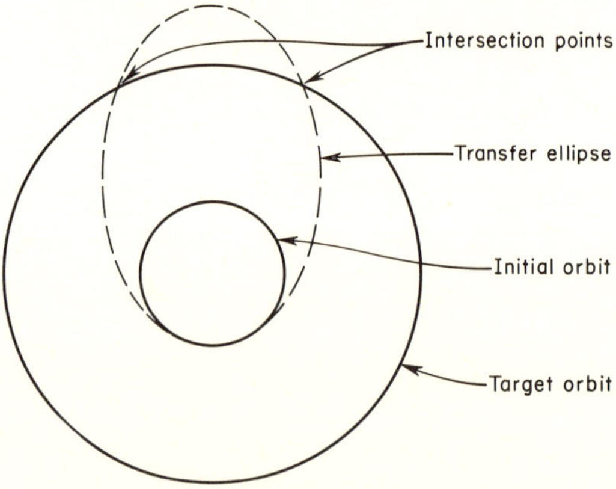

FIG. 3-20. Intersecting Transfer.

velocity is bounded by the circular velocities of the inner and outer circular orbits. That is, when the transfer ellipse is tangent to the inner circle, the velocity must be greater than the circular velocity of this orbit, and when the ellipse is tangent to the outer circle, the velocity must be less than the corresponding circular velocity. Thus, the ellipse velocity is bounded,

$$v_{c_1} < v_e < v_{c_2} \tag{3–93}$$

The intersecting coplanar transfer ellipse is not bounded by the circular orbit intersected twice, but may approach the circular velocity which is representative of its perigee or apogee (or apcynthion, aphelion, etc.) as the case may be. How nearly the particular tangential circular velocity is approached, is a function of the orbit (ellipse) eccentricity.

Thus, zero eccentricity will be a circle with no deviation from the circular velocity, while infinite eccentricity will be a straight line, radial from the field center, with no component of the circular velocity at apses. The true constant term is the orbital energy, h, which includes kinetic and potential energy

$$\mathrm{h} = \mathrm{v}^2 - \frac{2\mathrm{K}}{\mathrm{r}} \tag{3–83}$$

The equations generally used for velocity at apogee and perigee are,

$$v_a = v_{ca}\,(1 - e)^{1/2} = v_{ca}\left(1 - \frac{e}{2}\right) \tag{3–94}$$

and

$$v_p = v_{cp}\,(1 + e)^{1/2} = v_{pa}\left(1 + \frac{e}{2}\right) \tag{3–95}$$

where v_{ca} is the circular velocity at apogee and v_{cp} is the circular velocity at perigee. The elliptic eccentricity may be readily obtained from

$$e = \frac{a - 1}{1 + a} \tag{3–96}$$

where a is the ratio of the semi-major and semi-minor axes,

$$a = \frac{r_a}{r_p} \tag{3–97}$$

We may represent a ballistic space vehicle as leaving the surface of a planet on an elliptic path, with the planet center at one focal point of the ellipse, and show the space vehicle intersecting the planet surface where the ellipse and planet coplanar (with the ellipse) circle intersect. This is illustrated in Fig. 3–21, in which the effective range of the vehicle, D, is thus the great circle arc on the sphere which connects the two intersections of the elliptic trajectory. (In actual Earth-bound ballistic missile work, atmospheric effects, thrust programming prior to burn-out and Earth's turning rate at the particular latitudes involved must be considered in establishing the impact point.) The minimum energy required to fire between two points on a non-revolving globe may be observed to be a function of the firing elevation angle (or burn-out velocity vector) in a manner similar to traditional artillery ballistics, so that an optimum vector must be employed to achieve either a particular point with minimum energy or a maximum range with a fixed amount of energy. Of course, the normal operational problem must take into account effects of Earth's turning rate (velocity imparted on lift-off and target motion during flight period), and will commonly involve a fixed rocket which traverses either of two elliptic paths which are determined by aiming appropriately either above or below the maximum range elevation angle.

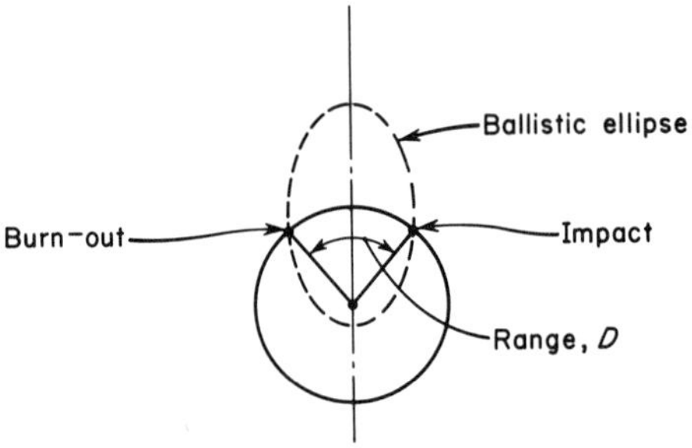

FIG. 3-21. Ballistic Trajectory.

Orbit-plane changes present another interesting problem. If we may assume that we merely wish to change orbit-plane with no change in orbit eccentricity, we may postulate the simple situation illustrated in Fig. 3–22. In this figure we assume that an instantaneous impulse velocity change will be imparted to effect the change in plane. Thus, the required velocity change is,

$$\Delta v = 2v \sin \frac{\alpha}{2} \tag{3–98}$$

where v is the velocity at the change position, and α is the angle between the orbit planes. Considering that the velocity at apses is the minimum velocity for any ellipse, this would be the minimum energy situation. However, this must require that the line of intersection between the two orbit planes is the axis of both ellipses. If such is not the case, one must consider the total energy required to realign both apses and the orbit plane. It can be inferred from the foregoing that there will be, in general, at least two solutions to the orbit-plane change problem and that the one occurring closest to apses will require the minimum energy.

3.3.3 Orbit Parameters

Before progressing further, it is desirable to recapitulate the essence of orbits and their parameters. Firstly, we have noted that in a central force field that varies radially as an inverse function of the radius, all orbits and trajectories will be contained in planes that pass through the center of the force field. Further, the orbits and trajectories will be conics in which the force field central is located at one of the focal points. Kepler's laws have stated that (1) the orbits of the planets are ellipses (a special case of the above); (2) the radius vector of each planet sweeps over equal areas in equal times (a property of a non-propelled body in an elliptic orbit); and (3) the ratio of the squares of the periods of two bodies is equal to the ratio of the cubes of the semi-major axes of their orbits. For circular orbits, the latter is quickly evident by taking two orbits of radii r_1 and r_2. The periods will be related to the circular velocities by,

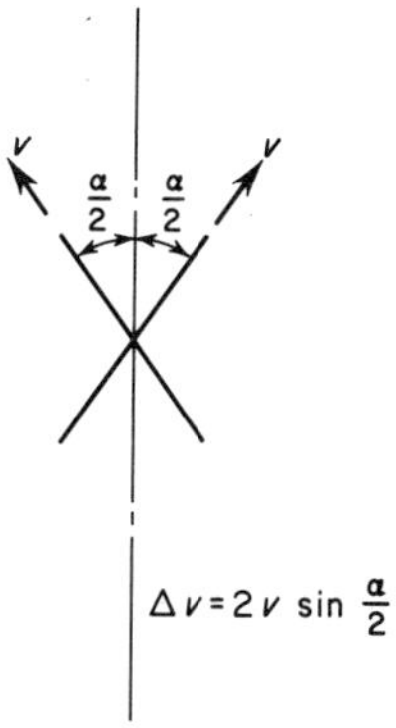

FIG. 3-22. End View of Orbit Plane Change.

$$T = \frac{2\pi}{\omega} = \frac{2\pi r}{v_c} = 2\pi r\sqrt{\frac{r}{k}} \tag{3-99}$$

Thus,

$$\left(\frac{T_1}{T_2}\right)^2 = \left(\frac{v_1}{v_2}\right)^3 \tag{3-100}$$

This relation holds also for ellipses if we substitute semi-major axes for radii,

$$\left(\frac{T_1}{T_2}\right)^2 = \left(\frac{a_1}{a_2}\right)^3 \tag{3-101}$$

If the bodies involved do not have negligible mass compared with the center mass, the square of the period of each body must be multiplied by the sum of the center body's mass, M, and the respective body mass,

$$\left(\frac{T_1}{T_2}\right)^2 \left(\frac{M + m_1}{M + m_2}\right) = \left(\frac{a_1}{a_2}\right)^3 \tag{3-102}$$

This equation defines the motion of the planets in the Sun's central force field very closely (and conveniently in terms of astronomical units). It may be expressed for a single orbit in terms of the gravitational constant, G, as,

$$\left(\frac{T}{2\pi}\right)^2 = \frac{a^3}{G(M + m)} \tag{3-103}$$

The *vis-viva* energy integral of Newton expresses the fact that the sum of the kinetic and potential energies is always unchanged in unpowered space (drag-free) flight. This may be expressed in velocity form as,

$$V^2 = G(M + m)\left(\frac{2}{r} - \frac{1}{a}\right) \tag{3-104}$$

The inclusion of higher order terms than $\frac{1}{r}$ (e.g. $\frac{1}{r^3}$, $\frac{1}{r^5}$, etc.) in the central force field will have some effect on the orbit parameters. Orbits may still be circular in this situation, but whether all orbits will be conics has not been determined clearly. The relative value of the coefficients for these terms in the Earth's case is 10^{-3} for $\frac{1}{r^3}$, and of the order of 10^{-6} for higher order terms in $\frac{1}{r^n}$ (see table in Chapter 2), so that for any considerable radial distance the $\frac{1}{r}$ term will predominate, and planar conic trajectories may be expected. The inclusion of latitude, longitude, and time-dependent terms in the central force field to account for oblateness, anomalies and tidal factors will also be important at extremely low altitudes. Where effective, these factors may be expected to cause some alterations in the orbit shape (magnitude of perigee and apogee), in the orbit phase (precession of perigee and apogee), the orbit period and even the orbit plane. These effects are treated in more advanced studies of celestial mechanics and the Electronic Systems Engineer will generally be concerned only with instrumenting a fire control problem for a particular defined orbit, performing in-flight measurements of the vehicle attitude and trajectory so as to initiate corrective maneuvers, or communicating with or between vehicles. In short, perturbative effects are left to more sophisticated classical studies.

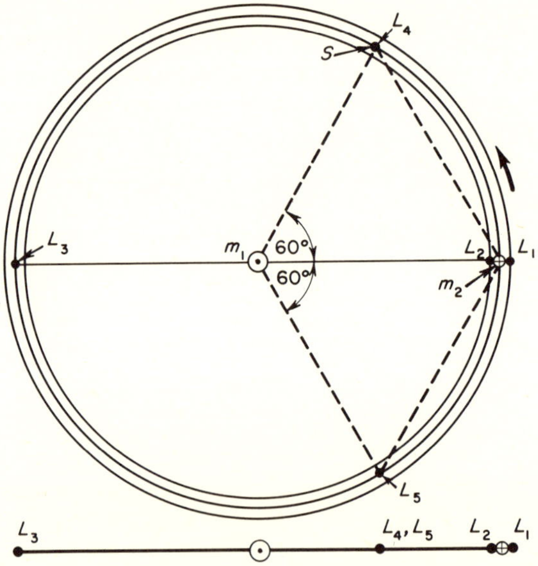

FIG. 3-23. Libration Points in the Sun-Earth System.

The principle of superposition applies to central force fields, so that when travelling from one central force field to another we may expect to encounter a "saddle" effect in the "net" or resultant gravitational force field. Treatment of orbits and trajectories in such a situation must fall back on the conservation of energy concept and must include both kinetic and potential energy. The charting of trajectories in such a force field is done by LaGrangian, Hamiltonian, and Jacobian techniques as illustrated by Ehricke (Ref. (*1*)). If we lump our Earth-Moon or the Sun-Earth system into one having a combined total mass, revolving about a "barycenter," we find that certain "libration" points possess peculiar properties in the common field. These equilateral points appear, mathematically, to be

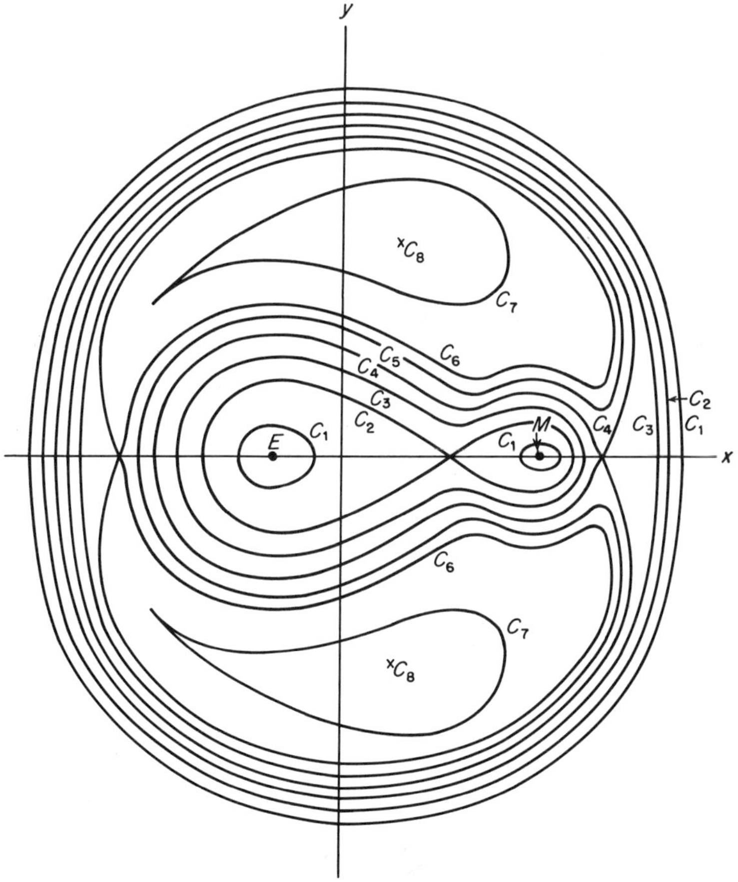

FIG. 3-24. Contours of Zero Relative Velocity in (x,y) Plane of Earth-Moon System.

stable unto themselves (Ref. (8)), that is, a body placed at these points with the appropriate velocity relative to the barycenter, will stay there indefinitely with, perhaps, only minor perturbations or oscillations about these libration points. Figure 3–23 illustrates the libration points for the Sun-Earth system. The two co-linear libration points, in close proximity to the Earth, are really 813,300 nautical miles away. The other co-linear libration point is in opposition to the Earth, and slightly closer to the Sun. Co-linear libration points are probably unstable (Ref. (8)). The equilateral libration points are interesting particularly, in that the Trojan Asteroids are clustered around two such points in the Sun-Jupiter system; this seems to verify the theory of stability of libration points.

Figure 3–24 illustrates contours of zero relative velocity in the (x, y) plane of the Earth-Moon system, and how these contours are related to the libration points. The counterpart of these curves in the simple central force field is a system of concentric circles. However, as the two central force fields merge, we may observe how the total field strength contours

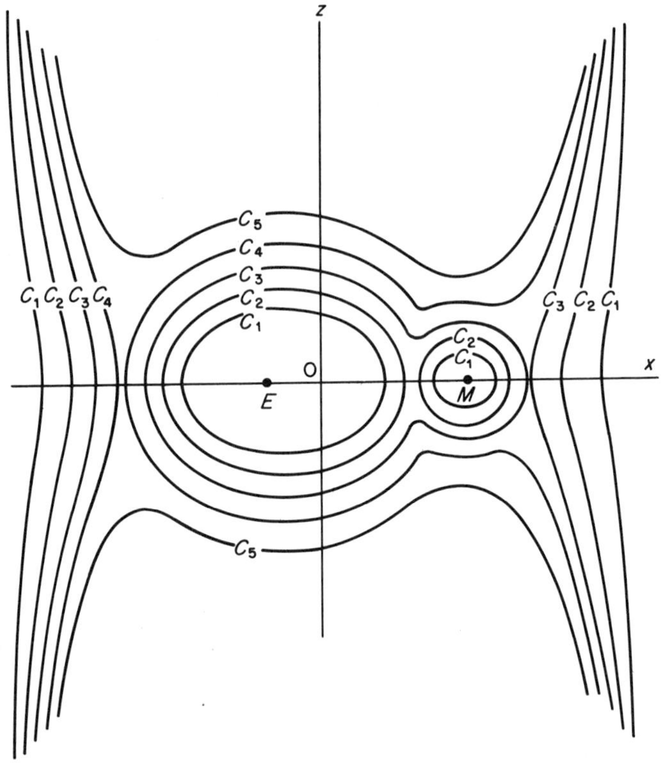

FIG. 3-25. Contours of Zero Relative Velocity in (x,z) Plane of Earth-Moon System.

develop three saddle patterns on the axis and two peaks at the equilateral points. (Note that these are constant field strength contours, not orbits.) Figures 3–25 and 3–26 illustrate the zero relative velocity contours in the (x, z) and (y, z) planes respectively. Motion in such a complexly contoured field will be quite complicated. Actual trajectories and orbits are generally planned for a specific flight by computer numerical integration techniques.

3.3.4 Nature of Trajectories

Trajectories in space are non-repetitive flight paths as opposed to orbits which are clearly repetitive. Trajectories do not have periods as such, but may possess original and terminal points. The ballistic missile travelling an elliptical path from one point on a globe to another describes a trajectory. A lunar landing or hard impact vehicle would describe another. All parabolic and hyperbolic paths (or the flight paths of those vehicles which possess escape velocities) may be considered to be trajectories rather than orbits. Trajectories in drag-free space without propulsive force will be conics. The addition of propulsive or corrective forces, atmospheric drag or a second (or more) force field will add perturbative effects. Many vehicles in the future will follow trajectories rather than orbits as transit

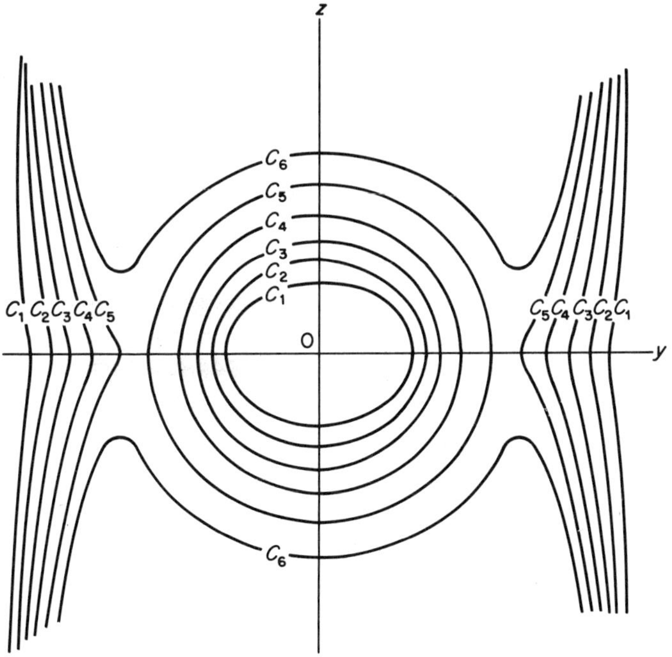

FIG. 3-26. Contours of Zero Relative Velocity in (y,z) Plane of Earth-Moon System.

times are minimized. The general vector equation for distance travelled will therefore be required:

$$\vec{s}(t) = \vec{s_0} + \int_0^t \vec{v_0}dt + \int_0^t \int_0^t \vec{a}(t)(dt)^2 \qquad (3\text{–}105)$$

This general equation is based on Newton's three fundamental laws of mechanics:

(1) A body not subject to a force continues in a state of rest or of straight line uniform motion.
(2) A body which is subject to a force undergoes an acceleration which is determined by the ratio of the force to the body mass.
(3) For each force exerted by one body on another, there exists an equal and oppositely directed force exerted by the latter on the former.

It is also premised on Newton's law of universal gravitation; the gravitational force holding two bodies together is proportional to the product of the two body masses and the reciprocal of the square of the distance between the bodies. Thus vector equation 3–105 is the basis for computation of all trajectories of powered, controlled, and aerodynamic flights. When velocities are approached which are significant percentages of the speed of light, relativity considerations become important and the equation must be modified.

3.4 RELATIVITY CONSIDERATIONS

Relativistic effects on orbits and trajectories depend on how fast a vehicle is going relative to the speed of light, c. The peculiar feature of relativity is that time is not considered to be progressing at a uniform rate in every location, but is considered to progress at a rate dependent on the observers' relative motion. Constancy of the speed of light is the key factor, and influences directly the validity of measurements of objects or distances while travelling at high speeds. Specifically, the three dimensional field-action potential equation for a Newtonian gravity field (or Maxwellian electromagnetic field) is, for empty space:

$$\Delta \mathrm{u} = 0 \qquad (3\text{–}106)$$

or with a source of density ρ,

$$\Delta \mathrm{u} = -(4\pi)\rho \qquad (3\text{–}107)$$

where Δ is the three dimensional Laplacian,

$$\Delta = \frac{\partial^2}{\partial \mathrm{x}^2} + \frac{\partial^2}{\partial \mathrm{v}^2} + \frac{\partial^2}{\partial \mathrm{z}^2} \qquad (3\text{–}108)$$

For a relativistic problem, a fourth coordinate, x_4, must be added, so that we have,

$$\Box u = 0 \qquad (3\text{–}109)$$

with $\Box$ the four-dimensional Laplacian,

$$\Box = \sum_{k=1}^{4} \frac{\partial^2}{\partial x^2 k} \tag{3–110}$$

x_1, x_2 and x_3 are the normal three-spatial coordinates, and,

$$x_4 = ict \tag{3–111}$$

an ortho-normal (with respect to the other three coordinates) time coordinate.

This set of equations may be employed in an analogous manner to Maxwell's equations for electromagnetic fields to show that gravity does not act instantaneously at a distance, but has a velocity of propagation equal to c, the speed of light. Forward (Ref. (*9*)) has done this, for a gravitational source of $GT_{\alpha\beta}$, arriving at a wave equation for gravitational potential:

$$\Box u = \Delta u - \frac{1}{c^2}\frac{\partial^2 u}{\partial t^2} = \frac{16\pi G}{c^4} T_{\alpha\beta} \tag{3–112}$$

in which $T_{\alpha\beta}$ is an energy-momentum source tensor. The significance of this expression is that the energy-momentum tensor may include gravity-source time variations, relative velocities, and rotations of the masses. Forward (Ref. (*9*)) has extended this further to indicate a relation between the acceleration of a particle in a curvilinear coordinate system and the various forces; coriolis, centrifugal, static gravitational attraction, gravitational induction, and a Lorentz-type velocity contraction force. The validity of this development of the general theory of relativity has not been established experimentally, and a specific experiment has been proposed (Ref. (*10*)) to measure the gravitational equivalent of the magnetic field which is implicit in the inductive effects term that results from the spinning of the two masses. This effect, if real, is expected to cause spin, or inductive effect, interactions which will cause gravity-like torques capable of precessing the spin axis of a satellite. The effect is so small in magnitude, that the problem of measuring it is beyond the current state-of-the-art in instrumentation.

The contraction factor, γ, resulting from the relative velocity difference, v, between observation points is,

$$\gamma = \frac{1}{\sqrt{\frac{1 - v^2}{c}}} = 1 + \frac{1}{2}\left(\frac{v}{c}\right)^2 + \ldots > 1.0 \tag{3–113}$$

This experimentally-proven relation is related to linear objects of velocity, v, relative to the observer at rest by,

$$L_v = \frac{L_o}{\gamma} \tag{3-114}$$

$$L_v < L_o$$

The contraction factor is related to incremental time measurements by,

$$\Delta t_v = \gamma t_0 \tag{3-115}$$
$$\Delta t_v > \Delta t_0$$

Also, the mass effect is

$$m_v = \gamma m_o \tag{3-116}$$

where m_0 is the rest mass.

The kinetic energy effect is,

$$\Delta E = m_v c^2 - m_0 c^2 = m_0 c^2 (\gamma - 1) \tag{3-117}$$

The momentum is,

$$\overrightarrow{m_v} = \gamma m_0 \overrightarrow{v} \tag{3-118}$$

Thus, whenever $v \ll c$, Newtonian mechanics are quite valid, and only when v is an appreciable percentage of c, does the contraction factor γ have a noticeable effect.

Electromagnetic radiation must have mass, energy, and momentum according to relativity. These are related by,

$$E_\nu = m_\nu c^2 = hf \tag{3-119}$$

$$\overrightarrow{m_\nu c} = E_\nu / \overrightarrow{c} = h/\lambda \tag{3-120}$$

However, radiation has no rest mass, m_0, and it is necessary to employ frequency, f, wavelength, λ, and Planck's constant, $h = 6.55 \times 10^{-34}$ Joule/sec to arrive at really useful equations (3–119 and 3–120). The Lorentz transformations apply to the contraction factor, γ and to transfer between coordinate systems. The transformations are,

$$x_1 = \gamma(x_2 + vt_2) \tag{3-121}$$

$$x_2 = \gamma(x_1 - vt_1) \tag{3-122}$$

$$t_1 = \gamma\left(t_2 + \frac{v}{c^2} x_2\right) \tag{3-123}$$

$$t_2 = \gamma\left(t_1 - \frac{v}{c^2} x_1\right) \tag{3-124}$$

where subscripts 1 and 2 denote the particular coordinate systems moving with a relative velocity, v.

Three significant experiments have validated relativity theory for general treatment of the field in empty space around a star out to the dimensions of the solar system. These experiments have been (Ref. (*11*)):

1. The advance of the perihelion of Mercury
2. The deflection of light near the Sun
3. The red shift or displacement of the Fraunhofer lines

These experiments required quite precise instrumentation, and are of interest primarily to the theoretical physicist. The extremely small magnitude of relativistic effects within the solar system may be taken as a good indication that these effects will be relatively insignificant in solar system exploration and development.

The extension of relativity theory to interstellar space with any accuracy or confidence requires that some new experiment be developed in validation thereof. Astronomical observations (Ref. (*12*)) have lent some credence to the non-universality of Newton's law. Specifically, there appears to be a finite limit on the size of clusters of galaxies, and there is no apparent clustering of the clusters of galaxies. Both of these phenomena are directly predictable by Newton's gravitational law, and their absence indicates a finite limit on the macroscopic side. Newton's law seems to break-down for D-critical $\cong$ 5 million light years. This break-down of Newton's law is also a break-down of the general theory of relativity at macroscopic distances (Ref. (*12*)), at least in its present form. The field equations used presently, in the limit for weak fields, basically adjust to Newton's law. Hypotheses to explain away the break-down of Newton's law and relativity theory include at least three: (1) that galaxies and clusters of galaxies of anti-matter may exist, thus thwarting our observations; (2) that shielding factors need to be introduced into the field equations of general relativity – the shielding to be a function of the mass sizes and their distribution; and (3) that bodies separated beyond D-critical simply do not interact at all. Further tests of Newtonian and field theory equations under macroscopic conditions may lead to an even better understanding of the true relativistic effects on space travel.

References to Chapter 3

1. K. Ehrike, *Space Flight*, Vol. 1, D. Van Nostrand Co., Inc., Princeton, N.J., 1960, pp. 421–441.
2. G. von Pahlen-Fedoroff, "An Accurate Determination of Attitude by Optical Means for Applications in Space Vehicles," Proceedings of the National Specialists Meeting on Guidance of Aerospace Vehicles, Boston, Mass., May 25–27, 1960, published by the Institute of Aeronautical Sciences (IAS), pp. 40–46.
3. P. L. Bender, "Atomic Clocks for Space Experiments," *Astronautics*, July 1960, p. 37.

4. B. Kovit, "Ultra-precise Magnetometer," *Space/Aeronautics*, July 1961, p. 119.
5. R. E. Fischell, "Magnetic Damping of the Angular Motions of Earth Satellites," *ARS Journal*, Vol. 31, No. 9, September 1961, p. 1210.
6. M. H. Arck & M. M. Merlen, "Horizon Sensors for Vertical Stabilization of Satellites and Space Vehicles," Proceedings of the National Specialists Meeting on Guidance of Aerospace Vehicles," Boston, Mass., May 25–27, 1960, Published by the Institute of Aeronautical Sciences (IAS), p. 34.
7. Povejsil, Raven & Waterman, *Airborne Radar*, D. Van Nostrand Co., Inc., Princeton, N.J., 1961., p. 725–759.
8. R. W. Bucheim, *Motion of a Small Body in Earth-Moon Space*," Rand Report RM-1726 of June 4, 1956.
9. R. L. Forward, "General Relativity for the Experimentalist," *Proceedings of the IRE*, May 1961, pp. 892–904.
10. R. L. Forward, "Satellite Might Be Used to Verify Einstein's Theory of Gravity," *Missiles and Rockets*, July 10, 1961, pp. 32–35.
11. H. A. Lang, *Some Elementary Facts of Relativity*, Rand Corp. Report No. P-1433 of February 24, 1958.
12. F. Zwicky, "Is Newton's Law of Gravitation Really Universal?", *Astronautics*, Jan. 1959, p. 18.

CHAPTER 4

COMMUNICATION

4.1 INTRODUCTION

Of prime interest in space electronic systems is the ability to communicate. Communication is taken to mean the one-way transfer of data or intelligence, thus embracing telemetry, television, and command links, as well as voice communications. The extreme ranges dictate unusual measures to make communication realizable; for example, selection of wavelength based on whether atmospheric effects are to be encountered or not. To minimize noise problems, information bandwidth must be considerably conserved.The transit times in space communication are several orders of magnitude larger than in Earth-bound systems; this means delayed response to commands or messages. Unusual antenna systems, in terms of gain and beam-width, will be required. In some environments, plasma sheaths or space charge effects will have considerable impact on the efficiency of the communication system. The noise background will be another unusual environment for space communications. Techniques employed will be based on the unusual propagation and environmental effects encountered. These techniques must include means for the acquisition, identification, and tracking of transmissions because of the eclipsing phenomena of the Sun and of planetary and satellite bodies. The long distances and eclipsing constellations will require further development of relay systems and diversity techniques. Minimum level signals must be enhanced by integration or correlation, and noise must be suppressed wherever possible. Modulation methods will vary somewhat according to information requirements and current equipment performance capabilities. However, the theoretical limits of information theory cannot be violated, and some standardization of modulation techniques is to be expected where communication links must operate with diverse facilities (e.g., as with a globe girdling receiver network). Communication equipment and techniques are expected to improve and change radically as a result of increasing demands for communications throughout space in the coming years.

4.2 PROPAGATION

The techniques to be selected and employed in any system are determined basically by two requirements: (1) the data to be transferred and

(2) the propagation environments. Fundamental to propagation analysis is the range equation which relates transmitted power to receiver capabilities. The application of the range equation to a true space environment is quite a treat for communication engineers because the absorption, scattering, refraction, multiple-path, and other such problems peculiar to Earth-bound communications will not be encountered, at least to the same degree. However, in Earth-to-space and reciprocal communications, some of these problems still occur, and will actually become unusually severe when one considers operating through the ionosphere, and trying to maintain communications with a re-entry vehicle which is under considerable thermal agitation as well as being surrounded by an ion sheath. So, we must re-examine these various parameters, at least for their significance.

4.2.1 Wave-lengths and absorption

Fundamental to propagation system design is the selection of wavelength or transmitted frequency based on various factors embodied in the range equation. In radar work, the range equation is for reception of one's own signal after reflection from some target object. In communication

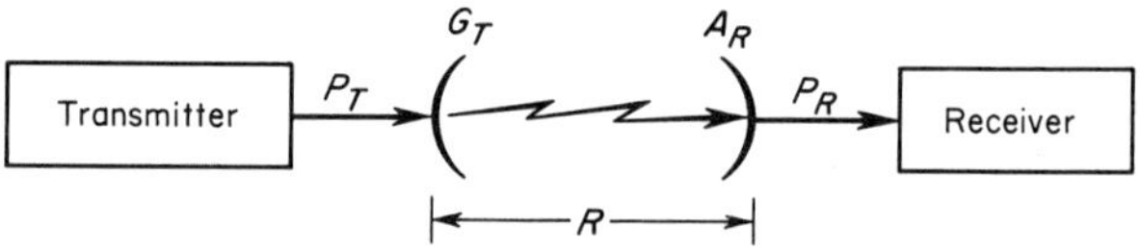

FIG. 4-1. Basic Communication Link.

systems the transmission is one-way as illustrated in Fig. 4–1. Thus, we may say that signal power available at the receiver input will be,

$$P_R = \frac{P_T\ G_T\ A_R}{4\pi\ R^2} \tag{4–1}$$

where P_T is the transmitter output power, G_T is the gain of the transmitting antenna in the propagation path direction, $4\pi R^2$ is the dilution of energy for the range (R), and A_R is the effective aperture area of the receiving antenna. By noting that the effective area of a parabolic antenna is,

$$A_R = \frac{\lambda^2}{4\pi} G_R \tag{4–2}$$

where λ is the carrier frequency wavelength and G_R is the maximum antenna gain; we may rewrite Eq. (4–1) as,

$$P_R = \frac{P_T\ G_T\ G_R\ \lambda^2}{(4\pi)^2\ R^2} \tag{4–3}$$

Now, if we can define some sort of minimum received signal power necessary to generate useful information within the receiver and its information processing system as $P_{R_{\min}}$, we may determine the maximum free space range for a particular communication system as,

$$R_{\max} = \frac{\lambda}{4\pi}\sqrt{\frac{P_T\ G_T\ G_R}{P_{R_{\min}}}} \tag{4–4}$$

where R and λ are usually expressed in different units, which requires that some dimension conversion factors be included in the foregoing equation. Also, P_T and P_R are usually expressed in decibels (db) relative to some wattage reference, and G_T and G_R in db. (Note, a power ratio P_1/P_2 is expressed in db as $10 \log_{10} \frac{P_1}{P_2}\Big)$. For more convenient units we sav,

$$10 \log P_R = -(36.6 + 20 \log f_{mc} + 20 \log R_{SM}) + 10 \log G_T + 10 \log G_R + 10 \log P_T \tag{4–5}$$

where 36.6 is the number of decibels necessary to account for the $\frac{1}{(4\pi)^2}$ factor and the use of frequency in megacycles f_{mc}, and range in statute miles, R_{SM}. G_T and G_R are still ratios. If we take P_T and P_R in db with respect to one watt (dbw) or one milliwatt (dbm), and G_T and G_R in db, we may now say,

$$P_R = -(36.6 + 20 \log f_{mc} + 20 \log R_{SM}) + G_T + G_R + P_T \tag{4–6}$$

or,

$$P_R = -L_{FS} + G_T + G_R + P_T \tag{4–7}$$

where L_{FS} is the free space loss.

If R is expressed in nautical miles,

$$L_{FS} = 37.8 + 2f_{mc} + 2R_{nm} \tag{4–8}$$

where 1.2 db has been added for the unit change from statute to nautical miles, and f_{mc} and R_{nm} are expressed directly in decibels relative to one megacycle (dbmc) and one nautical mile (dbnm) respectively.

From Eq. (4–1) we see that there is no particular advantage, from a propagation standpoint, in choosing any particular wavelength (λ) as a carrier for intelligence. The practical choice must include some assessment of the transmitting power available from output tubes at various wavelengths, reasonable antenna sizes, antenna beam coverage, intelligence band-widths, and the efficiency of the various components and circuits. However, it is generally held that all wavelengths will propagate equally well in free space. That is, the power density per steradian is frequency-invariant. When we introduce an atmosphere, plasma, ions, atoms, or molecular environment of any sort, the situation begins deteriorating; or, at least, we begin to encounter attenuation, absorption, scattering, reflection, and refraction effects so that we must become more restrictive in our wavelength selection. In Fig. 2–14, page 82, we depicted the absorption of radio energy by oxygen and water vapor in the air in db/km when operating through a fully saturated standard atmosphere at sea level. If we are concerned with communication between space and a sea level Earth station we must add an atmospheric loss figure ($-L_{ATM}$) to Equation 4–7,

$$P_R = -L_{ATM} - L_{FS} + G_T + G_R + P_T \tag{4–9}$$

The atmospheric loss factor caused by one-way transmission through the atmosphere may be determined from Fig. 2–15, page 83, which depicts the combined attenuation rates in db per km versus frequency. The equivalent atmospheric thickness as a function of ray path elevation above the horizon is portrayed in Fig. 2–16, page 83. Figure 2–17, page 84, illustrates the effect of oxygen absorption. Figure 2–18, page 85, portrays the combined water vapor plus oxygen absorption in db as a function of both frequency and ray path elevation from the horizontal while operating at sea level. Thus, one may read directly from Fig. 2–18, an L_{ATM} in db for inclusion in a propagation calculation using Eq. (4–9). From Fig. 2–18, it is immediately obvious that frequencies above 6 to 10 kilomegacycles will not be very useful if ray paths near the horizon are to be used and maximum range or minimum transmitted power are design requirements. With ray paths above 30 degrees elevation from the horizontal, frequencies up to 100 kmc may be useful. Thus, wavelength or frequency selection for a particular Earth-to-space communication link will depend somewhat on the system geometrical relations which may include duration of communication period, number of Earth-bound stations involved, etc., as with a network of transmitting/receiving stations on a rotating Earth, communicating in sequence with some deep-space probe or station (e.g. a fixed station on the Earth-oriented lunar surface).

Similar problems will no doubt occur in communication through Venutian, Martian, and other planetary atmospheres; however, it remains for the future to ascertain how much absorption will be experienced in specific atmospheres. The Martian atmosphere is well enough known that we can be reasonably assured that it provides less absorption than that of the Earth.

Of interest also is the effect of transmitting through the atmosphere twice, as with a communication relay system. Designating the two communication paths, 1 and 2, by subscripts, we may say, for a passive satellite reflection (e.g., Project ECHO satellite),

$$P_R = \frac{P_T\ G_T\ A_R\ l_1\ l_2\sigma}{(4\pi)^2\ R_1^2\ R_2^2} \tag{4–10}$$

where l_1 and l_2 represent the atmospheric path losses, and σ is the effective reflection cross-section which includes size and efficiency factors.

For an active or repeater satellite, the one-way propagation Eq. (4–9) should be used for each link. For radar detection of a passive satellite, Eq. (4–10) becomes the conventional radar equation,

$$P_R = \frac{P_T\ G_T\ A_R\ l^2\sigma}{(4\pi)^2\ R^4} \qquad (4\text{–}11)$$

And, for free space radar operation,

$$P_R = \frac{P_T\ G_T\ A_R\ \sigma}{(4\pi)^2\ R^4} \qquad (4\text{–}12a)$$

Another effect experienced when transmitting through the atmosphere is that the polarization of the transmitted wave will be rotated in direct proportion to the number of charged particles encountered in traversing the ionosphere. This is known as the Faraday effect, and is currently the subject of several experiments.

Of recent interest is the LASER (*l*ight *a*mplification by *s*timulated *e*mission of *r*adiation) technique which portends the development of modulated light beams for space communication applications. The short wavelength of light beams, the lack of atmospheric scattering and absorption in space, and the prospect of modulating the light beam at microwave frequencies indicate that a high potential exists for introducing the LASER as an extremely long-range communication device. The principal advantage is that the extremely short wavelength allows the generation of extremely narrow, hence high gain, beams with reasonable diameter apertures.

If we can concentrate our transmitted energy sufficiently so that the receiver collector can intercept all of the transmitted energy, we will have made a vast improvement in our power usage, and the one-way propagation equation will reduce to the form of a transmission line equation, or,

$$P_r = P_t - L_{\text{line}} \qquad (4\text{–}12b)$$

where L_{line} = line loss.

Practical achievement of this result is possible over relatively short ranges, as between satellites in an orbiting communication network. However, an attendant disadvantage is that such a tight beam results in extremely long search periods if it must be used to scan a significant volume of space in order to detect a transmission.

The concept of concentrating all of one's transmission on a passive reflector for dissemination to receivers which "look at" the reflector applies to large satellites or reflecting clouds, and large microwave transmitter antennas. Such a technique, if achieved, will modify the propagation Eq. (4–10) to,

$$P_R = \frac{P_T\ A_R\ l_1\ l_2\ k}{4\pi\ R_2^2} \qquad (4\text{–}12c)$$

where l_1 and l_2 represent path losses and k represents an efficiency factor for the reflector.

Similarly, a tightly beamed radar used as an altimeter has the propagation equation,

$$P_R = \frac{P_T\ A_R\ l^2\ k}{4\pi\ h^2} \tag{4–12d}$$

where h is the altitude, and the total beam energy is reflected with an efficiency ratio, k.

4.2.2 Bandwidths

The received signal power, P_R, must contain sufficient information in terms of frequency bandwidth to convey the desired message or data. This same signal power must be of sufficient magnitude, relative to the noise power, that the desired data may be recovered in the presence of the noise.

The noise present in a receiver output is a direct function of the receiver bandwidth, B (or Δf), the input circuit or antenna equivalent temperature, T, and the receiver noise figure, NF, which is the power ratio of noise generated within the receiver to noise generated at the antenna. The standard equation for noise power is,

$$P_N = KTB\ (NF) \tag{4–13}$$

where K is Boltzmann's constant, 1.33×10^{-23} joules per degree Kelvin. KT is often taken as 4×10^{-21} watts per cycle per second so that B may be taken directly in cycles per second.

The output display or data processing system will determine some useful minimum signal-to-noise ratio as,

$$S/N_{\min} = P_{R_{\min}}/P_N \tag{4–14}$$

where $P_{R_{\min}}$ may be related directly to the propagation Eq. (4–4) by,

$$P_{R_{\min}} = S/N_{\min} \times P_N = \frac{P_T\ G_T\ G_R\ \lambda^2}{(4\pi)^2\ R^2_{\max}} \tag{4–15}$$

or,

$$R_{\max} = \frac{\lambda}{4\pi} \sqrt{\frac{P_T\ G_T\ G_R}{KTB\ (NF)\ S/N_{\min}}} \tag{4–16}$$

For two-way transmission with simple reflection, the free space radar equation will be,

$$R_{\max} = \sqrt[4]{\frac{P_T\ G_T\ A_R \sigma}{(4\pi)^2\ KTB(NF)\ S/N_{\min}}} \tag{4–17}$$

The noise bandwidth will obviously be made as low as possible in any system. This has caused a considerable amount of study as to how to reduce the ratio of transmitted bandwidth to message bandwidth,

$$\gamma = B/M \tag{4–18}$$

This ratio will depend on the type of modulation employed, and may be improved somewhat by special signal processing techniques which take advantage of any *a priori* knowledge of the signal frequency content. These topics will be discussed more fully in Section 4.3 (Communication Techniques).

4.2.3 Directivity versus power

The foregoing sections have indicated antenna gains for both the transmitter and receiver, a typical expression being,

$$R_{\max} = \frac{\lambda}{4\pi}\sqrt{\frac{P_T\ G_T\ G_R}{KTB(NF)\ S/N_{\min}}} \tag{4–16}$$

From this expression, we may obviously work with smaller transmitter powers by increasing transmitter and/or receiver antenna gains. Figure 4–2 portrays the antenna gain and effective area for a variety of antenna types as a function of wavelength and antenna dimensions. For a uniformly illuminated circular aperture parabolic reflector, we may say the 1/2-power beamwidth in degrees is,

$$BW = \frac{70\lambda}{D} \tag{4–17}$$

where D is the diameter.

The actual antenna pattern for a parabolic reflector of this type is expressed by,

$$E = \frac{2J_1[(\pi D/\lambda)\ \sin\ \theta]}{(\pi D/\lambda)\ \sin\ \theta} \tag{4–18}$$

where θ is the angle relative to the aperture normal, and $J_1[(\pi D/\lambda)\sin\theta]$ is the Bessel function of the first order. Equation 4–18 is of the form, $E = 2\ J_1\,(x)/(x)$, which is plotted as an antenna pattern in Fig. 4–3. The illumination taper may be varied to suppress the side-lobes better than with the uniformly illuminated antenna, and at the same time providing additional beamwidth; effectively, a more efficient direction of the transmitted power. The beamwidths achievable as a function of aperture width and illumination taper are depicted in Fig. 4–4, including values for the first side lobe level relative to the antenna gain. In general, we may say that the maximum effective aperture is,

$$A_{em} = \frac{\lambda^2}{4\pi}\ G_m \tag{4–19}$$

where G_m is the directivity which is unity for an isotopic radiator, and must lie between unity and infinity. The ratio gain of an antenna is related to the directivity by an efficiency factor, k, of value between 0 and 1,

$$G = kG_m \tag{4-20}$$

Thus G may have any value from 0 to ∞.

Radiator	Gain above Isotropic Radiator	Effective Area
Isotropic radiator	1	$\lambda^2/4\pi$
Infinitesimal dipole or loop	1.5	$1.5\ \lambda^2/4\pi$
Half-wave dipole	1.64	$1.64\ \lambda^2/4\pi$
Optimum horn (mouth area = A)	$10\ A/\lambda^2$	0.81 A
Horn (maximum gain for fixed length) (mouth area = A)	$5.6\ A/\lambda^2$	0.45 A
Parabola or metal lens	6.3 to $7.5\ A/\lambda^2$	0.5 to 0.6 A
Broadside array (area = A)	$4\pi\ A/\lambda^2$ (max)	A (max)
Omnidirectional stacked array (length = L, stack interval $\leqslant \lambda$)	$\approx 2L/\lambda$	$\approx L\ \lambda/2\pi$
Turnstile	1.15	$1.15\ \lambda^2/4\pi$

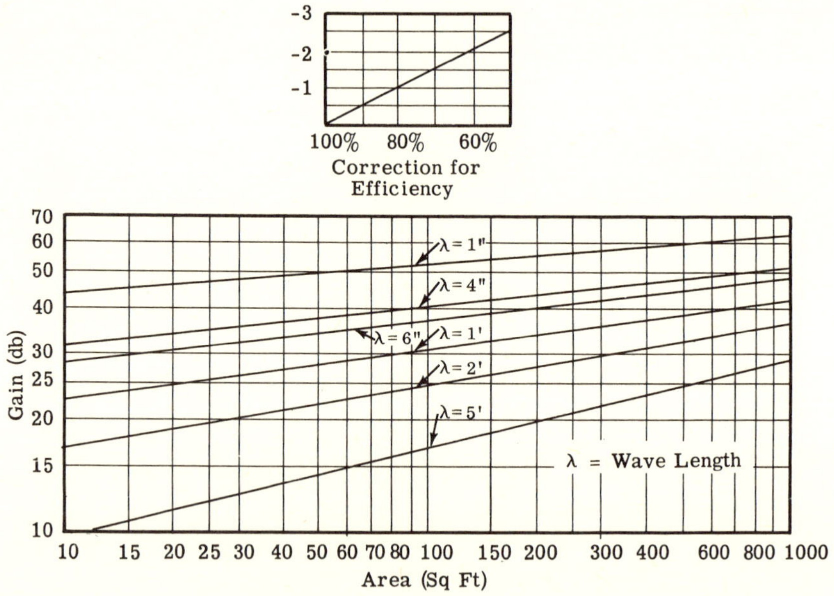

FIG. 4-2. Antenna Gain-Wavelength—Aperture Relations.

A general expression for effective aperture of an antenna is,

$$A_e = \frac{I^2 R_T}{P} \tag{4–21}$$

Where I is the rms terminal current in amperes, R_T is the terminal resistance in ohms, P is the power density (or Poynting vector) of the incident wave in watts per square meter, and A_e is the effective aperture in square meters. The fields of a short dipole of length, l, are

$$E_r = \frac{I_0 l e^{j\left(\omega t - \frac{2\pi r}{\lambda}\right)} \cos\theta}{2\pi E_0}\left(\frac{1}{cr^2} + \frac{1}{j\omega r^3}\right) \tag{4–22}$$

$$E_\theta = \frac{I_0 l e^{j\left(\omega t - \frac{2\pi r}{\lambda}\right)} \sin\theta}{4\pi E_0}\left(\frac{j\omega}{c^2 r} + \frac{1}{cr^2} + \frac{1}{j\omega r^3}\right) \tag{4–23}$$

$$H_\phi = \frac{I_0 l e^{j\left(\omega t - \frac{2\pi r}{\lambda}\right)} \sin\theta}{4\pi}\left(\frac{j\omega}{cr} + \frac{1}{r^2}\right) \tag{4–24}$$

$$E_\phi = H_r = H_\theta = 0$$

where, $r >> 1$ and $\lambda >> 1$,

E and H are the electric and magnetic field strengths. θ, ϕ and r are the orthogonal coordinates of the field, with r also the propagation distance. $\theta = 0°$ in-line with the dipole axis and 90° in the normal plane bisecting the dipole. E_r is taken radially in-line with r. E_θ is taken normal to E_r in

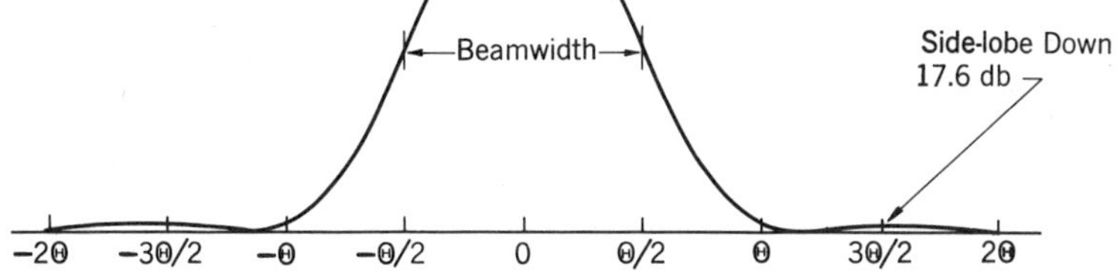

FIG. 4-3. Antenna Pattern.

the θ – plane which includes the dipole axis. E_ϕ is normal to E_r and E_θ. I_0 is the maximum current amplitude in the dipole. E_0 is the permittivity of free space. Equations (4–22), (4–23), and (4–24) include both near and far-field terms. For nearly all space applications, the near field terms may be deleted, and only the far-field terms considered. These terms are the

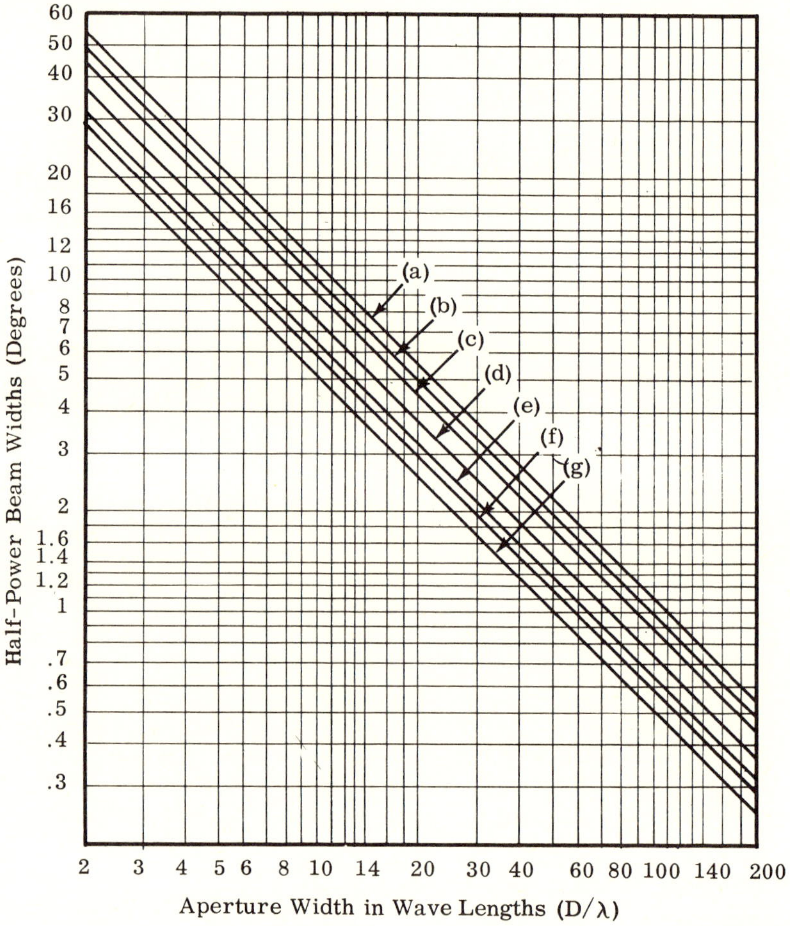

Curve	Illumination Taper	Theoretical First Side Lobe Level
(a)	(Cosine)^4	44.6 db
(b)	(Cosine)^3	39.0 db
(c)	(Cosine)^2	32.8 db
(d)	(Cosine)	25.8 db

Curve	Illumination Taper	Theoretical First Side Lobe Level
(e)	Average antenna	20.0 db
(f)	Uniform	17.6 db
(g)	Uniform	13.2 db

Curves (a) thru (f) are for circular or elliptical apertures. Curve (g) is for rectangular aperture.

FIG. 4-4. Antenna Beam width versus Aperture.

radiation field terms which vary with $1/r$. If we let $\omega \rightarrow 0$, we may obtain the quasi-stationary (d.c.) fields. The total field of any antenna is the summation of all the fields of the elemental dipoles composing the antenna. If one approaches the antenna aperture too closely, the $1/r^2$ and $1/r^3$ terms may become effective. Also, a large antenna aperture may include a near field zone in which the elemental dipole fields do not add in direct phase relation. For parabolic type apertures, the near field zone limit is generally taken as,

$$R_{NF} = \frac{D^2}{\lambda} \tag{4–26a}$$

This is illustrated in Fig. 4–5 as a break-point range separating the two field zones. In reality, a finely defined break-point does not exist. A common system design practice is to use twice the break-point range as a minimum range for antenna pattern measurements to insure that the far field is fully formed. Thus, one may encounter the relation,

$$R_{NF} = 2D^2/\lambda \tag{4–26b}$$

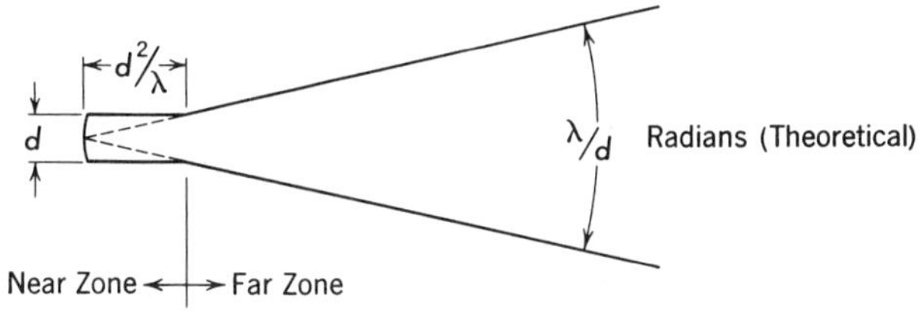

FIG. 4-5. Parabolic Antenna Field Zones.

As may be surmised, the use of directive antennas to communicate over longer ranges or with lower powers has some limitations. A directive antenna must be aimed at the desired receiver or transmitter within the tolerance of the beamwidth. Acceptable tolerance will be determined by the amount of gain needed and beamwidth limits that provide this gain. The determination of this vector can be quite difficult if very small beamwidth, high-gain antennas are employed. Tracking systems may be used which are really radiation line-of-sight direction finders.

Von Handel and Hoehndorf (Ref. (*1*)) have investigated the refractive effects of the atmosphere on microwave propagation. The basic equation for index of refraction, n, for microwaves penetrating the atmosphere is,

$$\nu = (n - 1) \times 10^6 = \frac{79}{T}\left(p + \frac{4800e}{T}\right) \tag{4–27}$$

where T is the temperature in degrees Kelvin, p is the air pressure in millibars, and e is the partial pressure of water vapor in millibars. From empirical data of refractive profiles, the relation between refraction index and altitude above mean sea level, h, has been established as (Ref. (*1*)),

$$\nu = \nu_{SL}e^{-h/H} = \nu_0 e^{-(h - h_0)/H} \tag{4–28}$$

where H is known to $\pm$ 1500 feet, and establishes a limit on the precision of angular and range measurements taken through the atmosphere. Subscripts 0 and SL refer to observation altitude and mean sea level, and h is the altitude. The general relation for refractive effects in a spherical system is,

$$\frac{\sin \alpha}{\sin \alpha_0} = \frac{R_{SL} + h_0}{R_{SL} + h} \times \frac{n_o}{n} \tag{4–29}$$

The refractive index is also defined by,

$$n^2 = \mu\epsilon \tag{4–30}$$

where μ and ϵ are the permeability and permittivity of the medium. The velocity of propagation through the medium is,

$$v_p = \frac{c}{\sqrt{\mu\epsilon}} = \frac{c}{n} \tag{4–31}$$

Thus, the refractive index will affect the apparent line-of-sight vector direction and any distance measurement based upon knowledge of the velocity of propagation. The correction in apparent range is,

$$\Delta r = [(R + h_0)^2 + (R + h_s)^2 - 2\,(R + h_0)\,(R + h_s) \cos \varphi_s]^{1/2} \tag{4–32}$$

where h_s is the altitude of the space vehicle above the nominal Earth's radius, R_0. φ_s is the angle subtended from Earth's center by the observation point and the space vehicle. The distance measurement, r_m, is, for a radar system,

$$r_m = \frac{\tau}{2} v_m \tag{4–33}$$

where τ is the echo time and v_m is the calibrated propagation velocity. The correction in apparent vector is,

$$\Delta\alpha = \sin^{-1}\left(\frac{R + h_s}{r_t} \sin \varphi_s\right) - \alpha_m \tag{4–34}$$

where α_m is the measured zenith angle and r_t is the true range. The interrelation of these equations, and the number of unknowns involved, require generally that some iterative procedure be followed in their application to specific measurements to refine the data. The space vehicle altitude, h_s, is generally determined on an electronic computer from (Ref. (*1*)),

$$\frac{T}{2} = \frac{1}{c} \int_{h_0}^{h_s} \frac{n^2 (R + h)}{[n^2 (R + h)^2 - n_0^2(R + h_0)^2 \sin^2 \alpha m]^{1/2}} \tag{4-35}$$

which is used iteratively with Eq. (4–28) to refine the accuracy of n.

4.2.4 Plasma sheaths

The entry of space vehicles into the Earth's atmosphere at hypersonic velocities causes these vehicles to be engulfed in a sheath of ions and free electrons known as a "plasma sheath." The reflection and attenuation effects of this sheath may cause a severe radio "black-out" which interrupts completely the communication services. In general, a plasma exhibits a resonant frequency which acts in a somewhat analogous manner to the cut-off frequency of a waveguide; a plasma will not support wave propagation at frequencies lower than this resonant frequency. The plasma resonant frequency is,

$$\omega_p = \left(\frac{ne^2}{\epsilon_0 m} \right)^{1/2} \tag{4-36}$$

where n is the electron density, e is the charge per electron, m is the electron mass, and ϵ_0 is the free-space permittivity. Above the plasma frequency, the plasma has dielectric properties. Immediately below the plasma frequency, a very high attenuation is experienced. Well below the plasma frequency, the plasma acts like a conductor, and the relative penetration depth of a signal varies in direct proportion with wavelength. The ionosphere and meteor trails are excellent examples of the latter case, in which an ionized plasma has been employed as a radio signal reflector because of its conductor-like properties. The electron density has a direct relationship to the resonant plasma frequency as illustrated by Eq. (4–36), and this relation has been employed to stratify the ionopshere layers by frequency soundings from the Earth. It is also being employed from above the ionosphere to sound the top-side strata from satellites and rocket probes.

The transmission of data through a plasma sheath is being investigated intensively at the present time, and the most promising approach appears to be the application of a static magnetic field (Ref. (*2*) and (*3*)).

The reception of data through the plasma sheath is encumbered similarly to the transmission. In addition to this, the receiving antenna experiences an increased ambient temperature, hence noise, from the vehicle body;

this is in addition to the noise generated from the plasma itself. The effective noise temperature of the antenna will be (Ref. (*4*))

$$T_n = A_\omega T_p + R_\omega T_v + T_\omega T_{ex}, \tag{4–37}$$

where A_ω is the power absorptivity, R_ω is the power reflectivity and T_ω is the power transmissivity of a plane wave incident on a plasma slab which blankets the receiving antenna.

Thus,

$$A_\omega + R_\omega + T_\omega = 1, \tag{4–38}$$

where T_p is the plasma temperature, T_v is the vehicle temperature, and T_{ex} is the temperature of any source external to both plasma sheath and vehicle. The principal noise sources will be from the plasma in the neighborhood of the resonant plasma frequency, from the vehicle at lower-than-resonant plasma frequencies, and from external sources at higher-than-resonant plasma frequencies. This would indicate shorter wavelengths as the best communication choice, except that resonant plasma frequencies are of the order of one centimeter wavelength which is generally a poor propagation region in the Earth's atmosphere. A static magnetic field, which will alter the plasma sheath characteristics so as to permit transmission at frequencies below the resonant plasma frequency, may be expected to decrease the noise power at these frequencies, thus providing some improvement in effective antenna noise temperature.

4.2.5 Galactic Noise

As communication techniques become more refined, internal noise sources will be diminished, and it is to be expected that in the "quiet" and "cold" of deep space, the galaxy, or Milky Way, will become a troublesome noise source. Superimposed on the galaxy and the celestial sphere are numerous radio stars which are discrete radio noise sources. The Sun is a considerable noise source with significant variations during solar storms. The planet Jupiter is known to produce radio noise sporadically, and there are some indications that Saturn and Venus similarly radiate. Thermal radiation by the Moon and the planets is another radio noise source. Smith, Ref. (5), has plotted brightness in watts per square meter per cycle per steradian versus celestial longitude and frequency as illustrated in Fig. 4–6. This figure shows a heavy noise concentration in the direction of the galactic center. Solar noise levels are illustrated versus wavelength in Fig. 4–7. The Sun is the greatest noise source providing 10^{-15} watt per square meter per cps in the 1 to 10 meter wavelength region on a sporadic basis. Galactic noise is constant, and is most intense in the 15 meter region reaching 10^{-19} watt-meter^{-2} cps^{-1} sterad^{-1}. The hydrogen line is a constant noise source at 21 cm with a maximum intensity of 6×10^{-20}

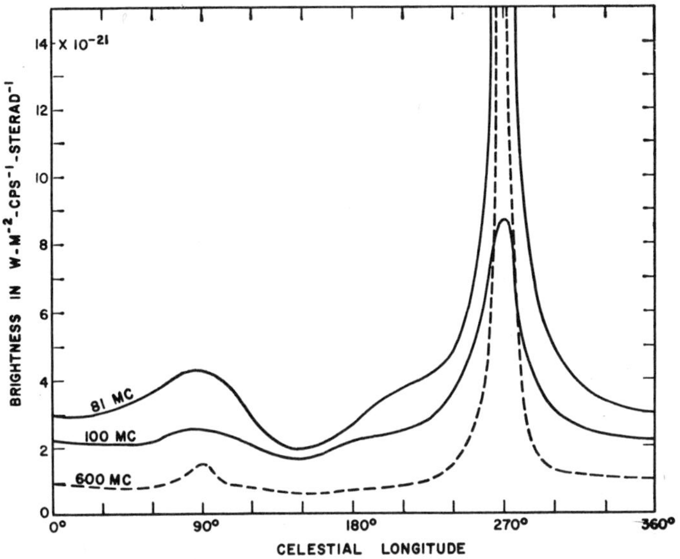

FIG. 4-6. Galactic Noise on the Ecliptic.

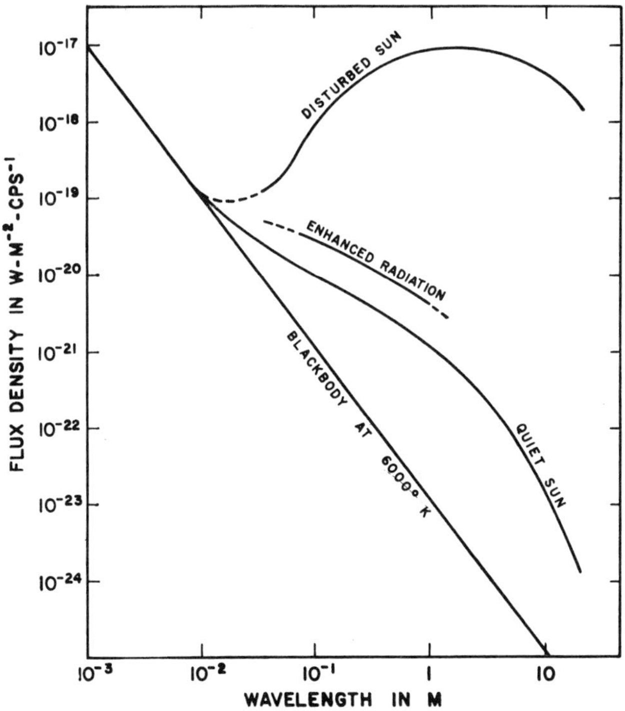

FIG. 4-7. Radio Noise from the Sun.

watt-meter^{-2} cps^{-1} sterad^{-1}. Jupiter exhibits sporadic bursts in the 15 meter region with peak intensities of 10^{-19} watt-meter^{-2} cps^{-1}.

All local bodies may be expected to exhibit thermal noise interference in their immediate vicinity. The Sun and galactic center would appear to be the primary natural radio noise sources in space.

4.2.6 Transit Times and Doppler Effects

Propagation transit times in space communications will be determined by the distances involved and the propagation velocity. Two-way communications, and radar-type distance measurements will suffer unusual transit time effects. The one-way transit time will be,

$$T = \frac{v}{c} \tag{4–39}$$

For an Earth-Moon distance of 400,000 km and a propagation velocity of 3×10^8 m/sec, the one-way transit time is 1.33 seconds which would cause some slight inconvenience in voice communication because a reply to a question would require in excess of 2.66 seconds. This time lag may be quite significant in terms of remote control of a lunar vehicle or some lunar operation. Of more consequence is the one-way transit time involved in direct communication with other planets (e.g., Venus between 3 and 15 minutes, Mars between 4 and 21 minutes, Pluto as much as 5.5 hours). Where relays are required, the transit times will be extended in proportion to the path increase and any delay in rebroadcast. A serious effect occurs when one considers the employment of very directive, high-gain antennas to communicate over the vast distances in space. To illuminate the lunar surface with energy is one thing; to communicate with a vehicle in orbit about Pluto may be impossible without some precomputation of when to transmit so that the vehicle will not be eclipsed by the planet when the transmission arrives, or else transmitting identical messages every half-Pluto-orbit period (if known) to ensure ability to receive and reconstruct the message.

Doppler effects must be contended with in communications and in tracking. The presence of high radial velocities will generate doppler frequency shifts in received signals in accordance with,

$$f_d = f_t \frac{v}{c} \tag{4–40}$$

Any noise rejection filtering system in the communication receiver must allow for the maximum doppler frequency content to come through, and hence the associated noise bandwidth, unless some special circuitry is devised to allow a receiver to lock-on and follow a very narrow band signal throughout the region of the expected doppler shifts.

The use of satellite-derived doppler reference data is subject to some errors because of the refractive effects and propagation velocity variations on the transmission as it passes through the Earth's ionosphere and atmosphere. Tischer (Ref. (*6*)) has considered these effects, and shown them to be directly accountable if the strata through which one is propagating is known. In the case of a non-homogeneous atmosphere and ionosphere, Tischer has indicated that a transverse (normal to radial) flight path doppler shift will exist where none should be apparent, and that this shift may be measurable.

4.3 COMMUNICATION TECHNIQUES

Techniques in communication will center generally in achieving maximum range, efficiency, or bandwidth (intelligence) over some particular geometric path or volume. The key equation is the one-way propagation expression,

$$R_{\max} = \frac{\lambda}{4\pi} \sqrt{\frac{P_T \; G_T \; G_R}{KTB \; (NF) \; S/N_{\min}}} \tag{4–16}$$

In space communication, the ranges are so great that it is almost axiomatic that some antenna gain must be employed unless the system is to have a very low pass band. In the use of very high gain antennas for either range extension or to obtain tracking accuracy a fundamental problem is the acquisition of the other end of the communication link. Once acquired, it must be identified as the proper or desired one, and then tracked. Relays will be used for range extension or horizon "hopping." Diversity links will be employed for bandwidth expansion or horizon extension. Special signal processing techniques will be used to reject as much noise as possible while retaining the signal's intelligence content.

4.3.1 Acquisition

The acquisition of communication transmissions by various receivers is a peculiar, many-faceted problem in space. The long ranges, and, hence, low signal strengths will make signal detection somewhat difficult. The general direction in which to listen is somewhat easier for non-maneuvering bodies provided some sort of ephemeris computer is available.

The need for acquisition stems from one or more of the following, depending on the geometrical and operational situation:

(1) Satellite eclipsing during planet-centered orbits.

(2) Probe, Moon and planet eclipsing because of Earth's rotation.

(3) The appearance of unknown or enemy vehicles that may require interrogation.

(4) Unscheduled maneuvers of spacecraft.
(5) Sequential use of communication relay facilities.
(6) Intermittent operational use of communication facilities.

The sequence of operations in acquisition requires that one search a pre-determined or assigned volume; detect any and all signals which have the characteristics of interest such as frequency, coding, unpredicted presence; select a particular, or the only, signal; and place tracking circuits on this signal to maintain lock-on or contact. This sequence is sketched in Fig. 4–8 which applies both to acquiring friendly or enemy transmissions and/or radar-type echos.

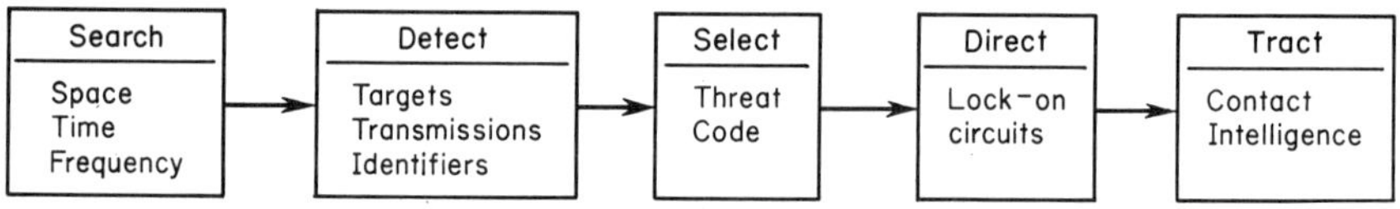

Fig. 4-8. Acquisition Sequence.

The search-function geometry is illustrated in Fig. 4–9. A search receiver or radar will listen through an antenna for transmissions in the volume of space defined by the antenna solid angle coverage. A search radar will interrogate and listen for echos or responses through a similar volume which will be restricted by both the transmitting and receiving antenna coverage. In order to increase the total spatial volume coverage, the antenna system may be scanned as illustrated.

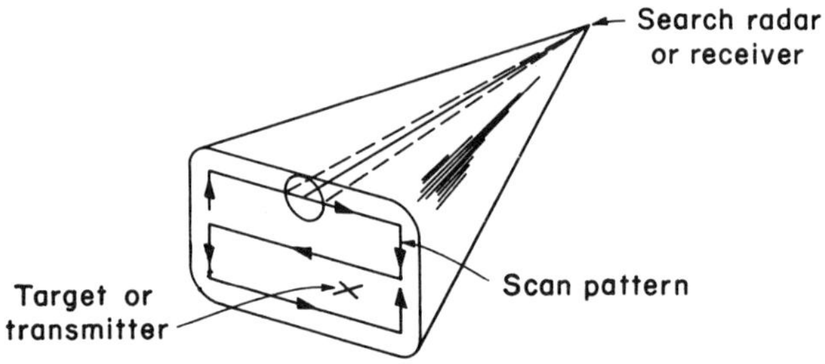

Fig. 4-9. Search Geometry.

The simple 3-bar scan pattern of Fig. 4–9 may be made much more complicated in order to cover larger volumes of space or to employ some available facility such as a conical scanning system which already is present in the equipment for use in subsequent tracking operations. Figure 4–10 illustrates a few scan pattern variations available to the system designer.

A detailed description of the various acquisition techniques employed in radar design will be found in a complementary volume of this series, Ref. (*7*). For communication link acquisition, we may assume a steady signal and a one-way transmission so that target fluctuations will not be the dominant factor in calculating detection probability. Also, the free space loss will be an R^2 function rather than R^4, and the power required to establish reliable communications for a similar (to radar) range will be on the order of the square root of the radar power assuming a common bandwidth requirement and similar antenna gains (and internal circuit losses). The detection probability ceases to be statistical if the transmitter and receiver antenna coverage are oriented in fixed positions, and the probability of detection may be expressed as,

$$P_D = 1 \text{ for } S/N > S/N_{\min} \tag{4–41}$$

and

$$P_D = 0 \text{ for } S/N < S/N_{\min} \tag{4–42}$$

where Eq. (4-16) is expressed in the form,

$$S/N = \frac{P_T G_T G_R \ \lambda^2}{KTB \ (NF) \ (4\pi R)^2} \tag{4–43a}$$

and $S/N_{\min}$ is the threshold detection level necessary for the communication receiver detection to operate.

The probability of detection then becomes the probability that the two antennas will be oriented with sufficient accuracy to provide the necessary gain, and the probability that the transmitter and receiver will be functioning in accordance with their design specifications.

Scanning will generally be necessary only at close ranges where sufficient signal strength will be provided easily, except for very wide bandwidth systems and the problem will devolve into having a detector whose response or integration time is less than the expected scan time interval during

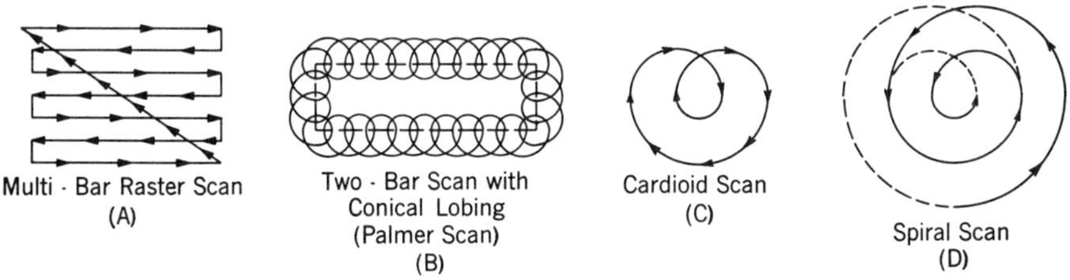

FIG. 4-10. Possible Pencil Beam Search Scan Patterns.

which the detector threshold will be exceeded. For lasers, scanning may be necessary at extreme ranges because of the tight beams involved.

Variable bandwidth/variable beamwidth systems can enhance detection probability and acquisition time considerably. Thus, a widebeam, narrow bandwidth system may be used to coarse-locate the transmitter, and to direct a high-gain, wide bandwidth system for fine-locating and communication. The detection and acquisition probabilities for communication satellite or ion cloud reflection systems will be similar to the bistatic radar problem in which the two one-way ranges are each squared and multiplied together to form an effective fourth power range term. The signal power will be degraded also by the effective cross-sectional area of the reflection which is usually somewhat less than the monostatic radar effective cross-sectional area and will have statistical variations. The signal-to-noise relation is,

$$S/N = \frac{P_T G_T \sigma_B A_R}{(4\pi)^2 \, (R_T)^2 \, (R_R)^2 \, KTB \, (NF)} \tag{4–43b}$$

where, σ_B is the effective bistatic reflectivity cross-section of the reflector and must consider how much is illuminated in common by the transmitting and receiving beams, scattering, and absorption.

R_T and R_R are the ranges to the reflector from the transmitter and receiver.

4.3.2 Identification

When more than one signal may be encountered, some means of identification will be required to assure that the proper signal is being followed. This identification may be by voice or visual contact and recognition in audio and television systems. Other communication systems (e.g., unmanned vehicles, telemetering, or beacon tracking links) require some more automatic identification means or coding.

Coding may be regarded as a noise rejection scheme. That is, only signals of predetermined characteristics will be accepted; all others are rejected. The effectiveness of coding is judged by the signal-to-noise ratio where noise is interpreted to include other man-made, and even similarly coded transmissions.

Coding may be accomplished by frequency, time, and space selectivity and combinations of these. The basic technique is to attempt correlation of the received signal with some predetermined coding plan. This is illustrated in schematic form in Fig. 4–11. The variety of identifying characteristics may be expanded by any and all combinations of those indicated as typical. The key point is that the stored data must have some memory of the anticipated "right" signal which will distinguish it from all "wrong" signals.

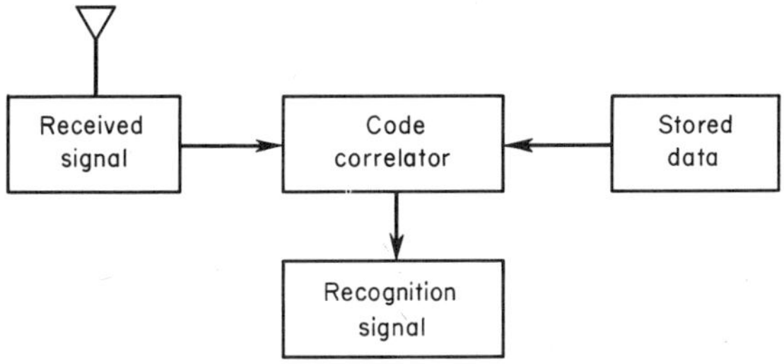

Identifying characteristics

Direction	Noise modulation
Frequency	Pulses
Side bands	Pulse length
Tone modulation	Pulse spacing
Frequency modulation	Multipulse sequence
Phase modulation	

FIG. 4-11. Code Correlation.

The basic mathematical function describing this process is the *cross correlation function*, which may be described in the time domain as,

$$\varphi_{12}(\tau) = \lim_{T\to\infty} \frac{1}{2T} \int_{-T}^{T} f_1(t) f_2 (t - \tau)\, dt \tag{4–44}$$

This function is thus a measure of the degree of matching between a stored or anticipated function, $f_1(t)$, and a received, time τ displaced function $f_2(t-\tau)$. Where sufficient correlation exists a recognition signal is generated which identifies the received signal.

For a radar system, storage of the transmitted signal frequency and modulation characteristics for comparison with the echo, yields the *auto-correlation function*,

$$\varphi_{11}(\tau) = \lim_{T\to\infty} \frac{1}{2T} \int_{-T}^{T} f_1(t) f_1 (t - \tau)\, dt \tag{4–45}$$

One might say that the purpose of any identification coding system is to make the stored signal identical with the "right" received signal which would cause $f_2(t)$ to approach identity with $f_1(t)$, and make the code correlation approach auto-correlation. The ability to accomplish this is known as *coherency*. That is, for complete identity of characteristics, the two signals must be coherent with each other in all significant characteristics.

If we take the identifying function, $f_1(t)$ as $e^{-i\omega t}$, we may say from Equation 4–44:

$$\varphi_{12}(\tau) = \lim_{T\to\infty} \frac{1}{2T} \int_{-T}^{T} e^{-i\omega t} f_2 (t - \tau)\, dt \tag{4–46}$$

Thus, the Fourier transform of $f_2(t)$ may be identifiable with the correlation function. This corroborates mathematically the intuitive idea that correlation may be in the frequency as well as time domain, as the memory function, $f_1(t)$, has been made a frequency-dependent term. Thus, we may correlate by frequency selection using predetermined filters or spectrum analyzers. Indeed, we may write a frequency correlation function as,

$$\varphi_{12}(\omega) = \lim_{\omega \to \infty} \frac{1}{2\omega} \int_{-\omega}^{\omega} f_1(\omega) f_2(\omega - \nu)\, d\omega \tag{4-47}$$

This frequency correlation function describes the degree of correlation one may expect through some spectrum-comparing device. Letting $f_1(\omega)$ be $e^{i\omega t}$ we may, of course, take an inverse Fourier transform and this expression reverts to the time correlation function. The significance is, however, that $f_1(\omega)$ need not be a continuous function of frequency, but would be a combination of positive "accept" filters, and negative "reject" filters such that any peculiar matrix "key" may be developed for code identification in the frequency domain just as a similar matrix "key" may be developed in the time domain by accepting pulses in certain spacing sequences as determined by passing the received signal through a tapped delay line and taking the product of the signals at the tapping points.

Examples of these two techniques are presented in Fig. 4–12 and Fig. 4–13. In Fig. 4–12, an illustration is made of acceptable and unacceptable signals being processed through a pre-set delay line. The delay line signals are multiplied together in a π-function operation to yield either a positive or negative (blank) indication of the signal acceptability. In Fig. 4–13, a similar process has been illustrated in frequency domain. False or unacceptable signals have been presented to illustrate how such signals may be rejected.

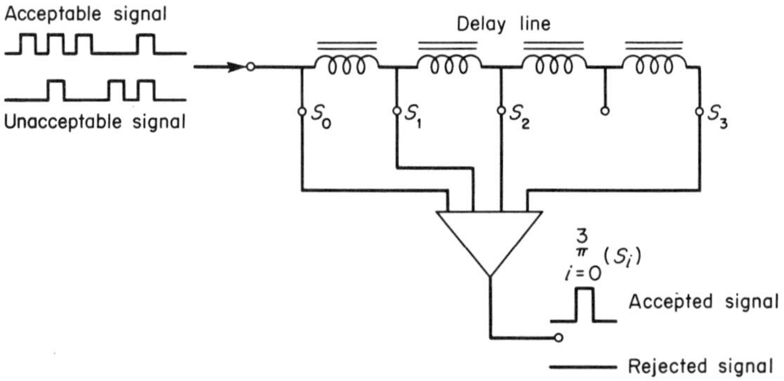

Fig. 4-12. Pulse Sequence Spacing Correlation.

The end output of the identification process is the presence or lack of recognition signal. To this, we may add the identification of false signals as an added feature.

Correlation by means of directivity appears practical only where antenna beams are sufficiently sharp as to make a signal cross-over the detection threshold level depending on whether the signal is being generated with sufficient power density within the received beam or not.

The end result of identification is taken to be the use of the recognition signal to operate appropriate tracking and data generation systems in space electronics. Military systems may include such techniques for determination of whether a transmitter is friendly or not, and the recognition signal may be used to initiate countermeasures, e.g., jamming action.

4.3.3 Tracking

Tracking will be required for two purposes in communication systems. (1) to maintain high-gain antennas oriented to the appropriate transmitter or reflector thus ensuring reliable communications, and (2) to generate data pertaining to the target position and its time derivatives. The first of these requirements is angular in nature, and involves the direction of an antenna beam through some physical device (e.g., gimbal drives, ferrite phase shifters). The generation of data to determine target position may be accomplished in three ways: (1) measuring the physical orientation of the directed beam or its mount gimbal system; (2) using the electrical measurement of the target in the beam directly; or (3) summing the physical measurement and electrical (pointing error) measurements. The

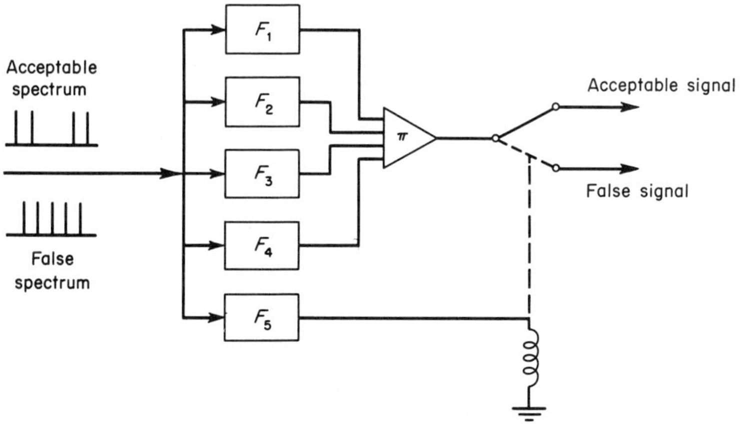

FIG. 4-13. Frequency Correlation.

determination of spatial position may require an electrical measurement of distance along the propagation path. This may be accomplished by radar echo time measurement, beacon response echo time measurement, or by employing synchronous clocks (oscillators) at each end of the communication link. More sophisticated methods are available when multi-station networks are considered. Trilateration from three diverse transmitters may identify a point in space by the intersection of three range spheres. Interferometric measurements of a signal phase relation may be used to determine a vector direction. Time difference measurements may be used to define a three-dimensional hyperboloid grid system of constant time differences, etc.

In measuring angular position by physical means, the beam must be directed towards the target by some means so that the antenna mount or gimbals represent, to instrument accuracy, the vector to the target. The vector is usually taken in two angles, azimuth and elevation, which may be referenced to vehicle frame coordinates, or through some stabilization system and/or computation to external or platform coordinates. A simplified block schematic of one axis of such a tracking system is depicted in Fig. 4–14. This figure illustrates the dependence of the antenna angle in space, A_{TL}, on various components and internal stabilization loop performance. A more detailed discussion of this tracking system is available in this book series, Ref. (7). However, one point of significance deserves elaboration. The basic error measuring sensor which determines the need for adjusting the antenna angle in space, A_{TL}, to a new orientation in order to follow more precisely the target line-of-sight angle, A_{LS}, must be some angle measuring device based in principle on an electrical measurement. We may simplify Fig. 4–14 further to a basic feed-back loop as illustrated in Fig. 4–15 where all the internal loops and components have been lumped into a single box whose performance is described by G. The error measuring device which determines,

$$E_{\mathrm{T}} = A_{LS} - A_{TL} \tag{4–48}$$

must be as accurate near the null cross-over as the overall system can hope to be, and must sense any error in position ahead of the subsequent corrective action of the feedback system.

Thus, in attempting to measure the line-of-sight vector, A_{LS}, we may:

1. Assume E_T is so small that A_{TL} and A_{LS} are coincident, and measure A_{TL} with physical sensors such as gimbal pick-offs.
2. Hold the antenna fixed, so that E_{T} equals A_{LS}, and the E_{T} sensor may then be used directly.
3. Measure A_{LS} as the sum of A_{TL} and E_T

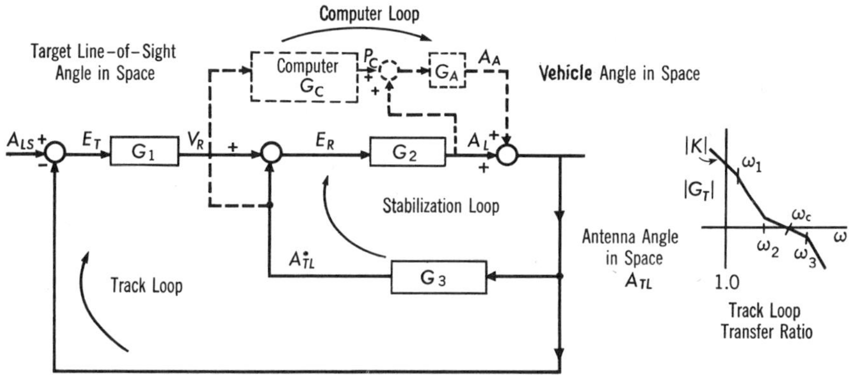

G_1 = radar receiver, amplifier, bandwidth shaping; G_2 = power amplifier, actuator, antenna structure; G_3 = rate gyro; E_T = radar angle error; E_R = rate loop error; $G_C G_A$ = computer, pilot, aircraft (or missile and autopilot only) (not easily described mathematically); V_R = rate command

TRANSFER FUNCTIONS

$$G_1 = \frac{K_1\left(1 + \frac{S}{\omega_2}\right)}{\left(1 + \frac{S}{\omega_3}\right)\left(1 + \frac{S}{\omega_3}\right)} \qquad \text{NOTE:} \quad \frac{G_2}{1 + G_2 G_3} \approx \frac{1}{K_3 S}$$

$$G_T = \frac{A_{TL}}{E_T} = \frac{K_1\left(1 + \frac{S}{\omega_2}\right)}{\left(1 + \frac{S}{\omega_1}\right)\left(1 + \frac{S}{\omega_3}\right)} \times \frac{1}{K_3 S} = \frac{K\left(1 + \frac{S}{\omega_2}\right)}{S\left(1 + \frac{S}{\omega_1}\right)\left(1 + \frac{S}{\omega_3}\right)}, \quad K = K_1$$

FIG. 4-14. Simplified Block Schematic of One Channel of Angle Tracking and Stabilization Loop.

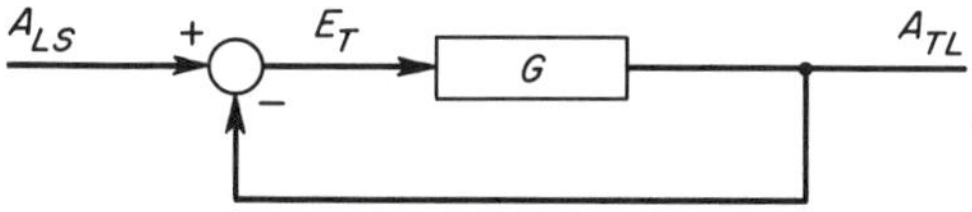

FIG. 4-15. Basic Feedback Loop.

The first of these approaches is commonly used in radar tracking systems. The second is generally impractical from signal strength considerations, since some significant antenna gain is generally required to maintain lock. Further, the typical antenna error sensitivity curve of such a system, Fig. 4–16, is not generally linear throughout the angular region of interest. Note that Fig. 4–16 is drawn for a particular "monopulse" radar. However, the monopulse lobe-on-receive-only (LORO) concept applies directly to one-way communication link angle tracking, and it really makes no difference in the static error sensitivity curve whether the received antenna lobing is simultaneous or sequential, and whether the transmission is pulsed or continuous wave.

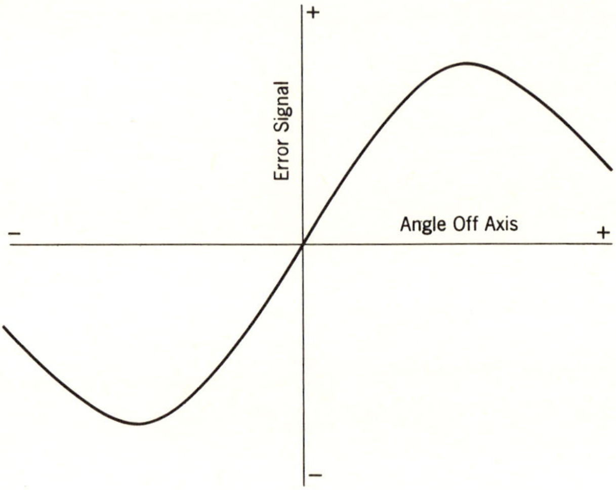

Fig. 4-16. Typical Antenna Error Sensitivity Curve.

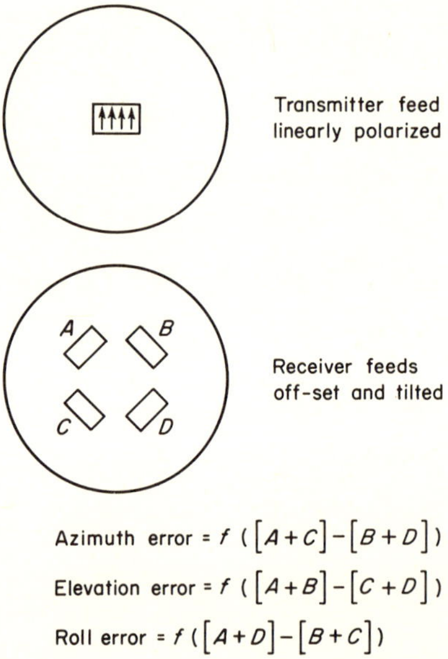

Fig. 4-17. Three-angle Tracking Technique.

The most accurate method of generating sight-line angle data is thus the third of the designated approaches. That is, use A_{TL} as coarse vector data, E_T as fine data, and sum the two signals directly for each tracking axis.

Roll attitude may be measured directly on the same communication link by polarization techniques provided no Faraday polarization rotation is experienced in the propagation path. The technique is to transmit a linearly polarized signal, and to orient the polarization of four received beams so that a polarization-dependent signal may be generated by comparison techniques. Figure 4–17 illustrates the technique. The error functions expressed relate the errors to some function determined by a comparison of one summation of signals with another summation of signals. Whether a difference, ratio, or other function should be employed depends on the geometry, the beam pattern, and whether the basic tracking is to be based on amplitude differences (a subtractive comparison) or phase differences (interferometric technique).

A fundamental limitation on the accuracy of angle measurement is the threshold or incremental sensitivity of the error measurement circuit. This sensitivity must be related to the antenna beam pattern gradient at the null or beam cross-over point. Thus, the beam off-set angle must be such that the pattern gradient at cross-over provides a sufficient difference signal for the desired angle sensitivity. As an illustration, if our angle detector minimum sensitivity is one db, and we wish to measure to 0.1 milliradian (mil) accuracy near bore-sight (or cross-over), we must have a pattern difference or gradient at cross-over of better than 10 db per mil. Thus, for any particular antenna pattern to be employed, we must offset the feed or beam axis until we reach this pattern gradient on the bore-sight axis.

Antenna systems that are phased arrays are basically interferometric in nature. That is, they determine vector by measuring the time or phase difference in receiving the same signal at two precisely known locations.

Angle rate data are commonly required in addition to angle measurement. Rate data may be obtained directly by tachometers or rate gyros mounted on the antenna drives or supports. Also, rate data may be obtained by differentiating the position data electrically or in a digital computer, if the rates encountered will permit this.

Range tracking is achieved by measuring the transit time of a signal. If the measurement is one-way, a requirement for synchronized clocks (oscillators) is implied. If the measurement is two-way, the echo time or beacon response time is merely measured. In any event, the measurement will be limited in accuracy by knowledge of the propagation velocity, time stability of the reference oscillator, stability or fidelity of the modulated signal, and stability of the measuring circuit. The fundamental equations are, of course,

$$R = T_{\text{one-way}} \times C \tag{4-49}$$

and

$$R = \frac{T_{\text{two-way}} \times C}{2} \tag{4-50}$$

Pulse measurements of range transit time is the most common technique. Such measurements suffer somewhat in accuracy or efficiency when compared with other techniques such as ramp modulated CW (continuous wave) and tone modulated CW. For a range-only measurement system, the most efficient modulation technique is to use the fine-grain carrier wavelength as an incremental yardstick, and a single tone modulation to set the maximum unambiguous range. The practical problems in instrumenting such a coherent system are considerable considering that carrier coherency requires a high order of stability of the carrier oscillator. To avoid such problems, multi-tone modulation is employed using longer wavelengths at which the stability can be better instrumented and controlled. The advantage of tone modulation over pulse and ramp-modulated CW is that the total transmitted power may be placed in the discrete carrier and sideband channels which establish the resolution and maximum range limits. These channels thus permit operating over the maximum possible range within transmitter power and receiver sensitivity limitations. Further, *a priori* knowledge of carrier and tone frequencies at the receiver permits the selection of these frequencies only by selective filters, hence more noise rejection, a better signal-to-noise ratio, and a better operating power margin. The primary limits on the noise rejection capability of the filter system will be instrumentation feasibility, and knowledge of the received frequencies. Doppler effects may be great enough to force a filter-design band-width widening since the receiver must track the signal through the velocity shifts. Once initial lock is established, it may be feasible to adapt the filter to a small band around the desired signals, and by doppler measurements on these signals cause the filters to track them in frequency.

Another method of great utility is the ICW (Interrupted-CW) modulation technique. This method employs a 50 per cent duty cycle pulse with the pulse duration "slaved" to the range measurement. Thus, time solution is available between the transmitter and receiver, and a maximum range measurement may be achieved (most efficient of power). This method provides coherent carrier data for "fine" range and doppler measurements as well. The actual doppler measurements made with the coherent carrier may be used to narrow the effective receiver band-width thereby providing maximum noise rejection and range. At low altitudes, the ICW system is generally slaved to the maximum practical PRF, and altitude is measured as a phase measurement on the fundamental or PRF frequency. This technique provides a capability of making altitude measurements down to where the rise time of the ICW "long pulse" becomes significant. In this mode of operation, the eclipsing action of the transmitter-receiver overlap provides discrimination against near targets, that is, the range equation for altimetry converts from a $1/R^2$ relation to $1/R$. This indicates a measureable improvement over other modulation techniques in ability to "see" through rain, dust, snow and fog. At extremely low altitudes, the ICW system generally is designed to operate as a conventional AM/CW altimeter using the pulse-transient overlap between transmit and receive signals, and detecting the overlap duration with a video receiver. The ICW technique also permits one to make the most efficient use of antenna aperture (hence weight) by generating several beams simultaneously through the same aperture because transmitter-to-receiver leakage is minimized (hence maximum range achieved) by time isolation. The ICW technique also permits one to obtain the best average power (hence maximum range with accuracy) from a peak power limited device such as current solid state transmitters.

Tracking in close proximity with another vehicle, as in space rendezvous docking, poses additional problems. If the antenna near fields are sensed, the angle and angle rate measurements will become meaningless and misleading. Range and range rate measurements should be valid in the near field zone provided CW techniques are employed. If angle tracking data are necessary right up to docking, the antennas must be off-set from the mating surfaces so that the near fields do not engage. This will generally mean erecting antennas external to the vehicle such that shadowing and shaping do not invalidate the measurements. An illustration of this is presented in Fig. 4–18.

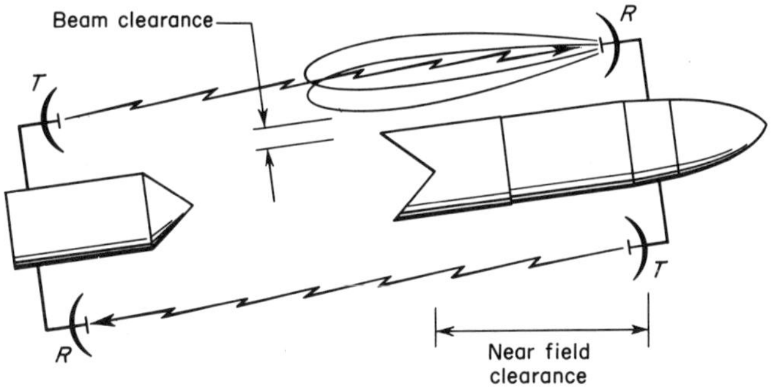

FIG. 4-18. Offset Antennas for Space Rendezvous Docking.

A final problem in contact docking is the extreme dynamic range of signals encountered by a receiver. The dynamic range of input signals will be

$$\frac{P_{R_{min}}}{P_{R_{max}}} = \left(\frac{R_{max}}{R_{min}}\right)^2 \tag{4–51}$$

If R_{min} approaches zero, the dynamic range will approach infinity, an impossible design goal for the receiver. The offsetting of antennas to avoid near-field degradation of angle measurements will assist in establishing a more useful R_{min}, hence dynamic range. A corollary technique is to program the transmitter power as a function of range. Another approach is to switch to a low gain "near-field" antenna at short ranges.

4.3.4 Relays

The purpose of communication relays in space is to (1) circumvent an eclipsing obstacle such as Earth-horizon or a large satellite, (2) maintain the security of a primary military broadcast location, or (3) serve as a noise rejection device. The first two of these objectives are merely solutions to geometric problems, and are not directly germane to communication techniques as such.

The use of a relay system as a noise rejection system may be illustrated by considering a noisy long-range voice communication link (Fig. 4–19). If the signal-to-noise ratio at the receiver is sufficiently poor, not all the intelligence will be transferred. If we interpose a similar capability, (i.e., manned receiver/transmitter station) somewhere between the original two, the relay operator may listen to the original transmission and rebroadcast it to the destination point. Thus, by shortening the individual link ranges, and having the relay operator correlate the message with his vocabularly and retransmit, we have improved our communication assurance measurably.

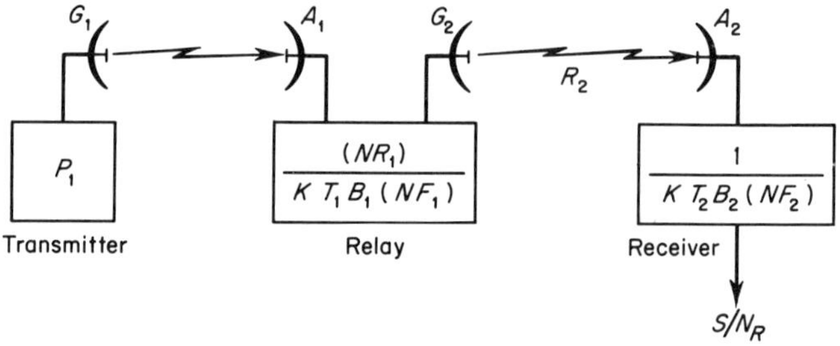

FIG. 4-19. Relay Geometric Situation.

In this illustration we have enhanced the overall signal-to-noise ratio by individual link shortening. The output signal-to-noise through a relay station will be,

$$S/N_R = \frac{P_1 G_1 A_1 G_2 \; (NR_1) \; A_2}{(4\pi)^2 \; (R_1)^2 \; KT_1B_1(NF_1) \; (R_2)^2 \; KT_2B_2(NF_2)} \tag{4–52}$$

where P_1 is the primary transmitter power,

G_1 is the primary antenna gain,

R_1 is the first link range,

A_1 is the effective receiver cross-section of the relay,

KT_1B_1 is the noise band-width of the relay receiver,

G_2 is the gain of the relay including second link transmitter and antenna gain,

(NR_1) is the noise rejection capability of the relay, and is dimensionally in KTB units,
A_2 is the final receiver effective cross-sectional area,
R_2 is the second link range,
KT_2B_2 is the final receiver noise band-width, and
NF_2 is the final receiver noise figure.

The geometric situation is depicted in Fig. 4–19. If we now write the direct transmission expression

$$S/N_D = \frac{P_1G_1A_2}{4\pi(R_1 + R_2)^2\ KT_2B_2\ (NF_2)}, \tag{4–53}$$

we may say that the relay has provided a margin of assurance or capability of

$$M_A = \frac{S/N_R}{S/N_D} \tag{4–54}$$

or

$$M_A = \frac{A_1G_2\ (NR_1)\ (R_1 + R_2)^2}{4(R_1)^2\ (R_2)^2\ KT_1B_1\ (NF_1)}. \tag{4–55}$$

This indicates that to be effective the relay station design must have sufficiently large values of $A_1G_2(NR_1)$ to more than offset the penalty of the remaining terms. This should certainly be feasible if useful size antennas can be employed in obtaining high values for A_1 and G_2. The noise rejection term (NR_1) is a measure of signal enhancement through correlation and noise rejection. This correlation and noise rejection may be accomplished by several means. As an example, a human operator can correlate a message with the vocabulary of words (or Morse code signals), and reject impossible words and combinations thereof; thus the operator performs a noise rejection function. Video and coded communications can be similarly improved if one can provide a logic and memory to ensure it is a reasonable message, and to reject illogical data, perhaps signalling for a repeat transmission if no logical content can be deduced.

4.3.5 Diversity

Diversity techniques are often employed to enhance communication assurance under unusual propagation conditions. Diversity may take three forms; frequency, time, and space.

In frequency diversity the same message content is broadcast simultaneously over two or more frequency channels or sidebands and the signals received in these channels are combined to provide a maximum signal strength. This is illustrated in Fig. 4–20. In addition to providing reliability or assurance through redundant transmission, the system has the added feature that the diversity combiner may recover the message from two signals, each of which may be too weak singly to provide the message.

A similar technique is available with time diversity as illustrated in Fig. 4–21. In this system two identical messages are sent with a time difference τ separating them. The receiver passes the first message through a delay circuit and the second message directly to the diversity combiner so that the two messages become coincident. The combiner then adjusts to the best signal-to-noise ratio combination of the two signals, and extracts the message.

Figure 4–22 illustrates the space diversity situation in which the same message is sent by different routes. As shown, two space-separated receivers are employed. Space-separated transmitters and relays may be introduced if desired. A critical item in space diversity is to have a common overall path length or time delay in the various channels to assure coincident arrival at the diversity combiner.

Various hybrids and multiple combinations of frequency, time, and space diversity are feasible to provide communication assurance under a wide variety of environmental conditions. The diversity combiner is basically a correlation device rather than a direct integrator. That is, rather than sum or multiply two signals of unknown quality, the combiner attempts a combination, and then seeks the best combination. A schematic of this is presented in Fig. 4–23.

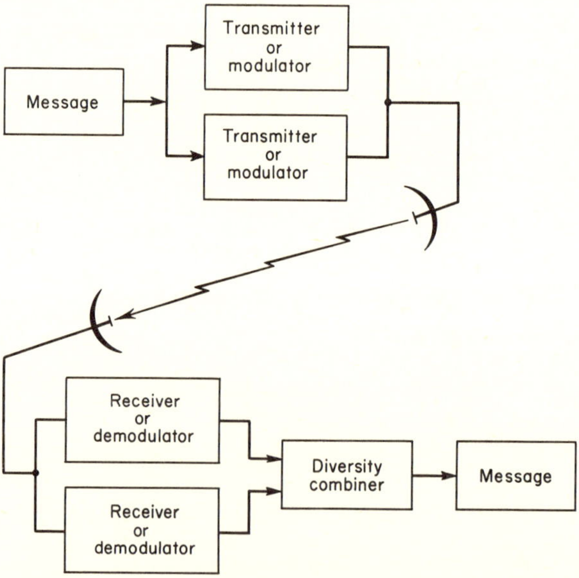

FIG. 4-20. Basic Frequency Diversity System.

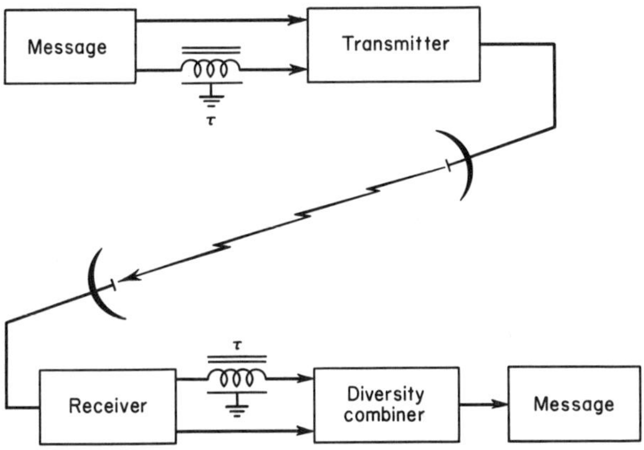

FIG. 4-21. Basic Time Diversity System.

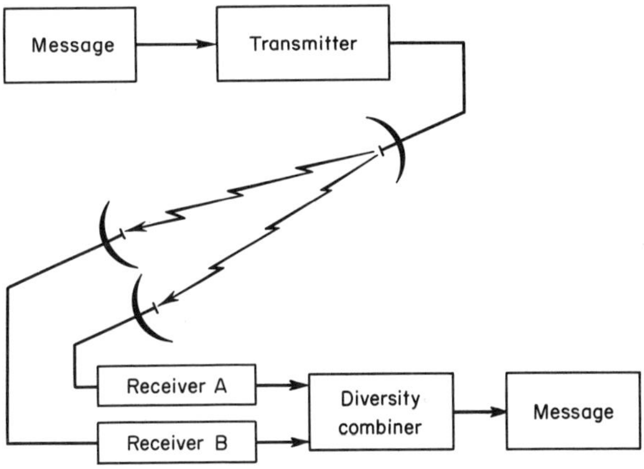

FIG. 4-22. Basic Space Diversity System.

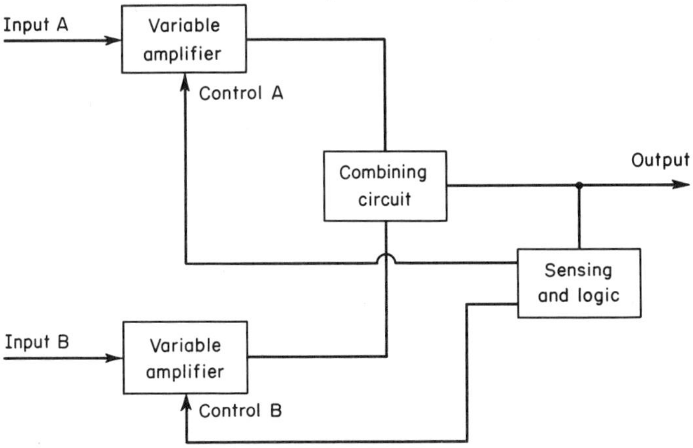

FIG. 4-23. Diversity Combiner.

Each signal is operated on independently in a variable amplifier prior to entering the actual combining circuit. The output of the combining circuit is monitored continuously, and whenever the signal or signal-to-noise drops below some pre-set level, the variable amplifiers are readjusted by a program in the logic box to find some new combination which produces a useful output signal.

4.3.6 Signal Enhancement and Noise Suppression

In pulse radar, and repetitive signal systems generally, a signal-to-noise ratio improvement is feasible by integrating successive signals with each other. That is, if we have an individual signal-to-noise ratio for each signal, s/n, we may integrate several pulses together for an overall improvement,

$$S/N = \int_0^T s/n \, dt \tag{4–45}$$

$$S/N > \frac{s}{n}$$

This arises because s is repetitive directly, and hence may be added consistently in phase, whereas n is non-repetitive, and merely integrates as the power density of random signals.

A second form of signal enhancement is correlation. This may be time, frequency, or space (directivity, range, or volume) correlation. The basic equations and processes have already been treated under identification. Fundamentally, correlation is a particular form of integration in which a signal auto-correlates directly with itself or with a stored memory, or cross-correlates with a pre-determined definition of what the signal should look like.

A fundamental purpose is signal enhancement or noise suppression. Thus, if any signal-to-noise ratio, S/N, may be operated on by some noise-rejecting or signal-enhancing function, an overall improvement in communication system performance may be anticipated. Thus, whenever we may narrow the frequency spread of our circuit acceptance, we reduce the noise level and S/N improves. This is permissible so long as the message content is not degraded. Low-noise input circuits are requisite to good system performance because the limiting S/N ratio will occur at receiver input circuits. If we can gate out noise by time gating we can make some significant improvements in S/N as is exploited in pulse circuits whenever feasible. If we can orient our antennas precisely and make them extremely directive, we can similarly improve S/N. Considering these factors we may look at the generalized communication equation,

$$S/N = \frac{P_T G_T \; A_R}{4\pi R^2 \; KTB \; (NF)} \tag{4–58}$$

where B represents the sum total effectiveness of all the bandwidth and time rejection features of a design, G_T and A_R the space gain or directivity rejection features, and NF the internal circuit rejection features.

The thermal noise factor, T, at the receiver input is of much concern in achieving maximum range and efficiency in communication and radar systems. McNaul, Ref. (*8*), has predicted the various capabilities inherent in such low noise input circuits as diode amplifiers, masers and travelling-wave tubes. These predictions are presented in Fig. 4–24, 4–25, and 4–26 together with cosmic and atmospheric noise temperatures and specific achievements. As illustrated, noise temperatures on the order of 100 degrees Kelvin are achievable currently. Continual research in the design of low input noise circuits should eventually produce techniques which approximate the galactic noise back-ground level for intra-space communications.

Of particular interest is the phase-locked detector which has applications for noise suppression by "locking-on" to a particular signal, and following it with a narrow band-width, or high Q, filter. The basic circuit is illustrated by Fig. 4–27. An input signal is mixed with a voltage-controlled oscillator to achieve a peak output signal strength at the desired frequency. Thus, the noise rejection capability (selectivity) of the filter is employed to enhance the signal-to-noise ratio while the voltage-controlled-oscillator (VCO) is used as an automatic phase (frequency) control device to keep the output of the mixer centered on the filter middle frequency. The technique is mathematically similar to conventional radar AFC (automatic frequency control) circuits, and is primarily unique in that it permits

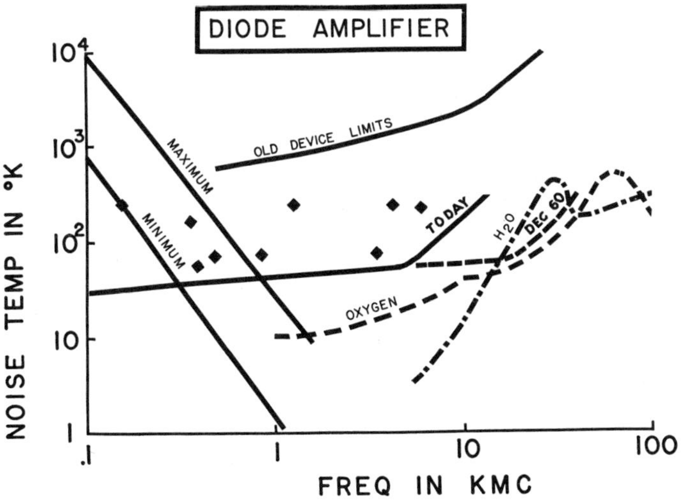

Fig. 4-24. Present and Predicted Diode Parametric Amplifier, Capabilities with Spotted Achievements and Cosmic and Atmospheric Noise.

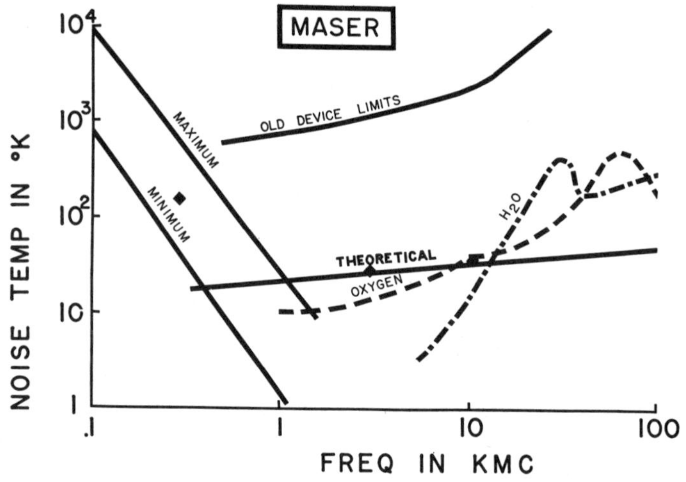

FIG. 4-25. Solid State Maser Capabilities with Spotted Achievements.

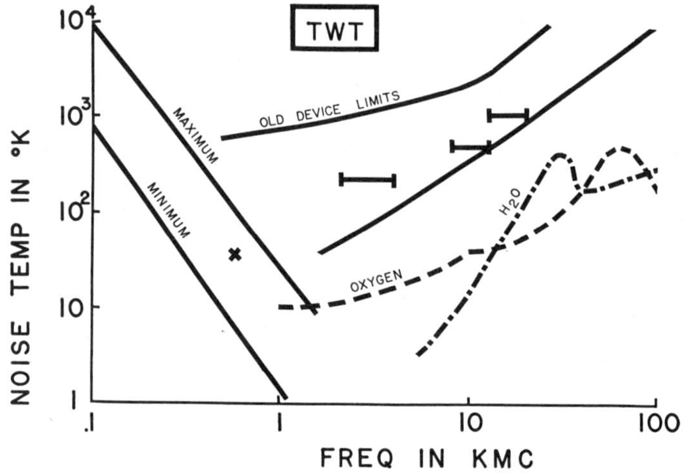

FIG. 4-26. Travelling-wave Tube Capabilities with 3 Achievement Bars and an Electron-Beam Paramp (x).

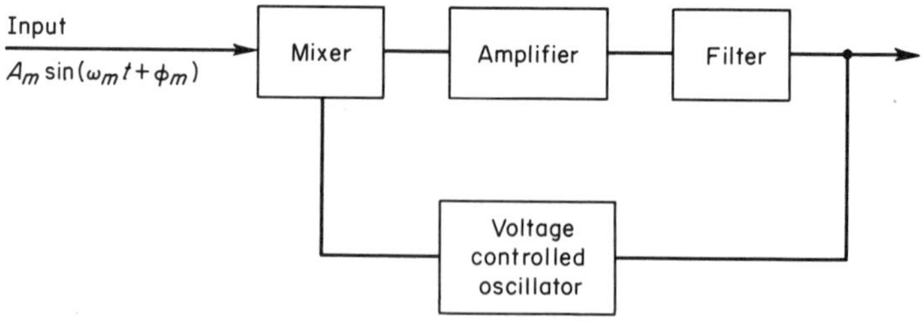

FIG. 4-27. Phase-Locked Detector.

post-second detector noise rejection through the use of a phase comparison technique.

4.3.7 Modulation Methods

The basic purpose of all communication systems is the transfer of information between two isolated points. The transfer of information depends on two significant parameters, power and bandwidth. We must have a sufficient amount of power at the reception end of the communication link to distinguish our information signal from the background noise, and we must have sufficient bandwidth to permit the transfer of the information content. As we have already seen, the signal-to-noise ratio at the receiver is bandwidth-dependent, i.e., the bandwidth entering the computation as a consequence of introducing the noise power. We may say that there will always be a trade-off between maximum communication range and bandwidth as is indicated in many equations in the foregoing sections of this chapter. In order to transfer a message of some information content, we must illuminate the receiver with a signal containing the message in some form or format. This method of impressing a message- or information-containing signal on a carrier wave form is known as modulation.

Basic to the concept of information transfer is the bandwidth efficiency of a modulation method:

$$\frac{1}{\gamma} = B_M/B_C. \tag{4–59}$$

Where B_C and B_M represent the bandwidths of the message, M, and carrier, C, respectively, γ, the ratio of transmitted bandwidth to message bandwidth will depend on the various modulation technique employed.

Black, Ref. (9), has defined the various modulation techniques presently in use, and explored their instrumentation and salient points in some detail. The purpose of this section is to provide a brief resume of these techniques so that the reader may be sufficiently well informed to understand their differences, and, to some extent major advantages and disadvantages.

The generic types of modulation are:

(1) Amplitude Modulation (AM)
- *a.* Amplitude Modulation (normal)
- *b.* Single-Sideband Modulation (SSB)
- *c.* Vestigial-Sideband Modulation (VSB)

(2) Angle Modulation
- *a.* Phase Modulation (PM)
- *b.* Frequency Modulation (FM)

(3) Pulse Modulation
- *a.* Pulse Amplitude Modulation (PAM)

b. Pulse Time Modulation (PTM)
  (*b*-1) Pulse Duration Modulation or Pulse Length Modulation: (PDM), (PLM)
  (*b*-2) Pulse Position Modulation (PPM)
c. Pulse Code Modulation (PCM)
  (*c*-1) Amplitude PCM
  (*c*-2) Frequency Shift Keying (FSK)
  (*c*-3) Phase Shift Keying (PSK)

If we consider the time-dependent function of a transmitted signal to be a continuous sinusoidal wave of a particular amplitude, frequency and phase characteristic, we may express the carrier as,

$$C = A_c \cos(\omega_c t + \phi_c). \tag{4–60}$$

Amplitude modulation implies that the carrier amplitude, A_c, is varied in some way with a time-function, $V(t)$, containing the message, or AM implies,

$$S = A_c[1 + kV(t)] \cos(\omega_c t + \phi_c) \tag{4–61}$$

where k represents the percentage modulation or the amplitude ratio of $V(t)$ and A_c. Typical waveforms for AM are illustrated in Fig. 4–28, which shows a voice modulating wave, a continuous wave carrier, and the resulting complex wave. If we consider that $V(t)$ may be handled as a spectrum or summation of various sinusoidal component frequency signals, or,

$$V(t) = \sum_{i=1}^{n} A_i \cos(\omega_i t + \phi_i) \tag{4–62}$$

we may say that,

$$S = A_c\left[1 + k\sum_{i=1}^{n} A_i \cos(\omega_i t + \phi_i)\right] \cos(\omega_c t + \phi_c) \tag{4–63}$$

From this expression we may observe that the actual signal content will be a spectrum centered around the carrier frequency, ω_c, and containing all the positive and negative sums of the ω_i frequencies with ω_c. Thus, the frequency spectrum of the signal will be symmetrical about the carrier frequency as illustrated in Fig. 4–29, which shows that the actual signal in AM is completely redundant in amplitude and phase characteristics.

Single-sideband (SSB) modulation implies the amplitude modulation of a carrier as just described, but the transmission of the frequency spectrum on only one side of the carrier frequency. The advantages of SSB over straight AM are that only half the frequency spectrum is necessary to transmit a particular bandwidth message, hence a more efficient spectrum usage and a theoretical two-to-one noise rejection potential. A further variation of AM and SSB is the suppressed carrier (SC) technique in which an even narrower band-spectrum may be employed if the receiver knows the carrier frequency *a priori*, and can introduce this frequency internally thus allowing the transmitter to concentrate all of its power directly in the side-band frequencies. An example would be the use of crystal-controlled carrier frequency references by the transmitter and receiver where the crystals are cut to resonate at the same frequency with an acceptable tolerance.

Vestigial Sideband Modulation (VSB) is a form of SSB-SC in which a small portion of the transmitter power is allocated to the sideband frequencies close to the carrier, so that the receiver may define the carrier frequency by recognizing the symmetry of the modulation around the carrier at one side of the SSB spectrum.

Angle Modulation implies introduction of the message, $V(t)$, to the argument of the carrier wave. Thus, phase modulation (PM) yields the expression,

$$S = A_c \cos [\omega_c t + \phi_c + K_p V(t)] \qquad (4\text{–}64)$$

where K_p represents the phase deviation, or peak difference between the modulation and carrier phase angles.

Frequency modulation implies the carrier frequency is operated on directly, or,

$$S = A_c \cos \{[K_F V(t) + \omega_c]t + \phi_c\} \qquad (4\text{–}65)$$

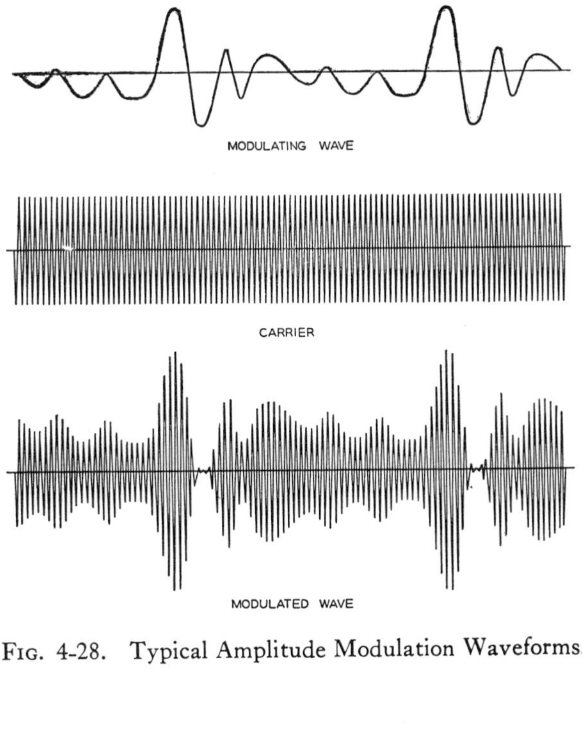

FIG. 4-28. Typical Amplitude Modulation Waveforms.

where K_F is now the frequency deviation or peak difference between modulation and carrier frequencies. The modulation index, M, is the ratio of frequency deviation and modulation frequency; or, for each modulating sine wave,

$$M_i = \frac{K_i}{\omega_i} \tag{4-66}$$

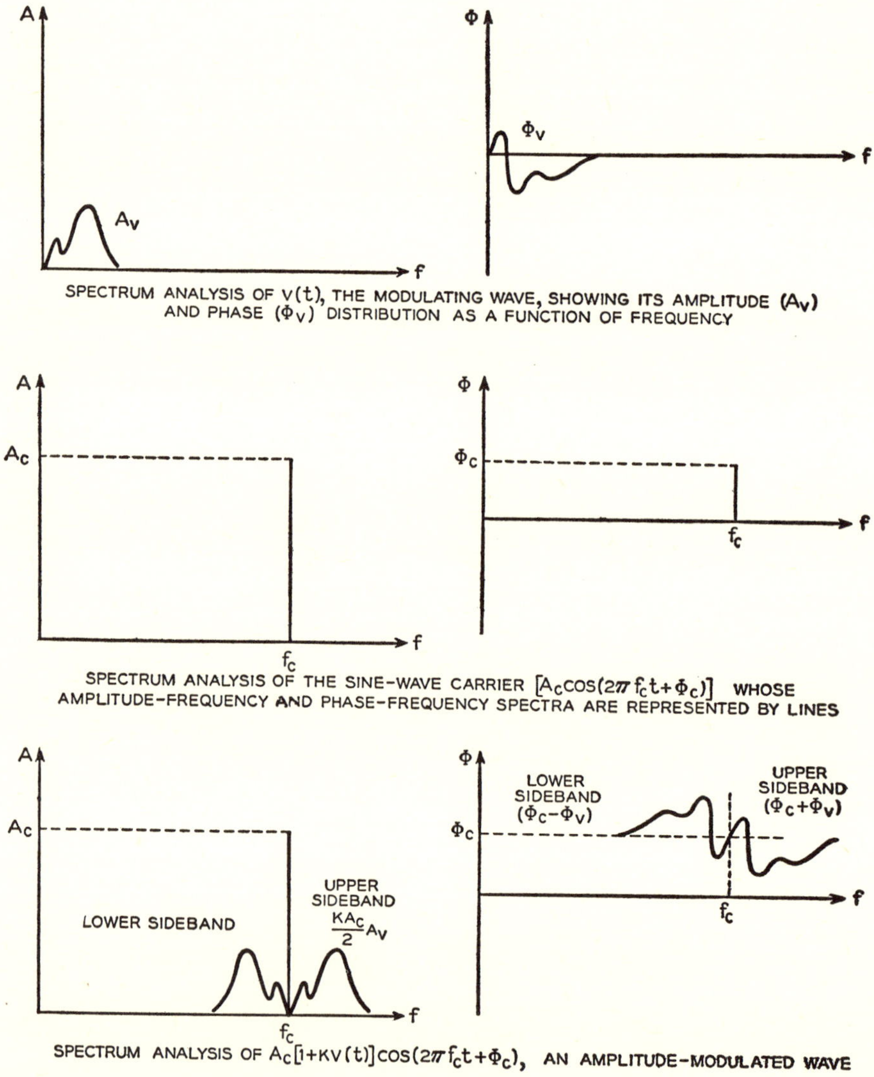

FIG. 4-29. Typical Frequency Spectrum of AM Signal.

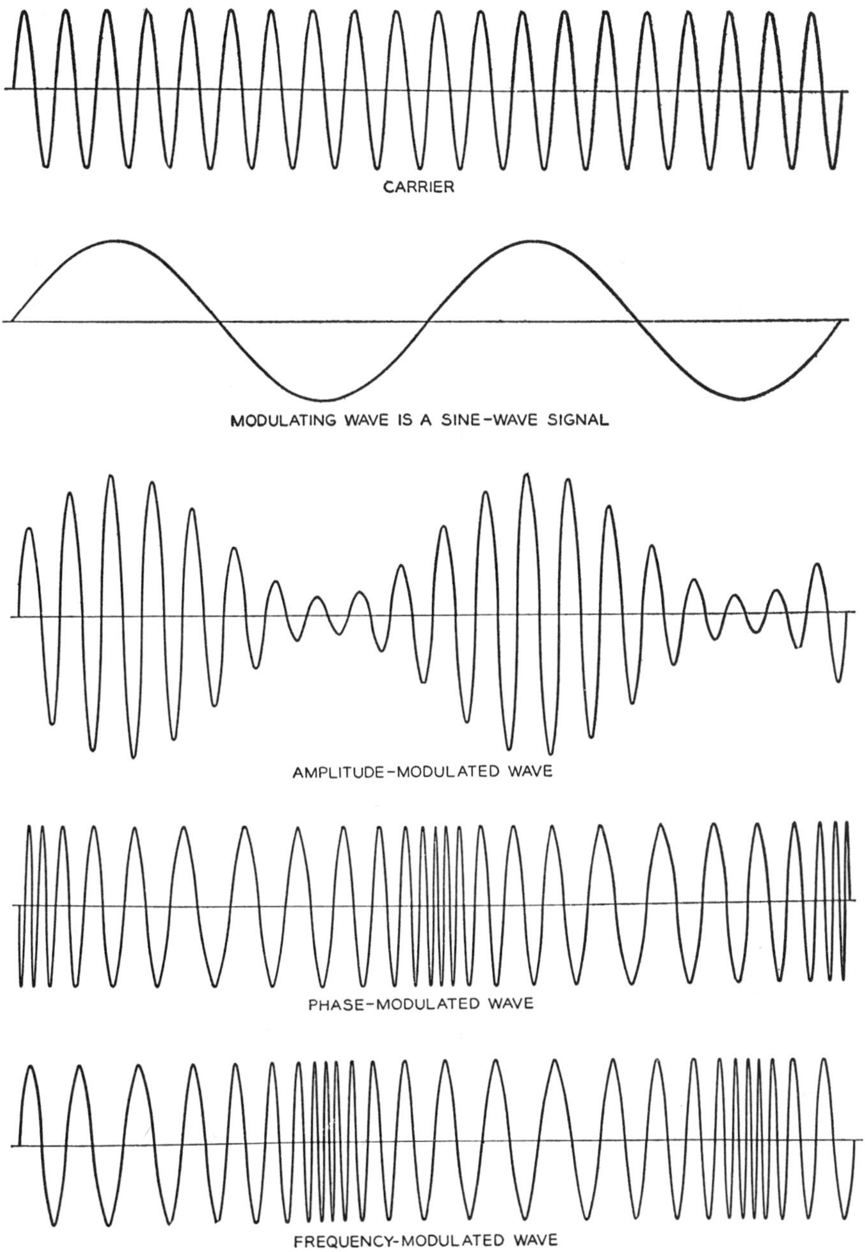

FIG. 4-30. Comparison of AM, PM, and FM.

Knowledge of modulation index, M, is particularly important for the band of frequencies being employed as the S/N for FM is related to S/N for AM by,

$$(S/N)_{FM} = 3M^2\,(S/N)_{AM} \tag{4-66a}$$

The (S/N) FM is related to (S/N) SSB by,

$$(S/N)_{FM} = \frac{3}{2}\,(S/N)_{SSB} \tag{4-66b}$$

The similarity between Eq. (4-64) and (4-65) indicates the close identity between phase and frequency modulation. By considering that,

$$\text{angular frequency} = \frac{d}{dt}\,(\text{angle}), \tag{4-67}$$

we may observe that an equivalent frequency modulation must exist for any phase modulation and vice-versa. Thus, $K_F V(t)$ is equivalent to $\frac{d}{dt} K_P V(t)$. Considering $V(t)$ as a spectrum of sine waves, the derivatives

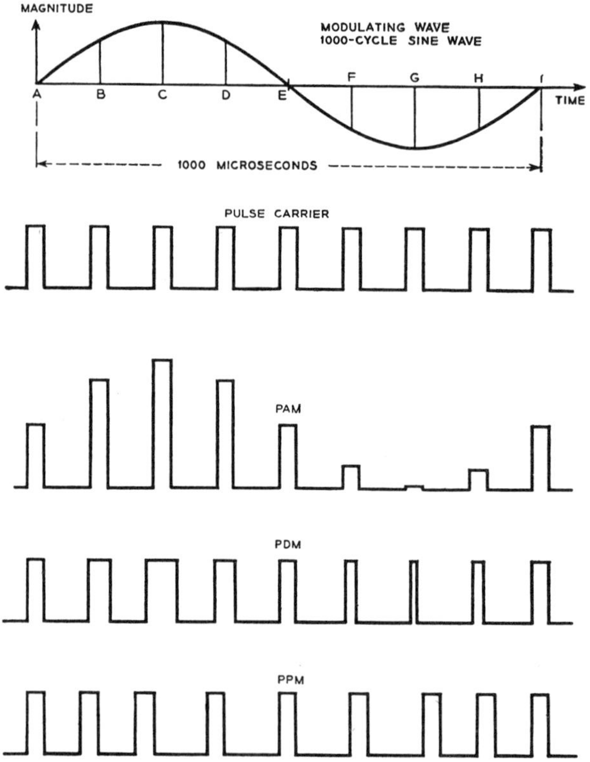

FIG. 4-31. Examples of Pulse Modulation.

are cosine waves, or sine waves phase shifted $\pi/2$ radians at the same frequencies with an amplitude change. Thus, phase and frequency modulation circuits will be found to be similar and will handle the same types of signals. Fig. 4–30 illustrates a comparison of AM, PM and FM using a single modulating tone on a continuous carrier. The $\pi/2$ radian relation between PM and FM is directly evident in this illustration.

Pulse modulation implies the use of a pulsed carrier frequency, and the conveyance of message intelligence by means of some variation of the pulse characteristics. The basic carrier frequency is generally pulsed at a constant rate with constant amplitude and length or duration pulses. A pulse modulation characteristic such as amplitude, duration or position is thus the means of carrying the information. Figure 4–31 illustrates three

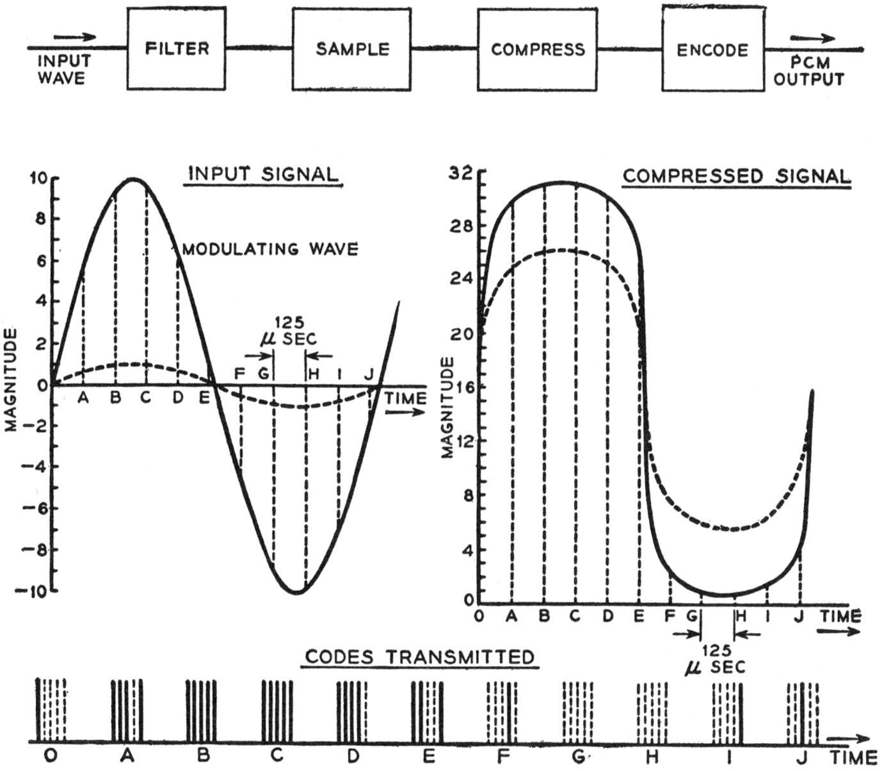

Transformation of voice to code pulses. The input signal, represented as a sine wave at the left above, is compressed logarithmically to give the curve shown at the right. Samples are taken every 125 microseconds, and each sample is converted to a five-pulse binary code as shown in the lower line. A solid vertical line is used to represent a pulse, and a dotted line, the absence of a pulse.

FIG. 4-32. Pulse Code Modulation; Sampling and Quantizing of a Voice-Frequency Wave.

basic types of pulse modulation waveforms, PAM, PDM, and PPM. Pulse modulation lends itself to time multiplexing of information. That is, several sets of information may be conveyed on one communication link by summing (in sequence) several pulse modulated signals in the transmitter, and separating the various signals in the receiver. A simple example used in telemetry is the use of a motor-driven commutator to sample a variety of input signals, and a stepping switch or synchronized motor to decommutate the information.

Pulse Code Modulation (PCM) requires the quantization of input data prior to transmission. A typical illustration of this technique is taken from Black, (Ref. (*9*)), and presented in Fig. 4–32. In this illustration, the input signal is operated on to generate a 5-pulse (or "bit") binary code "word" representing the signal amplitude at 125-microsecond sampling intervals. A major advantage of PCM is that so long as each binary code "bit" can be identified as 0 or 1, a new signal can be generated which reproduces it faithfully with considerably increased power. Thus, this technique is used extensively in communication repeater techniques and, through redundant circuit operation, may be employed for extended periods without circuit maintenance as in submarine cable signal boosters and satellite repeater stations.

PCM provides an additional advantage in that a summation of the binary bits transmitted in each "word" may be made to provide a "self-check" against a single additional bit provided in the transmission for this check. That is, a word grouping of 1011011 would be followed by a 1 to indicate an odd number of 1's was present. A 0 would be used to indicate that an even number of 1's should be counted. Thus, loss of an odd number of digits is used to reject a "word," and call for a repeat. This technique requires *a priori* knowledge of the number of bits per word, and of the self-check code being employed (i.e. significance of 0 or 1 as a check and the location of the self-check digit in the word—usually at end). This PCM/FM technique has been quoted (Ref. (*10*)) as capable of an 8.8 db improvement over conventional FM/FM, 5.9 db over PAM/FM and 6.0 db over PDM/FM in a comparison for use in missile telemetering.

Frequency Shift Keying (FSK) and Phase Shift Keying (PSK) are alternates to normal PCM. In normal PCM, one may think of transmitting the binary bits as either "on" (1) or "off" (0). FSK employs a singular frequency shift of a continuous power output to distinguish the bit state (1 or 0). PSK employs a phase shift (usually 180-degrees) on a single carrier to distinguish the bit state.

Viterbi, Ref. (*11*), has made a comparison of various detection methods for pulse modulations. This comparison is presented in Table 4–1, and illustrates some of the various hybrid forms pulse and PCM may take, the trade-offs potentially available in choice of detector, carrier pulse coherency, transmission time or period, number of code or quantization levels, etc.

TABLE 4–1 PULSE MODULATION DETECTION COMPARISON

Detection of Modulation	Telemetry System	Bandwidth, CPS: Adjacent channels not phase coherent	Bandwidth, CPS: Adjacent channels phase coherent
Linear synchronous detection	PAM	$1/T$	$1/2T$
	PSK	$1/T$	$1/T$
	PDM	$50/T$ for less than 1% channel cross-modulation	
Correlation detection	Quantized PPM	L/T	$L/2T$
	Quantized PSK	L/T	$L/2T$
	PCM, orthogonal codes	L/T	$L/2T$
	PCM, nonredundant codes	$\text{Log } L/T$	$\text{Log } L/2T$

T is the transmission time
L is the number of quantization levels
PAM: amplitude modulation
PSK: phase shift keying
PDM: pulse duration modulation
PPM: pulse position modulation
PCM: pulse code modulation

Of particular interest is the concept of transmission period. This concept may be employed in long-range space communication to reduce bandwidth (hence, noise) while conveying the same method.

To communicate, we need power, bandwidth, and time. By halving our bandwidth and transmitting the same power over twice the time, we may convey the same message with a much superior signal-to-noise ratio (3db). Thus, we have introduced a time-bandwidth product into the appropriate range equation. This is essentially the same concept employed in integrating a repetitive signal (as in a pulsed radar) over a period of time to improve the signal-to-noise ratio. Quantized pulse coding modulation techniques allow us to make direct use of this method of time-stretching our transmission in order to communicate over the vast distances of extra-terrestrial space.

While new equipment and techniques will be continually evolving, system designers should, whenever feasible, consider making the system compatible with the existing global instrumentation facility capabilities if communication with these facilities is envisioned. Such documents as Ref. (*12*) through (*15*) should be extremely useful in system test and operational planning. These and other related references will be found to contain the current operational standards for modulation formating to ensure compatibility with the various equipments in the global range complex.

References to Chapter 4

1. P. F. von Handel & F. Hoehndorf, "High-Accuracy Electronic Tracking of Space Vehicles," *IRE Transactions on Military Electronics*, Vol. MIL-3, No. 4, Oct. 1959, pp. 162–172.
2. T. P. Harley & G. Tyras, "Transmission of Electromagnetic Waves Through an Ionized Layer in the Presence of a Strong Magnetic Field," *Proceedings of the IRE*, Vol. 49, No. 12, Dec. 1961, pp. 1822–1824.
3. H. Hodara, "The Use of Magnetic Fields in the Elimination of the Re-Entry Radio Blackout," *Proceedings of the IRE*, Vol. 49, No. 12, Dec. 1961, pp. 1825–1830.
4. M. P. Bachynski, I. P. French and G. G. Cloutier, "Antenna Noise Temperature in Plasma Environment," *Proceedings of the IRE*, Vol. 49, No. 12, Dec. 1961, pp. 1846–1857.
5. A. G. Smith, "Extraterrestrial Noise as a Factor in Space Communications," *Proceedings of the IRE*, Vol. 48, No. 4, April 1960, pp. 593–599.
6. F. J. Tischer, "Propagation-Doppler Effects in Space Communication," *Proceedings of the IRE*, Vol. 48, No. 4, April 1960, pp. 570–574.
7. D. J. Povejsil, R. S. Raven, and P. Waterman, *Airborne Radar*, D. Van Nostrand Co., Inc., Princeton, N. J., 1961.
8. J. P. McNaul, "Electronic Components for Space Instrumentation," Transactions of the IRE Professional Group on Military Electronics, Vol. MIL-4 Nos. 2–3 of April–July 1960, pp. 308–312.
9. H. S. Black, *Modulation Theory*, D. Van Nostrand Co., Inc., Princeton, N. J., 1958.
10. "A Big Step in Digital Telemetry," *Missiles and Rockets*, July 13, 1959.
11. Andrew Viterbi, Jet Propulsion Laboratory, "Detection and Signal Processing," Proceedings of 1961 WesCon, IRE.
12. "Electronic Trajectory Systems, Catalog, Vol. II, Radar Beacons," Inter-Range Instrumentation Group (IRIG) No. 104–58 of June 1958.
13. "Radio Command Control Equipment Catalog, Vols. I and II," IRIG No. 105–59 of Nov. 1959.
14. "IRIG Standard Time Formats," IRIG No. 104–60 of July 1960.
15. "Telemetry Standards," IRIG No. 106–60 of November 1960.

CHAPTER 5

GUIDANCE

5.1 INTRODUCTION

The subject of guidance of space vehicles is similar to missile guidance in some respects, and yet, dissimilar in many ways. Missile guidance is generally oriented towards interdiction of some other missile or a target aircraft, or else detonating a warhead at a set of ground coordinates which may be predetermined as with free-flight, ballistic, loran, celestial, map-matching, etc., systems or determined enroute as with radiation seekers or air-to-surface command systems. In space vehicle guidance, these same problems exist, and the solutions generally run parallel. The companion book in this series, *Guidance*, by A. S. Locke treats these problems effectively and the purpose of this chapter will be to indicate and treat those areas peculiar to guidance in a space environment. The inclusion of a man as a sensor and/or a control device and the variations in reference characteristics are considered as well as the dominating effects of the gravitational fields, orbit characteristics and transit times. Special attention is given to rendezvous and docking techniques and problems, and the establishment and maintenance of stations in orbit and at libration points.

5.2 INFORMATION REQUIREMENTS

Fundamental to any guidance problem is the knowledge of the desired terminal location or destination in terms of present or enroute coordinates. The coordinate system is arbitrary, and generally may be selected in terms of the best available information source. Some means must be available to interpret these data into vehicle control signals allowing for transformation of coordinates to vehicle control coordinates and the effect of vehicle control forces on the ultimate destination including gravitational field effects. This implies both coordinate sensing and conversion, and optimization of control equations based on some criteria such as energy conservation, minimum transit time, impact softening, docking orientation, station-keeping orbit parameters, etc. Guidance information requirements thus will vary considerably, and will be dictated very much by the nature of the mission and the payload. Consideration must be given to the need for and requirements placed on (1) pre-launch guidance or fire control, (2) launch or early-flight guidance, (3) mid-course, and (4) terminal guidance. Some of these may be non-existent or combined for particular missions, while others may go through several time-sequenced stages or modes. An example of the latter would be individual guidance of each stage in a multi-stage rocket launching. Another example would be rendezvous and docking guidance that requires orbit plane transfer, orbit altitude transfer, orbit synchronization, docking alignment, closure control, and impact softening. Many of these problems can become quite sophisticated in their own right, and their treatment here will be devoted to fundamentals because of further developments needed in the state-of-the-art.

5.2.1 Position and Derivatives

Fundamental to guidance is the requirement that one know the vehicle position and time-derivatives of position with respect to some target or expected future target location. Knowledge of one's position, velocity, acceleration, etc. in some set of absolute or reference coordinates, and navigating or guiding a vehicle with respect to these coordinates (e.g. celestial latitude-longitude reference system) will suffice provided one knows the terminus of the guidance problem in the same coordinate system. Time derivatives also are required as well in predicting or ensuring orbit synchronism and "soft" rendezvous.

We may thus categorize information requirements in terms of knowledge of our own vehicle position and time-derivatives relative to either the target or the predominant central force field to some accuracy determinable by the terminus requirement and control-capabilities. Knowledge of one's position in terms of target coordinates will generally be useful in terminal guidance systems. Knowledge of one's position with respect to the predominant force field coordinates will generally be useful in all antecedant guidance systems, and will generally require knowledge of the target coordinates in terms of the same force field coordinant system. Thus, one may generally expect a launch through mid-course guidance system which requires knowledge of the local vertical position in the central force field coordinate system, and motion sensors which measure relative to these references which, in turn, must include some arbitrarily selected azimuth such as the Vernal Equinox. The target position may be known from its ephemeris in the same system, or from sighting it with a sensor and computing its position, successive sightings being employed to generate the required time derivatives.

Terminal guidance systems will generally rely on tracking sensors, and will be used to make up for the errors accumulated during the mid-course guidance period. Terminal guidance may employ a common sensor with the earlier guidance modes, perhaps using the sensor in some more accurate fashion.

Operation in bi-central force fields such as the Earth-Moon system, will present peculiar problems in that a vehicle's ephemeris may not be approximated by a simple conic orbit (or incremental conic if under powered flight). Similarly, the target's ephemeris will not be obtained easily. Knowledge of position and time derivatives of both vehicle and target in terms of the bi-central field will help in that "dead-reckoning" integration of these quantities can enable one to predict a trajectory, and, by an iterative search, obtain a set of suitably intersecting trajectories. Augmentation with a target-tracking sensor will simplify the operations. However, utilization of a minimum energy trajectory in such an interception will still require quite sophisticated path prediction and mating.

5.2.2 Path Prediction

The path of any free-flight vehicle may be predicted at any time based on knowledge of an infinite set of constants. That is, if one knows a vehicle's position over a short segment of its flight path, this knowledge will contain data on the position at the infinite set of points included in this path segment, and from knowledge of this infinite set of points, one may extrapolate to determine the total path, thus position at any time. Similarly, if one knows a free-flight vehicle's position at one point, and also knows all the time-derivatives (an infinite set) of the path at this point, the total path is completely defined.

From these considerations, one might assume that guidance in space will be relatively easy since the direct approach would be to measure a set of positions on a path segment, or a set of derivatives at a single point on a path. The difficulty in applying this concept lies in the requirement that an infinite set be measured, and that the measurement accuracy be perfect; two unattainable goals. The compromise required is that one make position and derivative measurements either continuously or periodically during a flight to some attainable instrument accuracy, and use these data to correct the vehicle flight path so as to reach the desired terminus or next "up-dating" point with a useful accuracy.

5.2.3 Destination or Future Location Prediction

As with predicting one's own future location, the ultimate destination location must be predicted in some manner unless a pursuit-type homing approach is to be followed. That is, to minimize energy, control, or transit time, one should follow a path which will extrapolate the target present position and path in such a way that the two paths will be coincident at some future time.

One may define a vehicle position in three dimensions relative to the destination as, in cartesian coordinates,

$$s = x, y, z \tag{5-1}$$

or, in polar coordinates,

$$s = \rho, \phi, \theta \tag{5-2}$$

If the target is considered stationary, and motion taken as relative to the target, then the vehicle may be navigated in terms of these coordinates. The overall transit time, T, may be used to express average values for the first derivatives,

$$\dot{s}_{\text{ave}} = \overrightarrow{v_{\text{ave}}} = \dot{x}_{\text{ave}}, \dot{y}_{\text{ave}}, \dot{z}_{\text{ave}} \tag{5-3}$$

which would be a good total as well as average relation for a constant speed, hard impact missile. For other vehicles, higher order time-derivatives are required so as to program the smoothness of contact and take-off, and to be able to outmaneuver the target if it displays acceleration capabilities. If both the space vehicle and the target are operating in some central force field, then knowledge of one's position and time-derivatives relative to the force field center and its gradient will be extremely important, for control purposes, in addition to knowledge relative to the target. Indeed, accurate knowledge of a cooperative target ephemeris in the central force field, plus knowledge of one's own vehicle ephemeris in the same field will be all that is required to effect suitable orbit changes in order to accomplish rendezvous. We may say that at any time, a vehicle's location is defined by the vector relation,

$$\overrightarrow{s_v} = \overrightarrow{s_{vo}} + \int_0^t \overrightarrow{v_{vo}}\, dt + \int_0^t \int_0^t \overrightarrow{a_v}\, (dt)^2 \tag{5-4}$$

For the target, we may similarly say,

$$\overrightarrow{s_t} = \overrightarrow{s_{to}} + \int_0^t \overrightarrow{v_{to}} dt + \int_0^t \int_0^t \overrightarrow{a_t} (dt)^2 \tag{5-5}$$

Now, intercept (hard) or rendezvous (soft) will occur when, and only when,

$$\overrightarrow{s_v} = \overrightarrow{s_t} \tag{5-6}$$

which therefore requires at intercept time, T,

$$\overrightarrow{s_{vo}} + \int_0^T \overrightarrow{v_{vo}} dt + \int_0^T \int_0^T \overrightarrow{a_v} (dt)^2 = \overrightarrow{s_{to}} + \int_0^T \overrightarrow{v_{to}} dt + \int_0^T \int_0^T \overrightarrow{a_t} (dt)^2 \tag{5-7}$$

or,

$$(\overrightarrow{s_{to}} - \overrightarrow{s_{vo}}) + \int_0^T (\overrightarrow{v_{to}} - \overrightarrow{v_{vo}}) dt + \int_0^T \int_0^T (\overrightarrow{a_t} - \overrightarrow{a_v}) (dt)^2 = 0 \tag{5-8}$$

Of these terms, the initial conditions will be relatively well known, or we should not be able to plan a flight. The target acceleration, a_t, may or may not be known, generally dependent on whether it is a friendly or hostile target. Our own acceleration, a_v, is under our control, and will be the primary determining factor in establishing the flight epoch, T. The programming of our acceleration, a_v, is thus of primary concern in determining whether an intercept time, T, can exist if we are energy-limited, or in minimizing T if we are time-limited in conducting our mission.

It should be noted that Equation 5–8 expresses the total relation in relative or difference terms as well as absolute. That is, knowledge of one's position, velocity, and acceleration relative to the target, or the target's trajectory parameters relative to one's own, will suffice to define an intercept.

5.2.4 Optimizing Energy Utilization versus Minimizing Transport Time

As indicated in the previous section (Eq. (5–8)), if we are to optimize our use of energy in a space flight we must program our acceleration accordingly. If we consider the relatively simple problem of "skeet-shooting" a ballistic missile (i.e., firing an anti-missile from somewhere near the ballistic missile launch point as might be done by a submarine chaser against a Polaris-type weapon), we may establish a common set of initial coordinates, and the initial relative velocity as zero, so that Eq. (5–8) reduces to,

$$\int_0^T \int_0^T a_t(dt)^2 = \int_0^T \int_0^T a_v(dt)^2 \tag{5–9}$$

A typical plot of this function for single-engine, free-flight missiles is illustrated in Fig. 5–1. The area under each acceleration curve will be the velocity achieved by each missile, and the product of these areas and the time to intercept from the area centroid, d, will be an acceleration moment about the intercept time, T. When the two acceleration moments are equal, we shall have established an intercept time, T. From this illustration, we may infer two very significant points. First, the earlier we impart acceleration to our vehicle (the earlier the acceleration centroid), the earlier intercept will occur.

The second significant point is that the total energy expenditure must be related to the time integral of the acceleration. Thus, to minimize energy expenditure in achiev-

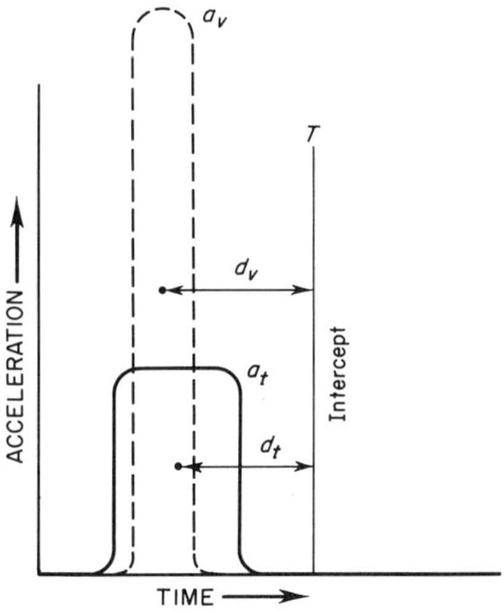

FIG. 5-1. Ballistic Missile Chase Intercept.

ing intercept by a certain time, T, we should minimize the acceleration area by obtaining as much moment arm, d_v, ahead of T as possible.

The two primary concerns in efficiency, energy and time, dictate the expenditure of as much thrust as possible as early as possible in the flight, the underlying difference being that minimizing energy expenditure will indicate a closely controlled or predictable thrust cut-off. Minimizing transport time will indicate a continued expenditure of thrust beyond the minimum energy requirement so as to build up an even greater acceleration moment, and move the intercept time to an earlier point.

5.2.5 Computational Requirements, Program Storage and Selection

In instrumenting Eq. 5–8 in our space vehicle, we must consider the computational requirements, program storage, and selection. If we assume our target or destination is a disabled spacecraft, we may postulate that we know its ephemeris to some precision, and that to effect a speedy rescue we must rendezvous as early as possible, and return to an acceptable Earth base. A simplified block schematic of the required guidance programming is illustrated in Fig. 5–2. A typical space vehicle might have several ephemeris programs available to it within a memory or storage bank. The programs would provide a prediction of orbit plane, perigee altitude and longitude, and apogee altitude, based on time of firing, fuel mixture, burn-out time, orbit injection impulse, maneuvering impulses, etc. An iterative ephemeris matching unit would compare the various programs with the target ephemeris, and, based on some selection criteria, establish a best fit. The selection criteria might be in the nature of best fit with blast-off in less than 2 hours, or with circular orbit matching target perigee (i.e., when apogee extends into a hazardous radiation belt). The "best-fit" program would then be modified to provide some better fit if possible, and then implemented. Implementation will consist of operation of vehicle controls, and presenting suitable data to the pilot so that he may operate such controls.

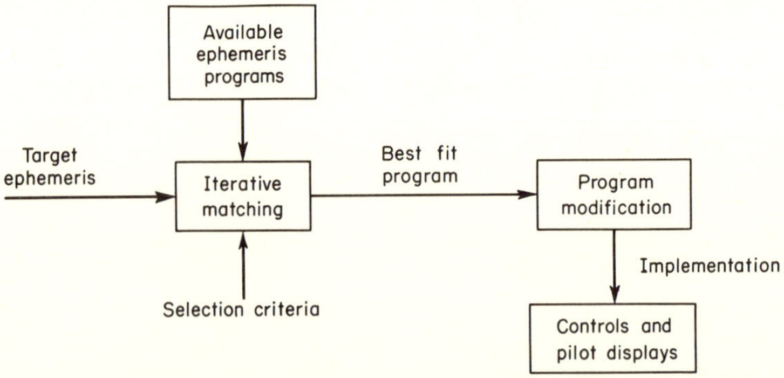

FIG. 5-2. Simplified Block Schematic—Rendezvous Guidance Programming.

Whether this scheme is ever instrumented is immaterial. The key item is that some computational elements are required to match the target ephemeris with the predicted vehicle ephemeris for some set of launch conditions. The computational elements must allow for any coordinate translation of data not already performed, and must either dictate a flight program and the launch conditions to an acceptable accuracy, or else compare the target ephemeris with predetermined programs to an acceptable accuracy. The magnitude of the computations to be performed will depend upon the number of operations necessary, and the accuracy. The latter will quite likely be determined by terminal guidance detection, acquisition and maneuvering capabilities, in addition to the error sensitivities of the computations.

G. Swanlund, in Ref. (*1*), has prepared a particularly good generalized analysis of guidance computational requirements for space vehicles. In this analysis, the differential equations of motion are taken as,

$$\frac{dx_i}{dt} = f_i\ (x, u, t) \qquad i = 1, \ldots, n \tag{5–10}$$

where $x(t)$ is an n-dimensional vector representing the actual vehicle state, and $u(t)$ is an ν-dimensional control state representing the guidance controls on the vehicle state. Note that the vehicle state may include spatial coordinates, time, vector velocity, mass, and heat (as in a re-entry problem). The control state may include vehicle attitude, thrust attitude, thrust magnitude, and thrust timing. In this technique, the guidance problem is to find a control function, $u(t)$, which takes $x(t)$ from an initial to a final point in the u-dimensional space while minimizing some function of the control process. The generalized technique is to form a Hamiltonian system of equations such that,

$$\dot{x}_i = \frac{\partial H\ (x, p, u)}{\partial p} \qquad i = 1, \ldots, n+1 \tag{5–11}$$

$$\dot{p}_i = \frac{-\partial H\ (x, p, u)}{\partial x} \qquad i = 1, \ldots, n+1 \tag{5–12}$$

where,

$$H = \sum_{i=1}^{n+1} p_i f_i \tag{5–13}$$

and p is the adjoint vector to x. The maximum of H with respect to $u(t)$ provides the optimum control function, u. Boundary conditions at $t=0$ and T (intercept time) are generally necessary to the solution. Reference (*1*) contains further generalized solution techniques for this problem (which can become quite complicated to solve) as the number and non-linearity of parameters increases.

5.2.6 Best Source Information Selection and Up-Dating Other References

The paramount problem in sophisticated Earth-oriented navigation systems, where a high order of accuracy is required, is the selection of the best source of reference information, and the use of these data to update the other navigational references. The same problem may be expected in space-craft, and therefore must be treated herein.

Decisions must be of two types, between instruments of similar nature, and between instruments of dissimilar nature. The outputs of instruments are taken as three: position, velocity, and acceleration measuring. Clocks are considered to be of a different class, although clock measurements are germane to the development of derivatives or integrals from other-order measurements of the same quantities.

The basic problem in determining which instrument to select between several of the same output may be resolved several ways.

1. The inherent accuracies of the various instruments may be known, and the one with superior accuracy employed.
2. The error sensitivity of the computations based on the input instrument accuracy may be used to judge whether the instrument is suitable.
3. Some self-check feature may be employed to establish or measure an instrument accuracy thus providing assurance of its performance.
4. Several similar instruments may be "averaged", so that the net error is always less than the worst individual instrument error.
5. Several similar instruments may be compared, and the median reading used.
6. Several similar instruments may be compared, and the most "popular" reading used.
7. Some combination of the foregoing may be employed.

The method used in any particular situation will depend on what is available, and what is required. However, the various techniques indicated provide several orders of sophistication and potential accuracy improvement if employed in various combinations.

The combination of data from various types of instruments is illustrated in Fig. 5–3. This may be an attitude stabilization control loop, a guidance control loop, or a navigation reference control loop. The design problem will be to assess the various accuracies available from the instruments, relate these to the overall accuracy requirement, and determine the relative weighting factor, K, and the complex frequency operation, $G(s)$, for each channel necessary to provide both inherent stability and suitable response to input or forcing functions. Internal and external noise sources must be minimized, and some analysis of these sources, their power spectrums, and the overall error sensitivity in terms of signal-to-noise ratio is generally required to ensure satisfactory system operation. The techniques described previously (such as averaging the measurements of similar instruments) are merely methods of obtaining a more accurate reading in substitution for one instrument. The gain or weighting factor, K, may be zero in some branch or for some period of flight, thus converting, for example, a stellar-doppler-inertial system into a stellar-inertial system. Also, a sampling term may be introduced in a lower-order path to indicate periodic up-dating of the higher order data sources.

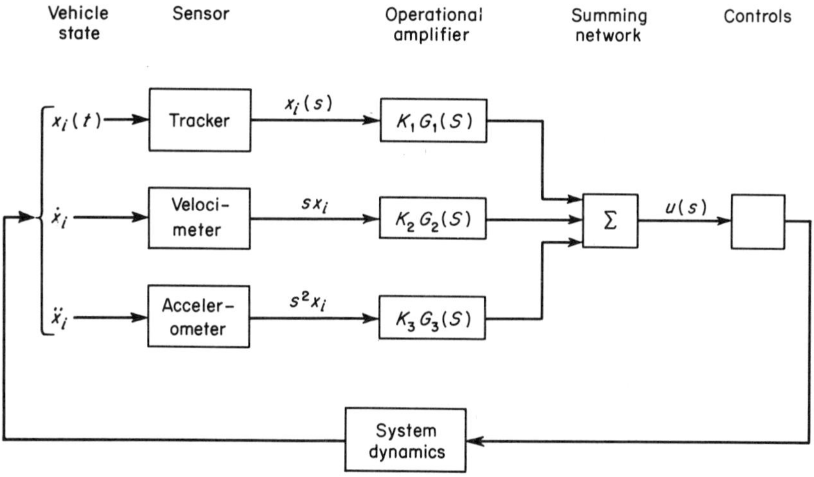

FIG. 5-3. One-axis Control Loop with Data Combination from Various Types of Sensors.

In general, one may think of the second-order measurement as having a superior short-term accuracy, and the position measurement a superior long-term accuracy, with the first-order measurement somewhere in between. This is illustrated in Fig. 5–4*a*. Thus, for any very long flight duration, it is obvious that some positional reference will be required, and that rate or acceleration measurements may be employed as interpolation devices if the position reference is available on a sampled basis only. It is also obvious that for relatively short duration flights, an all-inertial (acceleration) system will be the best choice for precision navigation, and for intermediate duration flights a rate measuring system will be best provided the rate measurement accuracy line in Fig. 5–4*a* does not cross the position measurement line prior to the acceleration measurement line crossing. The actual picture may be distorted somewhat from this. Position measurements may not always be available or of uniform accuracy with time. Thus, a rate system will often be superior for long duration flights. A space-borne, three-dimensional Loran system, for example, will have accuracy variations with distance and aspect relative to the base lines; also the typical Loran hyperbolic accuracy dilution factors will be encountered. Such things as Schuler-oscillations in inertial measurements, the ineffectiveness of orbital inertial measurements (by virtue of their lack of contact with terminal bodies), enemy countermeasures in a war situation, shadowing of constellation references, error build-ups in computer program control, statistical variations in measurements, etc. must be considered for any single design or mission problem.

Figure 5–4*b* illustrates the accuracy improvement achievable by combining rate and acceleration measurement systems. The rate sensor curve establishes an error rate maximum which can be translated to lower errors but not error rate (slope of dotted curve) by an acceleration sensitive system. Such a system is commonly called a D-I (Doppler-Inertial) system.

Figure 5–4*c* illustrates the system accuracy function for an acceleration-sensitive (inertial) system which is up-dated periodically by a position reference system. Figure 5–4*d* similarly shows a rate sensitive system up-dated periodically by a position reference system. The choice between up-dated inertial and up-dated rate systems will basically depend on whether the interval between up-datings is greater or less than the time interval at which the inertial and rate systems have equal error build-ups. For long free-running intervals, the rate system is obviously the superior choice.

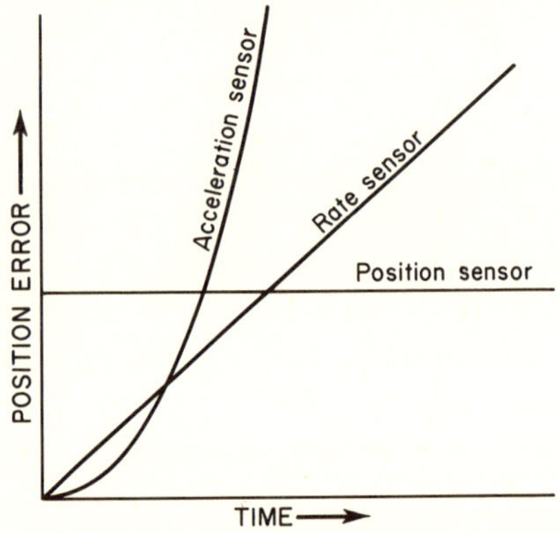

Fig. 5-4*a*. Typical Accuracy Relations of Various Sensor Types.

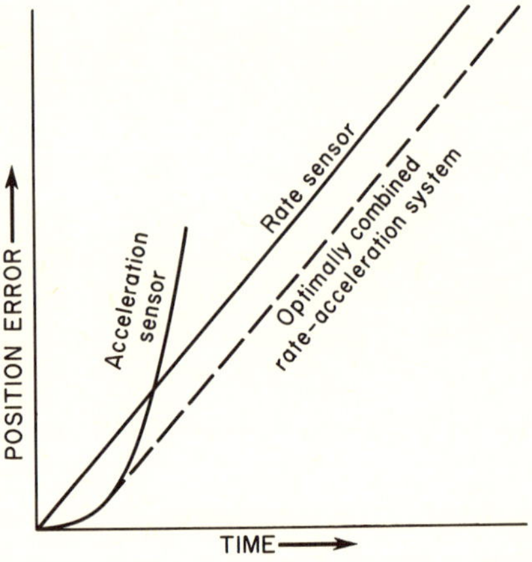

Fig. 5-4*b*. Accuracy Improvement Achievable by Combining Rate and Acceleration Systems.

5.3 ENERGY CONSERVATION METHODS

The two primary criteria for guidance and control optimization are energy conservation and time conservation. Energy conservation requires that the flight be accomplished with a minimum expenditure of energy. As

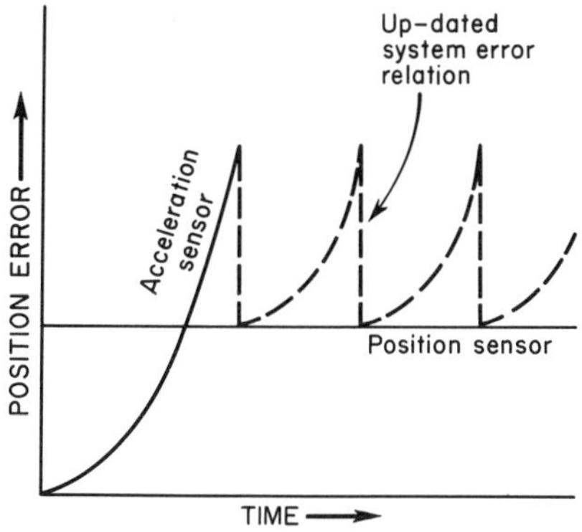

FIG. 5-4*c*. Accuracy of an Acceleration System Periodically Up-Dated by a Position System.

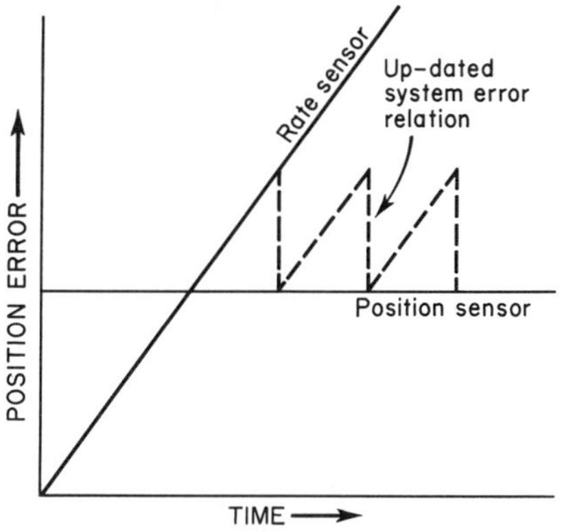

FIG. 5-4*d*. Accuracy of a Rate System Periodically Up-Dated by a Position System.

indicated previously in Section 5.2.4, early expenditure of the energy is fundamental to minimizing energy expenditure. The problem in guidance, under energy conservation restraints, is thus that of controlling the early impulse, mid-course, and terminal corrections, and the peculiar problems associated with orbit transfer and orbit plane charges.

5.3.1 Early Impulse Control

The fundamental property imparted to a space vehicle during guidance is a change in velocity. Rockets are rated on the velocity change, or impulse velocity, that they can achieve. Two control functions must be accomplished during the accompanying energy expenditure. The first is to orient the thrust vector, and the second is to control the burning or burn-out time so that the desired incremental or impulse velocity is achieved. If we look at the typical problem of placing a vehicle in a circular orbit we will have the situation illustrated in Fig. 5–5. In this figure we see an

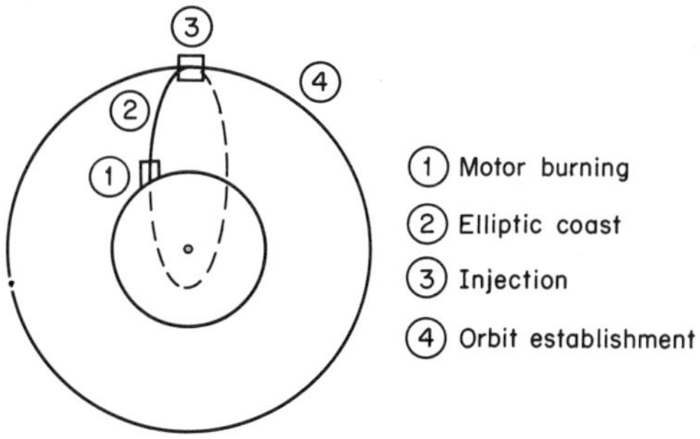

FIG. 5-5. Elliptic Ascent into Orbit.

early expenditure of energy during the motor burning period, an elliptical coasting trajectory to orbit altitude, and injection into orbit at apogee of the ellipse. Note that to establish the desired orbit, one must first reach the altitude and, for minimum energy expenditure, this indicates matching elliptic ascent apogee with the desired orbit altitude.

Precise control of the ascent ellipse must be established by proper aiming or thrust vector control, and by thrust impulse control. Vector control will generally be established by inertial references since the burning period is relatively short. Motor shut-off control is generally accomplished by integrating an accelerometer measurement to determine the impulse velocity. Ground tracking and radio command systems are commonly employed where the inertial instrument precision is inadequate. Also, ground tracking and radio command are commonly used to control the orbit injection altitude in conjunction with an internal inertial system for attitude control.

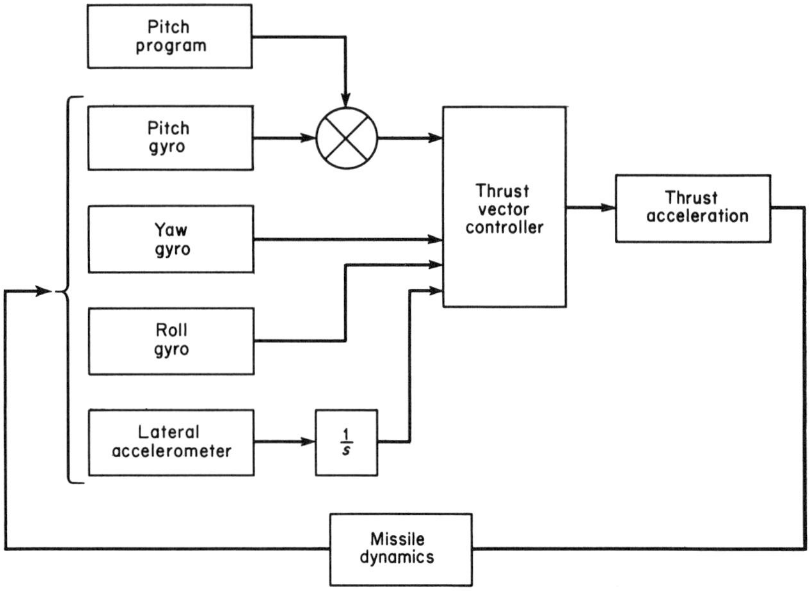

FIG. 5-6. Ballistic Missile—Attitude Control Systems.

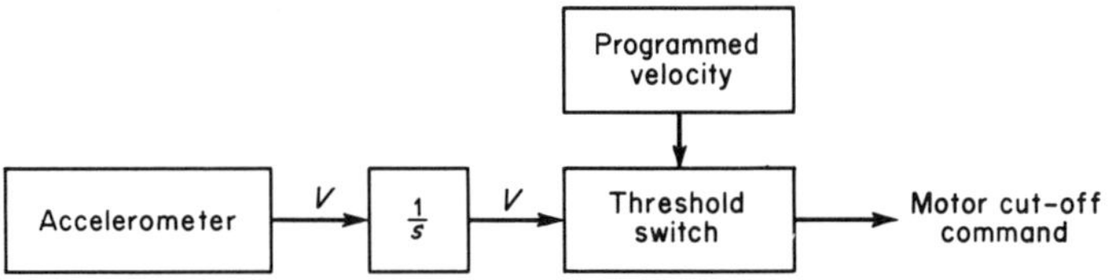

FIG. 5-7. Thrust Duration Control System.

Very sophisticated guidance schemes have been devised for controlling both thrust vector and burning time for ballistic missiles because it is desired to require no further guidance or control after motor burn-out. A typical attitude control system is shown in Fig. 5–6, and a thrust duration control computer in Fig. 5–7. In the attitudes control system, it will be noted that basically the yaw and roll attitude are held fixed, or at least near zero rate. Also, a lateral accelerometer is used to null out cross-course errors. Note that the acceleration measurement is integrated once to obtain a velocity measurement correction to control the thrust vector attitude. The basic attitude function determining impact position will be the pitch gyro measurement which is compared with a precomputed pitch attitude program to control the vehicle trajectory.

The thrust measuring accelerometer is axially mounted so as to measure the acceleration history of the vehicle along the flight path. The acceleration measurement is integrated once to obtain the impulse velocity imparted to the vehicle at any time. This velocity is compared with the programmed impulse velocity, and when the two velocities are equal the motor is cut-off. Naturally, various error sources exist in the instruments used, their calibrations, and the computations. In the event that these errors are excessive for the mission, and more accurate instruments and controls cannot be employed, some further guidance and control system may be used during the mid-course or terminal phase of the flight.

5.3.2 Mid-Course Control

The need for mid-course guidance arises basically from an inadequacy of the initial guidance or fire control system. Thus, one may think of a mid-course system as being of an "up-dating" nature. Although, mid-course techniques may be continuous in operation, thrust generation techniques and error data availability may dictate that mid-course corrections be impulsive by nature rather than continuous.

In general, we may expect mid-course corrections to be developed from some position-measuring system such as a remotely tracking radio-command system, or a stellar-sighting system. The addition of such data to acceleration data has been treated in Section 5.2.6. As indicated in Chapter 3, inertial systems do not measure effectively in free flight, and when used, must generally be up-dated from other sources unless their long-time error accumulation can be tolerated (e.g., corrected by the terminal guidance system).

The effect of an impulse correction on a trajectory is illustrated in Fig. 5–8. With each new impulse added vectorially to the vehicle's velocity and location on the prior ellipse, we start the vehicle on a new ellipse. One focal point of each ellipse will be the force field center. Thus, we may see that as significant errors are determined, corrections may be made (and should be made early) in order to achieve a minimum energy expenditure.

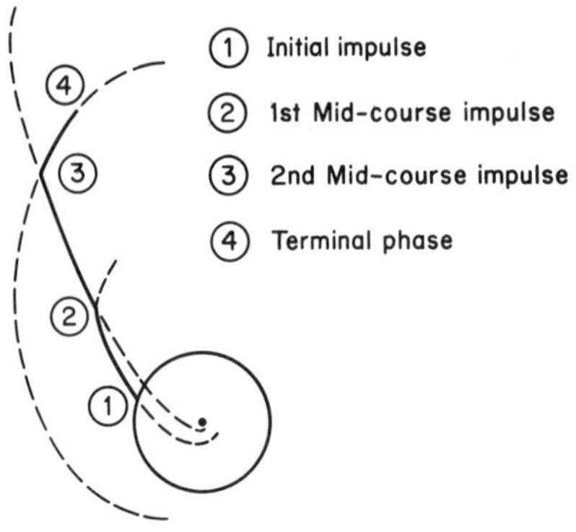

FIG. 5-8. Mid-Course Guidance Trajectory with Impulse Corrections.

If it were efficient from a thrust generation stand-point, a continuous correction system would be desirable for energy conservation because of the area-moment effect of early changes as discussed in Section 5.2.4. Off-plane errors may be corrected by normal impulse rockets, or by vector summation of impulse requirements and vector control of the motor impulse.

5.3.3 Terminal Corrections

The least efficient location for flight path corrections is the terminal location. Nevertheless, previous guidance will not always be sufficiently accurate to ensure complete and accurate achievement of the desired orbit or terminus. The primary purpose of terminal guidance will be to make up the deficiencies of previous guidance systems. The actual energy requirements will be relatively small if the previous systems have functioned reasonably well. The basis for corrections will generally be position and rate data generated with respect to the target or line-of-sight to the target. Homing and attitude mating techniques are particularly important in rendezvous.

Of particular interest is the direction of velocity impulse corrections in trying to establish a particular phase in orbit as well as altitude (or period). If we are trying to rendezvous with another vehicle, or achieve synchronism with other satellites as in a communication or meteorological network, we shall be attempting to match actual orbit position sequence (implying velocity plus a position or phase) with a target orbit. The situation is depicted in Fig. 5–9, which shows the target orbit, attained or vehicle orbit, altitude error, and phase error. As illustrated, the attained orbit lags by the phase error, ϕ, and is lower than the target orbit by the altitude error, Δh. To effect a correction of these errors would, at first glance, seem to require only an increase in the attained orbit speed, v_v. This will result in eliminating Δh provided the transfer impulses are attitude controlled to obtain the circularity indicated. However, the orbit period at the higher altitude will be longer, hence the phase error will not be eliminated by simply increasing the velocity.

If we make the correction between elliptical orbits, the problem will become even more severe because both apogee and perigee altitudes must be matched as well as their orbit phase angles.

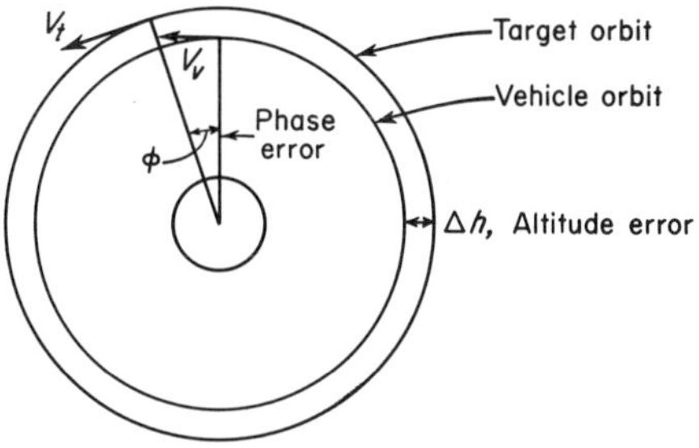

FIG. 5-9. Terminal Guidance Error Representation.

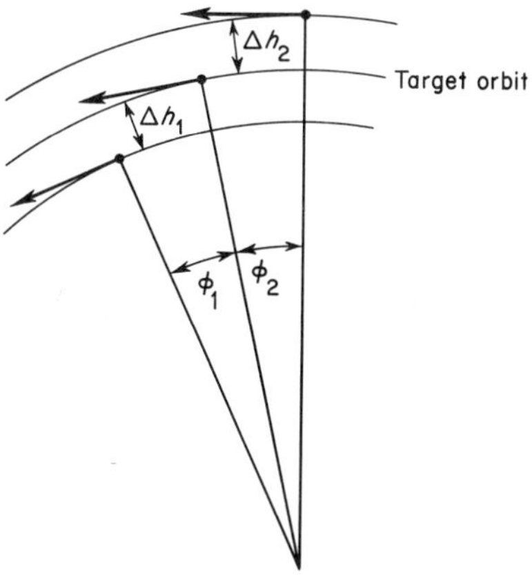

FIG. 5-10. Proper Phase Relation for Orbit Matching.

The procedure necessary in terminal correction to a target orbit sequence is to lead the target phase if below, and lag it if above. This is illustrated in Fig. 5–10. With this geometry, increasing the velocity of the lower altitude orbit position will reduce both the altitude and phase errors. Similarly, decreasing the velocity of the higher altitude orbit position will reduce both the altitude and phase errors. This is an anomalous relation from the normal "Earth-bound" feel for control systems. In space we accelerate, or "give it the gas" to catch a target behind us (in phase).

5.3.4 Orbit Transfer

The basic energy-conservative transfer between orbits is called the Hohman transfer, which is cotangential with both originating and target orbits. Of particular importance in this transfer is that the apogee and perigee (or aphelion and perihelion) of the transfer be 180 degrees separated in orbit phase relation. This requires the prediction of the apses in advance of perises (or vice versa). If the orbit periods are short, as in low altitude planet-centered orbits, this will merely require some means of determining the two orbits, computing their relations, and waiting for a suitable transfer initiation period to occur (assuming adequate instrumentation accuracy).

When the periods are large, the need for precise instrumentation and scheduling will become even more important in order to avoid missing an opportunity for such a transfer. This is illustrated by Fig. 5–11 which shows the distances and periods of Mercury, Venus, Mars, and Jupiter relative to Earth. From the relative orbit periods of the various planets we can determine the frequency of minimum-energy transfer opportunity to the planets. Note that the nearest planets, Mars and Venus, have longer relative orbit periods because their absolute orbit periods are closer to the Earth's absolute period. Note also that the orbit of Venus generally takes it farther away from the Earth than the (approximately) four month point of closest approach by Mercury. Thus, to get to Venus by MET (Minimum Energy Trajectory), we will have to wait better than 18 months if we miss an opportunity, and we may expect 4 more favorable possibilities for transfer to Mercury during the waiting period. Also, the MET requirement for transfer to Mars occurs only once in approximately 2 years. The MET to Jupiter and planets further out will occur with periods of slightly more than one per year.

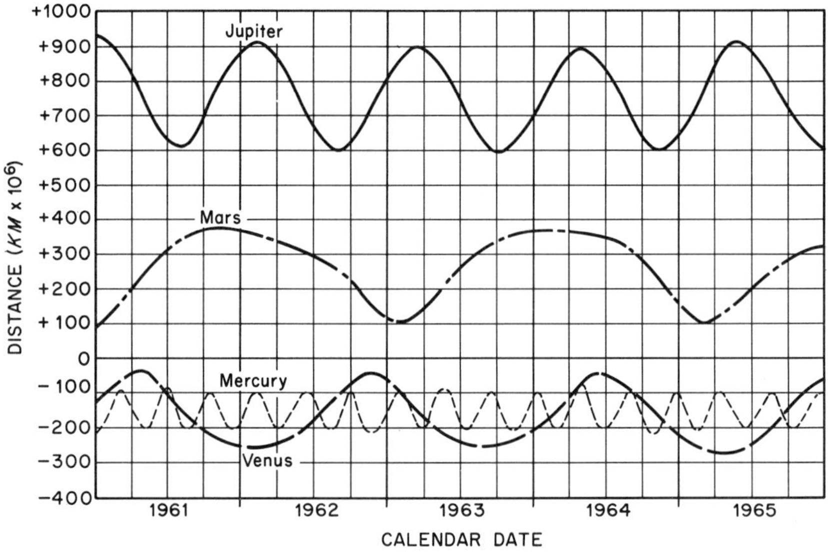

FIG. 5-11. Relative Distances and Periods of Planets.

The actual mission profiles for Earth-Mars and Earth-Venus transfers have been presented by Ruppe in Ref. (*2*). These profiles are illustrated in Fig. 5–12 and Fig. 5–13. As indicated in Fig. 5–12, the MET transfer to Mars requires that the Earth lag Mars in orbit phase by 44 degrees at launch. Two hundred and sixty Earth-days will be spent in transit, at which time the Earth will lead Mars by 76 degrees. The return MET trip must be a mirror reversal of this trip, so 449 days is the required waiting period to achieve the necessary orbit phase lag of 76 degrees. Thus the total round trip will take 970 days. As illustrated in Fig. 5–13, a Hohman transfer to Venus will take 146 days each way plus a 460 day waiting

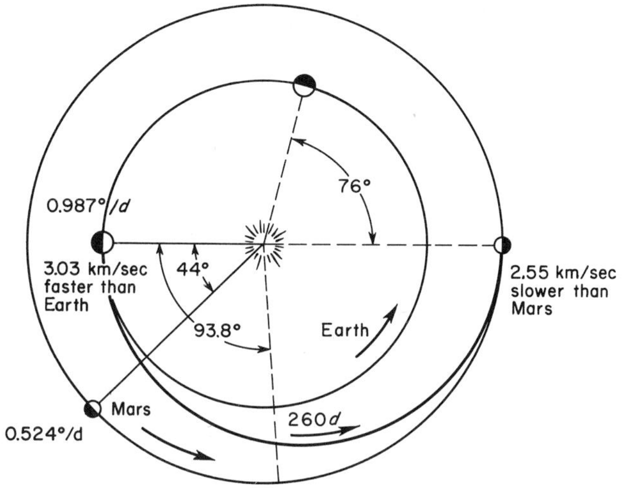

FIG. 5-12. Earth-Mars Mission Profile (Hohmann Ellipse).

Opposition occurs (44)/(0.937 – 0.524) = 95 days = 93.8° after launch. The waiting time is in this case (360 – 2.76)/(0.987 – 0.524) = 449 days. So a return mission lasts (260 + 449 + 260) = 970 days.

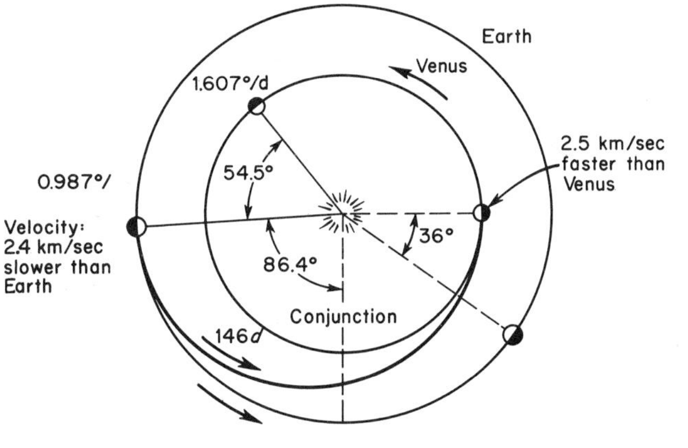

FIG. 5-13. Earth-Venus Mission Profile (Hohmann Ellipse).

Inferior conjunction is (54.5)/(1.607 – 0.987) = 87.6 days after launch; the angle is 87.6 – 0.987 = 86.4°. From this follows the rule: Launch 86.4° = 87.6 days before inferior conjunction.

For the return flight, we need to have Earth leading by 36°. (In above picture, we simply reverse all directions for the return flight.) So Venus has to catch up 360 – 2.36 = 288°, which takes (288)/(1.607 – 0.987) ≈ 460 days. So a return mission lasts 146 + 460 + 146 = 750 days.

period for a total time of 750 days. Figure 5–14 and Fig. 5–15 indicate the optimum injection dates for the Earth-Mars and Earth-Venus transfers. These dates are directly related to the opposition of Mars (closest approach to Earth) and the inferior conjunction of Venus (closest approach to Earth) as indicated in Fig. 5–12 and 5–13.

Of interest also is the possible Earth-Venus-Mars-Earth trip illustrated in Fig. 5–16. The indicated trip would take one by both planets and return to Earth with a total flight time of 547 days which is more than the round trip flight time to either planet singly, but considerably less than the combined round trip flight times. This round trip will require a peculiar constellation at injection, but should require less energy if accomplished without stopping and waiting at the planets.

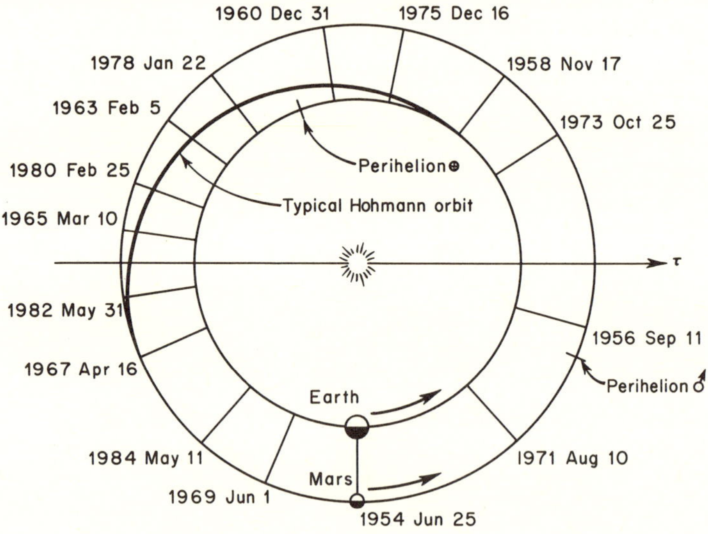

FIG. 5-14. Opposition of Mars and Hohmann Ellipse Injection Date.

Hohmann Ellipse Injection Dates:

23 Aug. 1958	23 Feb. 1969	7 Oct. 1977
25 Sep. 1960	21 May 1971	29 Oct. 1979
16 Oct. 1962	7 Aug. 1973	4 Dec. 1981
9 Nov. 1964	12 Sep. 1975	28 Jan. 1984
25 Dec. 1966		

NOTES:

Maximum error in injection dates: 1958–1971, 3 days; 1971–1984, 5 days. The opposition dates were obtained by extending the ephemeris of the Earth and Mars by a numerical integration method. This ephemeris was extended through 1984 by use of a graph showing synodic period of Mars versus longitude of opposition, prepared from the American Ephemeris and Nautical Almanac. The injection dates were obtained by use of a graph showing opposition time minus injection time versus longitude of opposition. This graph was prepared by choosing a multiple of conjunction times of the Hohmann ellipse and the orbit of Mars and from this, calculating the period and the theoretical opposition times and longitudes. These were then plotted to obtain the graph.

Another possible Earth-Mars-Venus-Earth trip is illustrated in Fig. 5–17. In this trip, the total trip is made on a single ellipse, the Mars-Venus Hohmann transfer. Injection is required to match this ellipse where it crosses the Earth's orbit. By this technique, impulses are not necessary at Mars and Venus, and the total flight may be accomplished in 12.5 to 13.5 months, which is an even shorter period (approximately 400 days) than the previously described circuit trip (547 days). This circuit trip does not allow for any waiting period, and requires a certain constellation at injection. June, 1971 is a possible injection time for this orbit.

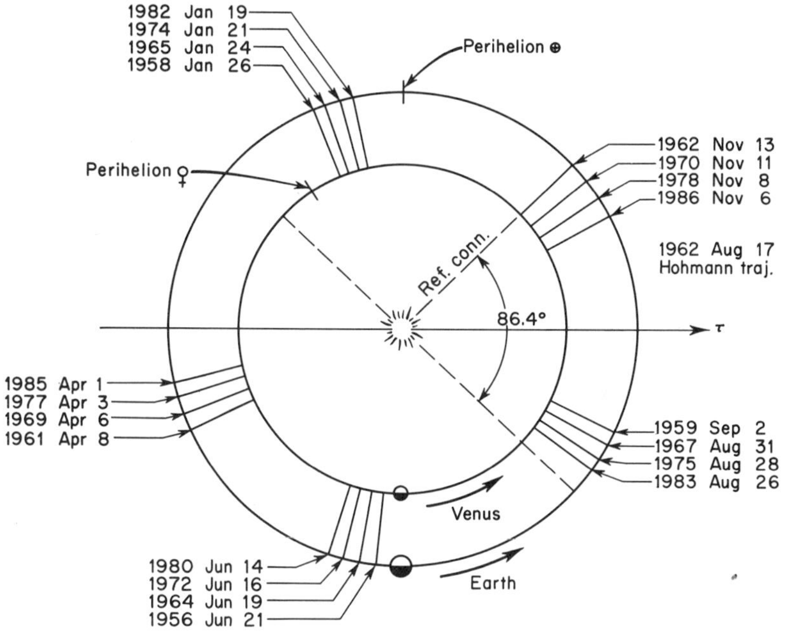

FIG. 5-15. Inferior Conjunction of Venus and Hohmann Ellipse Injection Date.

Hohmann Ellipse Injection Dates:

1956 Mar. 26	1967 Jun. 4	1978 Aug. 13
1957 Oct. 31	1969 Jan. 8	1980 Mar. 18
1959 Jun. 6	1970 Aug. 15	1981 Oct. 23
1961 Jan. 10	1972 Mar. 21	1983 May 30
1962 Aug. 17	1973 Oct. 26	1985 Jan. 3
1964 Mar. 23	1975 Jun. 2	1986 Aug. 10
1965 Oct. 58	1977 Jan. 6	

NOTES: (1) Estimated error of a conjunction or a Hohmann injection date is 4 days maximum.

(2) Orbits and circles; r(0) = 1.00 au, r(+) = 0.723 au.

(3) Dates and longitudes of inferior conjunctions are: t = JD 243 5062.3 + 583.931N (days); and λ = 575°.5180N; Reference date = 1954 Nov. 15.29 UT.

(4) Constant synodic period = 513.921 (days).

(5) Hohmann date and longitude; t − 87.6 days, λ − 86°.4.

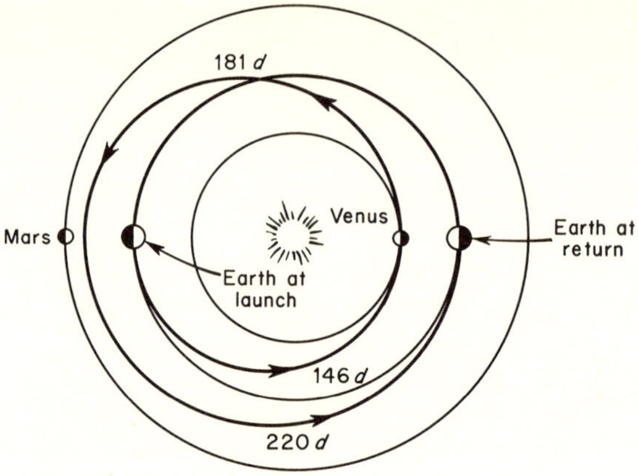

FIG. 5-16. Hohmann's Round Trip Trajectory.

Maneuvers:

1.	Brake original 29.8 km/sec of Earth by	2.4 km/sec
2.	Increase on passing Venus to 39.4 km/sec by	1.8 km/sec
3.	Increase on passing Mars to 24.8 km/sec by	2.3 km/sec
4.	Brake 31.5 km/sec arrival at Earth by	1.8 km/sec
	Total Velocity Requirement	17.6 km/sec
	Total Flight Time	547 days

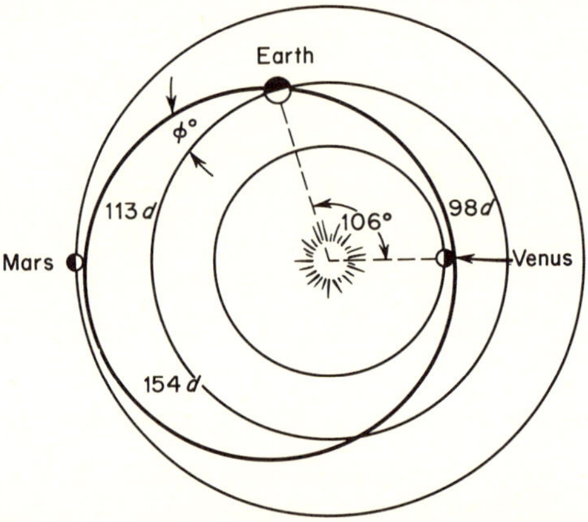

FIG. 5-17. Crocco's Round Trip Trajectory.

Total Velocity Requirement	15.968 km/sec
Total Flight Time	12.5 to 13.5 months
Possible Launch Time	June 1971
ϕ	16.04°

5.3.5 Orbit Plane Changes

The previous section on orbit transfer has considered the orbits to be co-planar. This is unlikely in any real situation, and some orbit plane changes will be necessary. The basic velocity increment required has been illustrated in Fig. 3–22, page 203, to be $2V \sin \phi$ where V is the velocity at the orbit-plane intersection, and ϕ is the half angle between the two planes. The critical guidance problem in effecting such a change successfully will be to predict the initial and target orbits with sufficient precision to calculate their intersection point and apply the necessary velocity increment accurately.

Some studies, such as Ref. (*3*) and (*4*), are being made relative to impulsive thrust transfers between non co-planar, non-congruent circular orbits. In some cases, it is indicated that transfers are feasible which require less energy than a Hohman-type with apogee at apses. Ling, Ref. (*4*), demonstrates that three impulses may or may not be more efficient than two impulses in orbit changes depending on the orbit geometry. Ling also concludes that the total impulse required can never be reduced by employing more than three impulses. The three impulse transfer is only more efficient than a two or four (or more) impulse transfer when the intermediate impulse is imparted at some radius from the force field center which is greater than the apses of both orbits. Ehricke, Ref. (*5*) discusses this problem in some detail as well.

5.4 TIME CONSERVATION METHODS

Fundamental to the use of time conservation methods of flight is the requirement that we have a surplus of energy over the minimum required for the flight, and that we wish to trade some of this surplus energy to save some time.

If we had infinite energy available, could accelerate (and decelerate) at an infinite rate, and travel at an infinite velocity, we could take a straight line path in zero time between any two points in Euclidean space. Because none of these requirements can be met, we must consider the prediction of terminus location, and the optimum energy use to effect a minimum transit or the earliest arrival at the destination (two distinctly different operation requirements).

5.4.1 Early Impulse Control

Figure 5–18 illustrates a solar probe orbit which is a current proposal for exploring the nature of the solar corona and solar flares. A particularly interesting feature of this orbit is that it is selected with a six-month period so that after two orbits the aphelion will coincide with the Earth's perihelion thereby ensuring recovery of data recorded while the probe is beyond communication range. This orbit is not a minimum time transfer, but does bring out a side issue which may be requisite to such transfers; that should the transfer be missed, it may be required that the transfer orbit period be synchronous with the launch orbit to effect recovery of the transfer vehicle occupants.

By comparing Fig. 5–18 with Fig. 5–13 we can see that by the initial expenditure of more energy to obtain a more eccentric ellipse we have shortened the transit time to the Venutian orbit from the Hohman 146 days to 54 days. The same tangential injection with more energy could be used to intercept Mars in less time also. Similarly, if we consider Fig. 5–18 as a generalized satellite orbit illustration, a non-tangential transfer between two orbiting vehicles represented by the Mercurial and Venutian

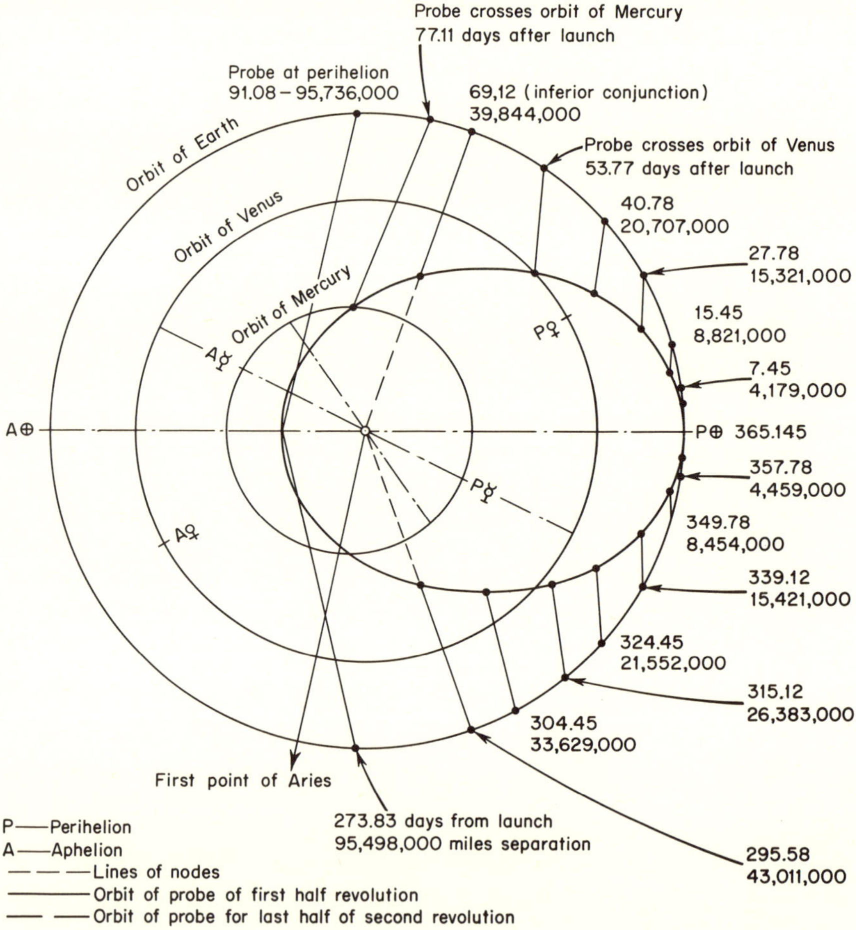

Fig. 5-18. Solar Probe Orbit. (Reprinted from MISSILES AND ROCKETS magazine.)

orbits is feasible. These transfer techniques are directly applicable to space vehicle-to-space vehicle missilery.

If a soft rendezvous is required, the transfer represented from the Venutian orbit to Earth in Fig. 5–18 would be desirable. This transfer orbit would make maximum use of the energy expended in the early phase of flight, and provide near synchronism of altitude at apses for the rendezvous system.

5.4.2 Continuously Powered Flight

The guidance of a vehicle in a continuously powered flight presents special guidance problems. The prediction of the trajectory becomes more complicated because it will no longer be a conic or the summation of a few impulse-generated free-flight conics. Very few initial attempts have been made to reduce continuously powered flights to analysis (e.g., Ref. (*5*) and (*6*)). The basic approach has been to define a set of differential equations which must be satisfied along the flight path. These equations may relate velocities, accelerations, positions, forces, mass, flow of propellant, cooling media flow, etc. The performance quantity to be minimized is expressed as a function of the initial and final variables of these quantities and time. Kelley (Ref. (*7*)) has studied the application of this procedure to the use of a "solar sail" (a sail that collects solar radiation pressure) to minimize the transit time to Mars for both orbit crossing and orbit matching intercepts. Using 10^{-4} g acceleration from the sail, he has developed an "optimum" sail angle program by successive approximations to the flight path integration.

The usual equations of motion employed are,

$$\text{Radial acceleration: } \dot{V}_R = \frac{\dot{V}_C^2}{R} - A_o\left(\frac{R_o}{R}\right)^2 + A_R(t) \qquad (5\text{–}14)$$

$$\text{Circumferential acceleration: } \dot{V}_C = \frac{-V_R V_C}{R} - A_C(t) \qquad (5\text{–}15)$$

$$\text{Radial velocity: } \dot{R} = V_R \qquad (5\text{–}16)$$

$$\text{Circumferential angular velocity: } \dot{\psi} = \frac{V_C}{R} \qquad (5\text{–}17)$$

The radial and circumferential quantities are always orthogonal. The radial and circumferential acceleration inputs to the system, A_R and A_C, are taken as functions of time. A_R and A_C may include functions of distance and altitude as well. In the solar sail case, Kelley programs the sail attitude, θ, as a function of time, and obtains radial accelerations of

$$A_R(t) = \alpha\left(\frac{R_o}{R}\right)^2 \left|\cos^3\theta\right| \qquad (5\text{–}18)$$

and circumferential accelerations of

$$A_C(t) = \alpha \left(\frac{R_o}{R} \right)^2 \sin \theta \cos^2 \theta \tag{5–19}$$

Other continuous functions of thrust, or thrust programs may be used depending on the energy supply available. The computation generally becomes iterative and sufficiently complex to require computer services to successively approximate the trajectories, and to minimize the control variable.

Another interesting special case has been studied by Faulders (Ref.(8)). In this study the thrust is taken as always in-line with the tangential velocity. The vector relation used is,

$$\dot{v} = a + \frac{dU}{ds} \tag{5–20}$$

where v is the tangential velocity, a is the thrust acceleration, and $\frac{dU}{ds}$ the gravitational gradient in the direction of the flight path, s. Propellant mass rate for the electric rocket is related to the acceleration by,

$$c\dot{m} = a(M_o - m) \tag{5–21}$$

the jet power is,

$$p = \left(\frac{c^2}{2} \right) \dot{m} \tag{5–22}$$

It is possible to eliminate c by combining these two equations, and propellant mass may be related to power by a new function,

$$\psi = \frac{2p}{M_o - m} = \frac{1}{M_o} + \int_0^t a^2 dt \tag{5–23}$$

From this,

$$\dot{\psi} = a^2 \tag{5–24}$$

The problem then becomes one of determining $a(t)$ such that the terminal values of the variables are satisfied. Using Euler-Lagrangian analytic techniques, Faulders shows that,

$$\ddot{a} = \left(\frac{d^2U}{ds^2} \right) a \tag{5–25}$$

and then studies three cases of solutions.

These cases are for

I. The boundary condition that velocity, v_T, is to be a maximum, or Δv a maximum at propellant expenditure in time T, m_T.

II. The terminal acceleration, $a_T = 0$ for propellant expenditure in time T, m_T.

III. The velocity, v_T, to be a maximum at a specified range, s_T, when the propellant is expelled in time T, m_T.

From this study, Faulders concludes that superior velocities and ranges can be obtained by optimum programming as opposed to a constant thrust program, and suggests some equations for first iterations in a computer study by numerical integration.

The state-of-the-art in this field is such that approximations by iterative procedures would seem to be with us for some time to come. Eventually, if and when thrust becomes more munificent, we may be able to go to some long-range lead-pursuit type homing procedure for inter-planetary transfer. Such navigation would be a coordinate transformation of the techniques covered in Ref. (*9*).

5.5 COLLISION WARNING AND AVOIDANCE

Warnings of imminent collision and avoiding such collisions in space seem academic subjects at this writing as they did in the very early days

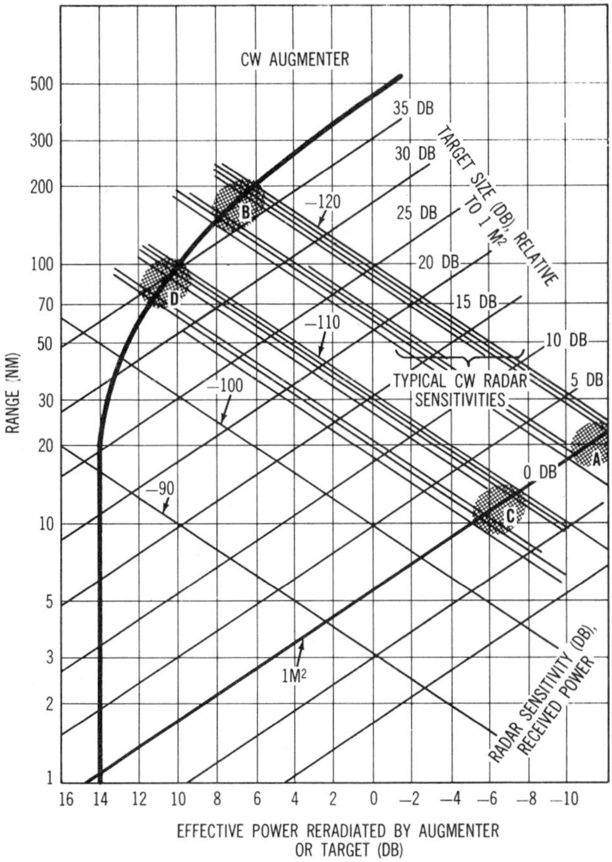

FIG. 5-19. CW Detection Ranges. (Courtesy, *Space/Aeronautics*, September, 1960.)

of aviation. Yet we have had many serious aircraft collisions, and even aircraft destruction from collision with birds.

In space, collisions are possible with both natural and man-made objects. The latter may be passive or actively radiating depending on whether they are hostile or friendly, and perhaps disabled. To determine collision imminence it is necessary to detect the object or its radiation, measure its relative position and velocities, and compute intercept time therefrom. To avoid such a collision will require some sort of maneuver, and a return to the original course if the mission is not to be aborted.

5.5.1 Object Detection

Objects may be detected by any of several media: ultra-violet, visual light, infra-red, microwave, or radio. The passive detection of the object will be a function of the amount of radiation emitting from the object and the range. *A priori* knowledge of the radiated spectrum and relative velocities may be used to narrow the receiver band-width and increase the detection range as discussed in Section 4.3. Figure 5–19 illustrates the relation between detection range and effective radiated power of a target. Also illustrated are the effects of target cross-section and of having a signal augmenter aboard the target. The volume *Airborne Radar* of this series, (Ref. (*10*)), covers in detail the problem of detecting and tracking targets in three-dimensional space.

5.5.2 Distance, Vector and Derivative Measurements

In order to predict imminence of collision it will be necessary to measure some or all of these vector quantities; line-of-sight range, range rate, azimuth and elevation angles and angular rates of the line-of-sight in radar or vehicle coordinates. Techniques for accomplishing these measurements have been discussed in Section 4.3.3.

If we assume that the velocity vectors of two colliding bodies are coplanar, we may write the relative motion equations in differential form as,

$$\ddot{r} - \dot{\sigma} r = \frac{F_r}{M_0} \qquad (5\text{–}26)$$

$$r\ddot{\sigma} + 2\dot{\sigma}\dot{r} = \frac{F_\sigma}{M_0} \qquad (5\text{–}27)$$

where r is the line-of-sight or range, $\dot{\sigma}$ is the sight-line (angular) rate, F_r is the force on the line-of-sight, F_σ is the force normal to the line-of-sight and M_0 is the vehicle mass. Note that F_r and F_σ include gravitational and coriolis forces. If powered flight is attempted, M_0 may be changed to $M_0 + \dot{m}t$ to allow for any significant mass expenditure. Nason, Ref. (*11*), has expanded on these equations to develop a terminal guidance concept which achieves collision by constraining the sight-line rate, $\dot{\sigma}$, to a constant. From this, we may infer that if the co-planar sight-line rate is a constant, collision will be imminent. If we have satisfied this requirement, we may compute an estimated collision time, T_c, as

$$T_c = \frac{r}{\dot{r}} \tag{5–28}$$

By differentiating this equation we obtain

$$\dot{T}_c = 1 - \frac{r\ddot{r}}{(\dot{r})^2} \tag{5–29}$$

If $\dot{T}_c$ is a constant we have the situation where T_c is a linear function of time,

$$T_c = kt \tag{5–30}$$

From the compution of $\dfrac{r\ddot{r}}{(\dot{r})^2}$ we may thus predict when the actual collision may be expected. If $\dfrac{r\ddot{r}}{(\dot{r})^2}$ is not constant, then we may be approaching either a softer or harder impact depending on whether $\dot{T}_c$ is decreasing or increasing. If $\dot{T}_c$ approaches zero, we will achieve a soft rendezvous.

5.5.3 Clearance and Maneuvers

Two general types of maneuvers appear feasible; to maneuver out of the collision plane, or to cross the other orbit at a point out-of-coincidence with the other object. Both of these techniques require in general that the sight-line rate be other than constant. Thus, to avoid collision with another vehicle or a large meteor, we need merely to apply some velocity change which will give us an appreciable sight-line rate, $\dot{\sigma}$. The computation of collision time, T_c, and its first derivative, $\dot{T}_c$, will provide a measure of the maneuvering period available.

5.5.4 Return to Course

The return to course may take several forms: a simple return to the original orbit, a return to the original position sequence, or a new orbit based on some optimization criteria such as fuel expenditure. If we are operating an orbiting observatory, and have maneuvered to avoid a meteor, we may thus re-establish our orbit plane, altitude, and velocity to return to course. If we are attempting to maintain a synchronous orbit with other satellites, as with a communication relay system, then we must achieve a rendezvous with the original path sequence. If we are concerned with transport to another planet or space station, we may wish to chart a new orbit based on some criteria such as fuel expenditure or transit time optimization. The treatment of these problems has been discussed under Sections 5.3 and 5.4.

5.6 RENDEZVOUS

The rendezvous of two vehicles in space may occur for many reasons, and hence may take many forms. A "hard" rendezvous or intercept of a vehicle by an interceptor missile may be required. An inspection and interrogation missile may require orbit synchronism and maneuvers about a potentially dangerous vehicle at various separation distances. A logistics or rescue mission may require actual orientation mating, closure, and soft impact without disturbing the orbit of one vehicle. All of these missions may be achieved automatically, and the inspection and logistics mission may be accomplished under some manual control as well, depending on the hazards of the situation.

Ascent trajectories are of concern in that they determine the search volume and orientation required of the rendezvous acquisition system. The purpose of the acquisition system is to place some measuring device on the target so that appropriate sensor, orientation and orbit matching controls may operate. Docking requires orientation matching of the two vehicles, and impact softening is necessary so that the orbits of the vehicles are not disturbed appreciably.

5.6.1 Ascent Trajectory

Several ascent trajectories have been considered for orbital rendezvous achievement. If we consider that the launch point may or may not be in the target orbit plane, we may postulate the methods illustrated in Fig. 5–20: (1) a launch timing system that waits for the target orbit plane to approach the launch point (breaks down for orbit planes which do not reach the launch latitude); (2) a direct ascent to target altitude on an intersecting plane with injection and orbit plane change accomplished simultaneously; or (3) establishment of a parking orbit at an altitude lower than the target, changing orbit plane, and transfer to target altitude in sequence. The first of these methods has some obvious energy savings, but requires launch time synchronism for both orbit plane and target position in the

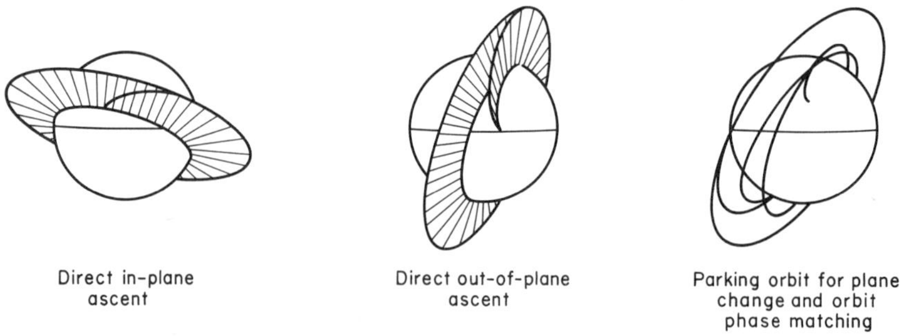

FIG. 5-20. Orbital Rendezvous Ascents.

orbit. Such synchronism may require waiting periods of many months for fulfillment. The second method has obvious time-saving advantages, and may have energy advantages in some applications. A critical period for simultaneous orbit plane change and injection into target orbit exists which requires precise launch fire control prediction to be sure both events can be accomplished coincidentally. Note that an interceptor weapon would most likely use this direct approach, but would merely detonate its warhead at the orbit intersection point. The third technique permits a greater flexibility of operation and solution of the intercept problem; it is expected that this technique will be employed generally for logistics and inspection missions. The requirement for orbit phase and perigee matching is most easily solved by placing the ascending vehicle in a low altitude, high phase velocity (short period) orbit.

5.6.2 Acquisition System

Typical rendezvous systems will require some sort of acquisition system to account for the inaccuracies of the primary flight control system, and to place the terminal guidance measuring devices in operation. As illustrated in Fig. 5–21, several error sources contribute to the mis-match of orbits between the ascending and target vehicles. The need for an acquisition system then becomes quite apparent.

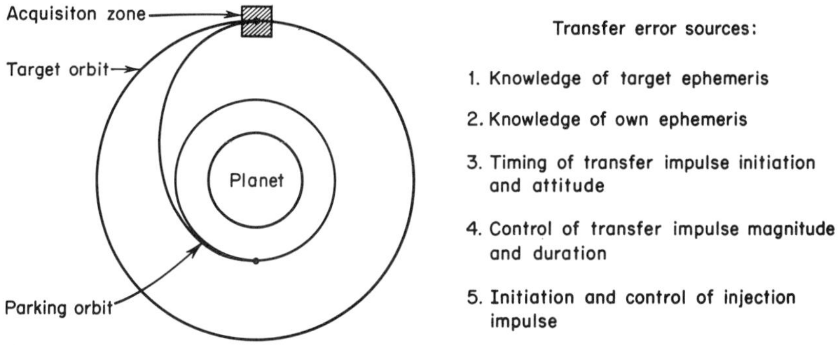

FIG. 5-21. Rendezvous Acquisition.

Detection devices may employ any wavelength in the electromagnetic spectrum as discussed in Section 5.5.1. However, the detection probability may be enhanced considerably if we take advantage of our prior knowledge of the target ephemeris, and plan our flight accordingly. Thus instead of the full spherical detection volume desired for collision warning (as for meteor or weapon detection), the acquisition volume may be reduced, and the detection probability should increase as a function of the reciprocal of the percentage of spherical coverage necessary.

A typical acquisition volume might be as illustrated in Fig. 5-22. The transfer ellipse is taken as a reference in the orbit plane, and the acquisition volume may be defined as $\pm\alpha$ degrees normal to the transfer plane, and $+\gamma$ and $-\beta$ degrees in the transfer plane. Some typical current values for the Gemini and Apollo programs are $\alpha = 10-25$ degrees, $\beta = 10-25$ degrees, and $\gamma = 25-60$ degrees. The selection of these values will depend on estimates of the accuracy with which the transfer can be controlled and predicted for some particular time and mission usage. For inspection of potentially hostile and maneuverable targets, the acquisition volume may have to be a full sphere.

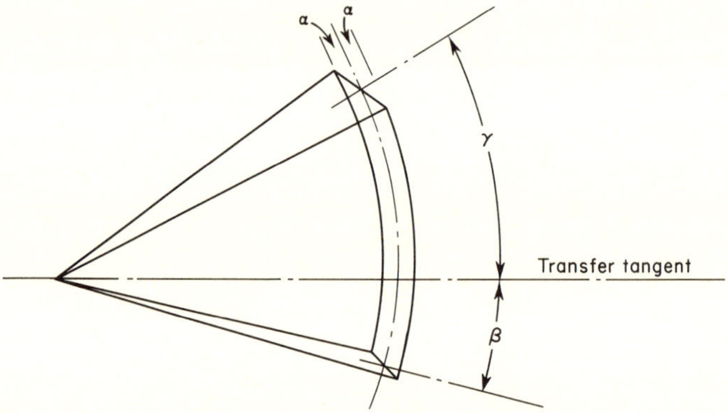

FIG. 5-22. Typical Acquisition Volume.

5.6.3 Distance, Vector and Derivative Measurement

The techniques for measurement of the range vector and its derivatives during rendezvous will be quite similar to those required for object warning and evasion in the earlier phases (discussed in Section 5.5.2). The docking phase will be somewhat more analgous to space refueling engagements under astronaut control. The lack of aerodynamic forces should make the maneuver somewhat easier.

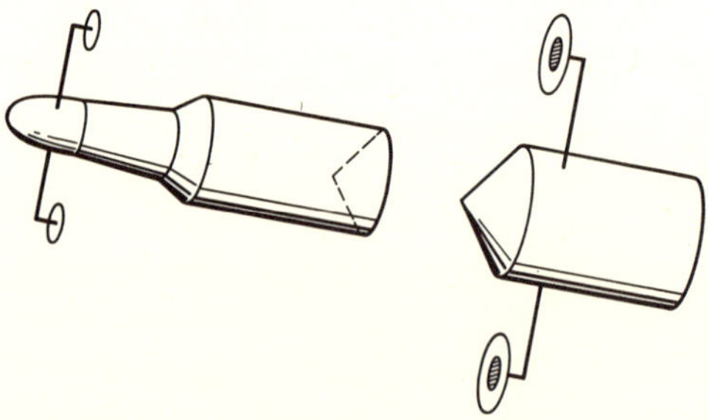

FIG. 5-23. Typical Docking.

A typical docking is illustrated in Fig. 5–23. In this situation the astronaut is mating a space vehicle to a final rocket stage or a tanker vehicle as proposed for the Apollo program. In such a mating, the astronaut is required to judge the docking situation from his command vehicle station through periscope devices. The target should carry identifying lights (e.g., two colors) to assist in establishing the correct mating orientation. The lights and periscopes have been erected beyond the vehicle surface to minimize body shadowing in establishing and maintaining contact.

The requirement placed on the astronaut is to "back" his vehicle into the other one gently until some interlocking coupler can engage (e.g., a funnel-shaped equivalent of railroad car couplers).

The requirements placed on the astronaut in rendezvous, and his ability to maneuver are the subject of many current investigations. The maintenance of a constant line-of-sight vector during the approach has been investigated, Ref. (*12*), including the determination of range and range rate from timed, free-flight optical measurements. Nason, Ref. (*13*), has studied the use of a restartable, constant thrust motor to achieve both intercept and soft rendezvous. Many other studies exist (e.g., Ref. (*14*)) which are pertinent. However, a closed-form optimum analytic solution is not generally available, except perhaps for some extremely constrained problems, and one must generally start with some set of initial conditions, instrument, and control accuracies, and perform numerical integration assuming some guidance system. The procedure is reiterated to bring one within an acceptable tolerance.

5.6.4 Docking

Docking, or the terminal phase of soft rendezvous, has been illustrated in Fig. 5–23. The general approach to rendezvous at docking has not been solved in closed form. However, Ref. (*11*) through (*14*) have indicated the merits of a zero sight-line rate approach in the terminal phase. Studies, Ref. (*15*) and (*16*) have been made of automatic rendezvous using a variable thrust motor. These studies have indicated the need for zero sight-line rotation. Reference (*16*) has indicated the requirement for achieving zero range and range rate simultaneously at docking. If we apply control and braking thrust in line-of-sight coordinates, or through a vector resolution system in terms of these coordinates we should thus arrive at a relatively direct computation and control system. That is, by controlling sight-line rate and sight-line acceleration directly we should have a useful docking system, provided our orientation is controlled to mate the coupling devices. Some measuring techniques for achieving these functions have been described in Section 4.3.3. Orientation may also be accomplished by horizon and magnetic field sensors.

5.6.5 Impact Softening

The final impact softening may be expected from some mechanical device associated with the coupler system. Some deformable structure or inflated cushion may be employed. The control of range and range rate to approach zero simultaneously appears to be extremely desirable in the final approach to docking or "stand-off" inspection.

We may satisfy the requirement that range, r, and range rate, $\dot{r}$, equal zero simultaneously by making $\ddot{r}$ equal to a constant, say A. This provides a square-law asymptotic closure of range which should be quite useful in impact softening. The fundamental relation will be of the form,

$$\ddot{r} = A \qquad (5\text{–}31)$$

$$\dot{r} = At + \dot{r}_0 \qquad (5\text{–}32)$$

$$r = {}^1/_2 At^2 + \dot{r}_0 t + r_0 \qquad (5\text{–}33)$$

r_0 and $\dot{r}_0$ may be measured at acquistion or some arbitrary range (a pre-determined or controlled r_0) between acquisition and impact. We may eliminate time from these equations and solve for A at the initial conditions to obtain,

$$\mathrm{A} = \frac{(\dot{r}_0)^2}{2r_0} \qquad (5\text{–}34)$$

This control law may be used to soften impact by modulating a variable thrust engine in a feedback system control. Constant thrust engines would have to be turned on and off based on instrumentation which measures range and range rate continuously, and some permissible percentage deviation from A in a switching logic circuit.

It is interesting to note that the use of a constant value for $\ddot{r}$ provides, from Eq. (5–29), a constant closure rate, $\dot{T}_c$. If more cushioning effect is desired, we might assume a constant $\dddot{r}$ (or higher order) and use a cube-law (or higher) asymptotic approach to impact. This will cause a time variation in estimated closure time rate computations. Considering the demonstrated feasibility of in-flight refueling of aircraft under pilot control and the probability that docking in space will be very similar, it may turn out that automatic guidance in the final phases will be unnecessary except for un-manned operations.

5.7 SATELLITE ORBIT MAINTENANCE

Of prime interest in space vehicle guidance is the need for satellite orbit maintenance. The accuracy with which a satellite orbit must be maintained will vary depending on the missions. The two extremes would be free-flight environmental exploration and geodetic surveying reference missions.

Orbit parameters may be measured or computed from measurements taken from the earth's surface and/or internally. The choice of system will depend on what techniques are available in the surface range complex, and whether their capabilities are suitable and can be made available. Internal measurements and on-board computers may be expected to reduce reliance on surface instrumentation.

Another requirement in orbit maintenance is orbit synchronism. This requirement may exist to assure continuous global coverage (or a limited portion thereof) by a network of satellites for communication, surveillance, meteorological, navigational, and geodetic purposes. Navigational and geodetic references may require precise synchronism for specific measurement reference purposes.

5.7.1 Surface Measurements and Command Corrections

The measurement of a satellite's ephemeris from surface instrumentation may be accomplished in many ways. Fundamental to the measurement is the requirement that the surface stations which sight the satellite must be accurately located on the geoid. Measurements taken at some "known" geodetic coordinate set must be transferred to a geocentric coordinate system to yield the satellite ephemeris properties. Note that the longitude in both coordinate systems are identical. Purcell and Cowan, Ref. (*17*), have developed an approximate relation which appears to be good to better than 10^{-9} radian in latitude and one foot in radius. Using the nomenclature defined by Fig. 5–24, the conversion from geodetic altitude, h, and latitude, λ, to geocentric radius, R, and latitude, μ, requires the following equations,

$$e^2 = 1 - \frac{b^2}{a^2} \tag{5–35}$$

$$\delta = \frac{e^2 \sin \lambda \cos \lambda}{1 - e^2 \sin^2 \lambda} \tag{5–36}$$

$$\mu_e = \lambda - \delta \tag{5–37}$$

$$R_e = b\left(1 + \frac{e^2}{2}\cos^2 \mu_e + \frac{3}{8} e^4 \cos^4 \mu_e\right) \tag{5–38}$$

$$\xi = \frac{R_e}{R_e + h} \tag{5–39}$$

$$R = R_e + h\left(1 - \frac{\delta \xi}{2}\right) \tag{5–40}$$

$$\mu = \lambda - \xi \tag{5–41}$$

Such equations are mechanized for real-time translation of satellite tracking data to geocentric coordinates. From a set of orbit points the satellite ephemeris may be determined. Command corrections to a satellite ephemeris require the comparison of the flight path with a desired flight path to yield errors and the command of the satellite propulsion system to reorient itself and emit corrective impulses of the appropriate magnitudes and/or durations.

The extent of the existing world-wide (American) instrumentation network is illustrated by Fig. 5–25 and 5–26. This network of stations provides orbit-determining measurements from several sources, parabolic antennas, telemetering antennas, the Minitrack system, and the Mercury

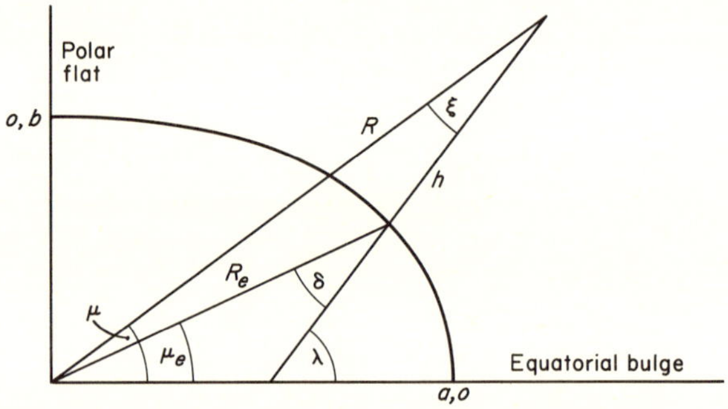

FIG. 5-24. Geometry of Latitude Conversion to Geometric Coordinates.

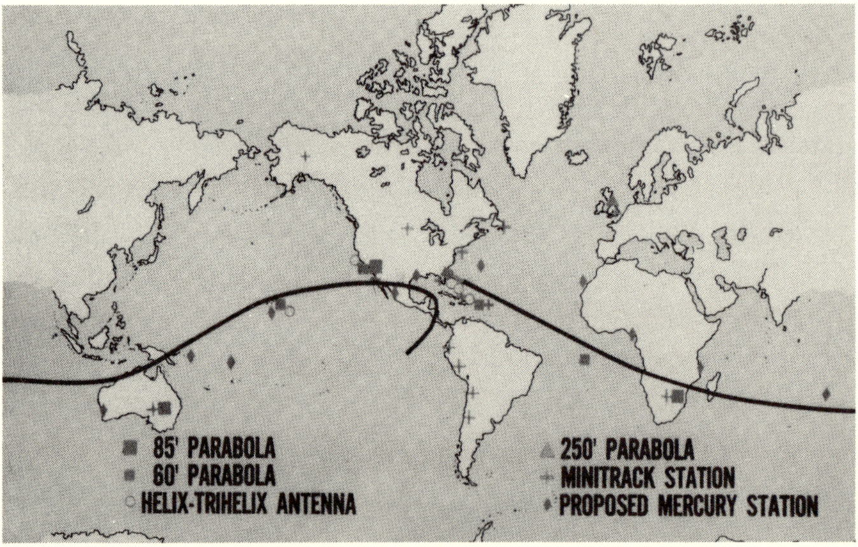

FIG. 5-25. World-Wide Ground Receiving Networks.

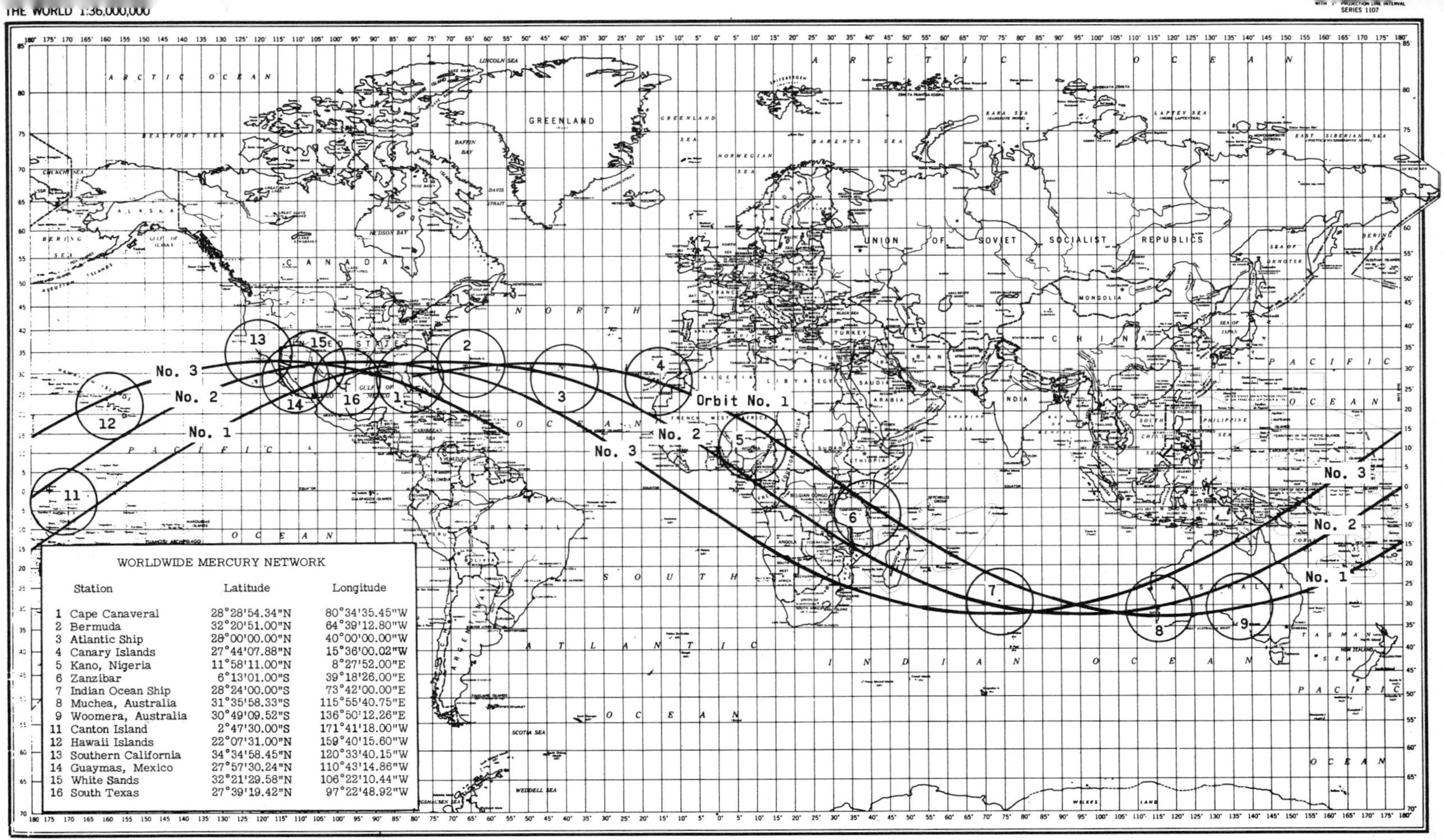

Station		Latitude	Longitude
1	Cape Canaveral	28°28'54.34"N	80°34'35.45"W
2	Bermuda	32°20'51.00"N	64°39'12.80"W
3	Atlantic Ship	28°00'00.00"N	40°00'00.00"W
4	Canary Islands	27°44'07.88"N	15°36'00.02"W
5	Kano, Nigeria	11°58'11.00"N	8°27'52.00"E
6	Zanzibar	6°13'01.00"S	39°18'26.00"E
7	Indian Ocean Ship	28°24'00.00"S	73°42'00.00"E
8	Muchea, Australia	31°35'58.33"S	115°55'40.75"E
9	Woomera, Australia	30°49'09.52"S	136°50'12.26"E
11	Canton Island	2°47'30.00"S	171°41'18.00"W
12	Hawaii Islands	22°07'31.00"N	159°40'15.60"W
13	Southern California	34°34'58.45"N	120°33'40.15"W
14	Guaymas, Mexico	27°57'30.24"N	110°43'14.86"W
15	White Sands	32°21'29.58"N	106°22'10.44"W
16	South Texas	27°39'19.42"N	97°22'48.92"W

FIG. 5-26. Mercury 3-Orbit Track Showing Location of Network Stations.

Network. The Mercury Network is detailed further in Fig. 5–27 as a variety of stations which include various combinations of radar, telemetry, communication and command links with the (Mercury) satellite.

The determination of satellite orbits and trajectories from a single set of data or a single radar is, of course, feasible. More accurate determinations may be made from several measurements and/or computations, and the average, or most consistent agreement, of the measurement used with greater confidence.

To provide an example of a tracking radar system, we may turn to Renzetti, Ref. (*18*). The 85-foot diameter Goldstone, California antenna is illustrated in Fig. 5–28. A block diagram of a typical space vehicle tracking system is shown in Fig. 5–29. This diagram shows the tracking loop used to point the antenna at the space vehicles. The target vector angles are taken from the antenna servo system. Range and doppler (range rate) data are available directly. In addition, the system is used to encode commands to the space vehicle, and to decode telemetered data. Figures 5–30 and 5–31 illustrate some typical range capabilities for such a

STA ABB	STA NO.	NAME	COVERAGE PASSES	RADAR		TELEMETRY RECEPTION	COMMUNICATION (CAPSULE)	COMMAND CONTROL	ACQUISITION			GROUND COMM			TIMING
				S	C				FA	SA	M	VOICE	TTY	SSB RADIO	
CNV	1A	CANAVERAL	1, 2, & 3	(X)	X	X	X	X	X	(X)		X	X		X
GBI	1B	GRAND BAHAMA	1, 2, & 3		(X)	X	X				X	X	X		AMR
GTI	1C	GRAND TURK	1, 2, & 3			X	X				X	X	X		AMR
BDA	2	BERMUDA	1, 2, & 3	X	X	X	X	X	X			X	X		X
ATS	3	ATLANTIC SHIP	1, 2, & 3			X	X		X				X	X	X
CYI	4	GRAND CANARY ISLAND	1 & 2	X		X	X		X				X	X	X
KNO	5	KANO, NIGERIA	1 & 2			X	X			X			X	X	X
ZZB	6	ZANZIBAR	1 & 2			X	X			X			X	X	X
IOS	7	INDIAN OCEAN SHIP	1, 2, & 3			X	X		X				X	X	X
MUC	8	MUCHEA, AUSTRALIA	1, 2, & 3	X		X	X	X	X			X	X	X	X
WOM	9	WOOMERA, AUSTRALIA	1 & 2		X	X	X		X			X	X		X
CTN	11	CANTON ISLAND	1 & 2			X	X			X			X		X
HAW	12	KAUAI ISLAND, HAWAII	2 & 3	X	X	X	X	X	X			X	X		X
CAL	13	PT. ARGUELLO, CALIFORNIA	2 & 3	X	X	X	X	X	X			X	X		X
GYM	14	GUAYMAS, MEXICO	1, 2, & 3	X		X	X	X	X			X	X		X
WHS	15	WHITE SANDS, NEW MEXICO	1, 2, & 3		X				X			X	X		X
TEX	16	CORPUS CHRISTI, TEXAS	1, 2, & 3	X		X	X		X			X	X		X
EGL	17	EGLIN, FLORIDA	1, 2, & 3	*	X				X			X	X		X
		GODDARD SFC		GROUND COMMUNICATIONS											X

SITE FUNCTIONS: FA = FULLY AUTOMATIC SA = SEMI AUTOMATIC M = MANUAL SSB = SINGLE SIDE BAND

*MPQ-31

FIG. 5-27. Mercury Network Site Capabilities.

station when used with the Ranger lunar probe vehicle. The various communication ranges are functions of antennas, transmitters and receivers used in the individual links.

Of particular interest is the horizon visibility of a surface mounted radar. This is illustrated in Fig. 5–32 as a function of orbit altitude for the three Deep Space Instrumentation Facility (DSIF) radars.

The means for obtaining tracking data from surface radar and optical measurements have been discussed in Section 4.3.3. Barrar and Deutsch have studied the determination of satellite trajectories with track-while-scan radars, Ref. (*19*). As indicated, optical devices provide excellent

FIG. 5-28. The 85-Foot Antenna at Goldstone

angle measurements, whereas radar measurements generally provide good range and range rate measurements but poorer angle measurements. The position vector of a satellite can be determined three ways:

1. By a radar that measures range, azimuth angle, and elevation angle.
2. By a radar that measures angles and ranges separately (such as an interferometer).
3. By a system that measures three ranges from three ground stations.

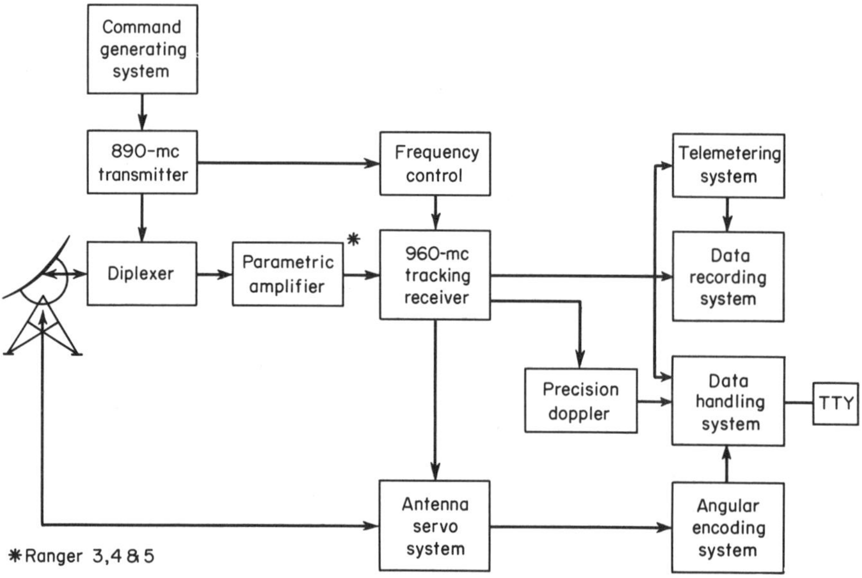

FIG. 5-29. Typical DSIF Station Block Diagram

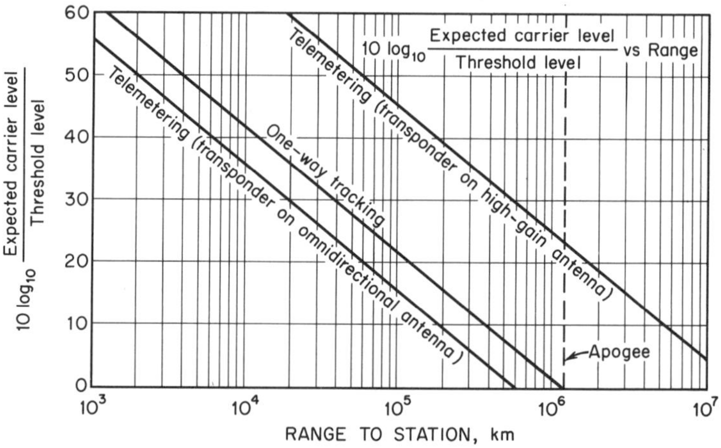

FIG. 5-30. Typical Spacecraft-to-Earth Communication Capability of Ranger 1 and 2

If a set of initial coordinates is accurately known, an orbit may be continuously computed from range and range rate data taken by a single station. To compute orbit positions from doppler frequency shift data

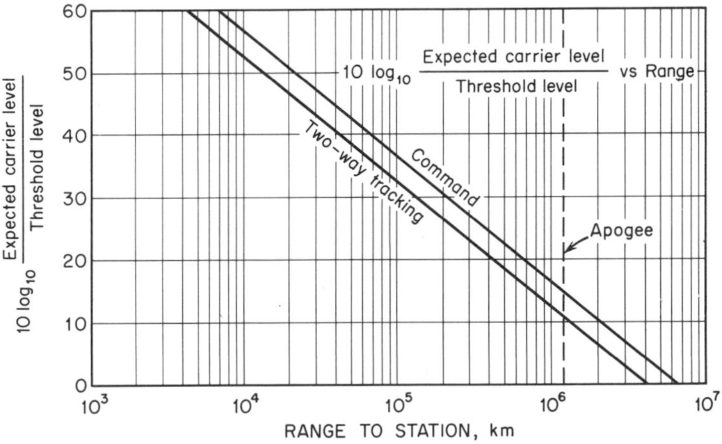

FIG. 5-31. Typical Earth-to-Spacecraft Communication Capability of Ranger 1 and 2

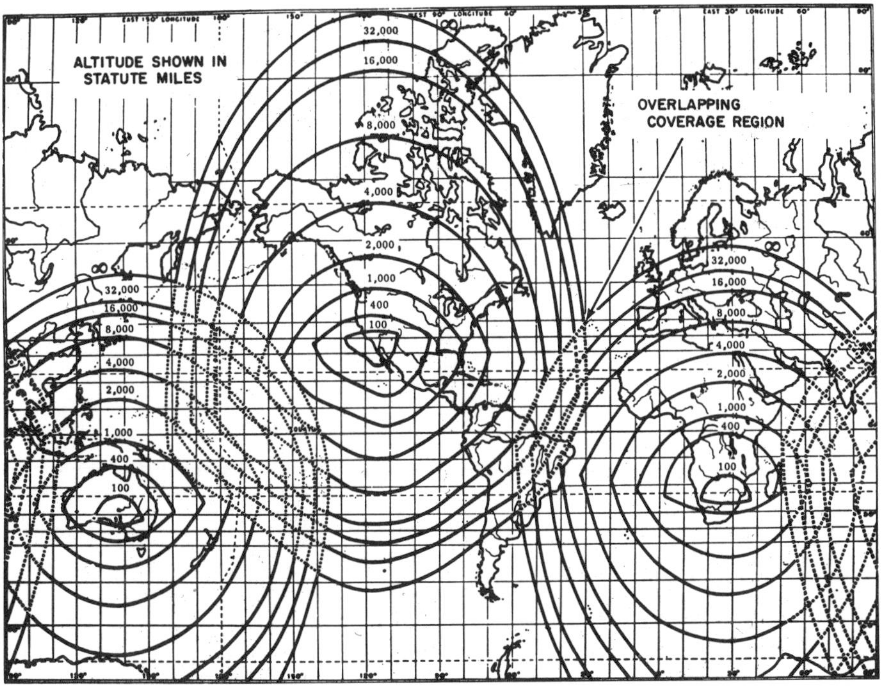

FIG. 5-32. DSIF Loci and Sub-Vehicle Points for Horizon Visibility.

alone requires several sets of stations. The relation that obtains when the ground-based transmitter and receiver are at separate locations is,

$$\dot{r}_T + \dot{r}_R = \left(\frac{c}{f_T}\right)(f_T - f_R) \tag{5-42}$$

where $\dot{r}_T$ and $\dot{r}_R$ are the range rates of the satellite from the transmitting and receiving stations, and f_T and f_R are the transmitted and received frequencies. By allowing for ground station movement, and using iterative procedures, a satellite orbit may be established from a series of such measurements (Ref. (*20*)).

The Minitrack system has been shown in Fig. 5–25 (page 292) to have several stations in the global instrumentation network. The Minitrack system utilizes a 108 Mcs. transmitter in the satellite, and a set of ground receivers oriented with fan-shaped beams on an orthogonal coordinate system. These receivers are operated in sets to make interferometric measurements of the satellite vector as it crosses each of the fan shaped beam systems. The technique requires very precise antenna beams and world-wide timing references to yield precision orbit data directly. Further accuracy is obtained by iterative orbit calculations and matching.

5.7.2 Internal Measurements and Corrections

Satellite orbits may be measured and maintained solely by internal measurements if desired. For orbiting other planets and their moons such instrumentation is quite likely. A vehicle returning from such a mission may employ the same equipment to establish an Earth-orbit prior to re-entry. An orbit may be established in terms of altitude information by direct radar altimeter measurements to the accuracy of the radar, and consistency or smoothness of the planet surface. If an initial altitude is known, doppler measurements on the nadir path may be integrated to establish the orbit and using a clock as a reference, the integrated doppler accuracy may be improved upon by knowing the altitude history, which should be repetitive every orbit period.

Knowledge of the gravitational parameter of the planet permits one to calculate ground velocity directly from altimeter measurements. Typical calculations indicate that a one-mile altimeter error is equivalent to a 3 to 4 feet-per-second ground velocity measurement error when orbiting the Earth or the Moon in circular orbits on the order of one-hundred to 600 miles altitude.

The satellite period may be determined directly by sighting a "fixed" star, and taking each eclipse of the star as one period. Self-contained attitude references while in orbit include those fixed in inertial space, celestial sightings, and horizon references.

Orbit corrections should be applied by vector resolution of the required changes in attitude and velocity or altitude. Inertial measurement units may be used to damp corrective impulses, but are insensitive to orbit parameters and can *not* measure ephemerides.

5.7.3 Orbit Synchronization

The synchronization of satellites into a specifically related orbit system requires some specification of the type of synchronism required. That is, are the orbits to be co-planar or not, co-altitude or not, of common eccentricity, perigee phase, etc? Based on the requirement for orbit synchronism, some measurement of the individual orbits is required, and the data on these orbits must be translated to some common coordinate system, usually geocentric. Once the data on the various orbits in the system are available for comparison with each other and with some Earth-reference point (e.g. Greenwich meridian and equator), a decision must be made as to which orbit(s) to correct, how much, and in what manner. Corrections to satellite orbits will, in general, require two impulses or more pairs if additional vernier corrections are required. That is, a satellite must be brought to the proper altitude, and then injected to the proper orbit speed with the correct orientation. If a satellite is at the proper orbit, but out of the required geocentric phase relation with the other satellites in the system, two impulses are required, one to advance or retard the phase relation, and the second to return to proper orbit velocity for the altitude.

Communication, meteorological, and surveillance systems will generally require only a particular orbit relation between the satellites. A geodetic or navigational reference system may require some specific relation be maintained with geodetic coordinates as well to avoid the error inherent in making the additional coordinate-correcting computations.

5.8 LIBRATION POINT STATION MAINTENANCE

In Section 2.3.2 we discussed the peculiar nature of the two equilateral libration points in the rotating Earth-Moon system. The peculiar advantages of maintaining a space station at these points for communication beyond the visible rim of the moon indicate that some considerations should be given to maintaining such a station.

The problem of arriving at the libration point involves operation of a vehicle in the bi-central cislunar force field. It appears desirable to select a trajectory to the libration point which will intersect the point with a remaining velocity component which will be useful in establishing the the required orbit around the complex system barycenter. The proper approaches to these points are illustrated in Fig. 5–33. A normal transfer somewhat similar to a Hohman ellipse has been depicted for P_1. Such a

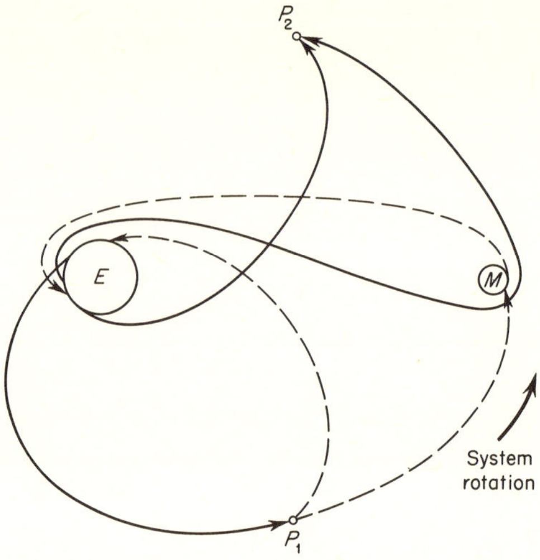

FIG. 5-33. Proper Approaches to Libration Points.

transfer may take advantage of the Earth's rotational velocity, and will be influenced only slightly by the lunar gravitational field. A similar transfer to P_2 might be accomplished as indicated by passing between the Earth and the Moon. Such a transfer will be more difficult because the lunar gravitational attraction will have more effect. As an alternate, another transfer is depicted which leaves the Earth (or Earth-staging orbit) in retrograde direction, approaches close to the Moon, and reverses its angular moment argument under the lunar influence. This latter approach will be extremely difficult.

The return from the libration point stations to an Earth-base may be the mirror-image reversal of the ascent trajectories. Returns are illustrated as dashed lines for P_1 only. An alternate return (and ascent as well) may be to stop at or approach close to the Moon without attempting the momentum argument reversal.

5.8.1 Libration Determination

As indicated by the Trojan asteroids which occupy the libration points of the Solar-Jovian system, we may expect the Earth-lunar equilateral points to be stable. It will be necessary, however, upon reaching these points, to determine that we are indeed there in order to inject our vehicle into the required orbit about the system barycenter.

The peculiar thing about these points is that they are equilateral with the Earth and Moon. This indicates that one may determine he has "arrived" when he can measure ranges to both the Earth and the Moon that are equal to Earth-Moon distance. Thus, one might attempt to approach a 240,000-mile orbit about the Earth, and maneuver in this orbit in the Earth-lunar plane until a 60-degree angle is measureable between the Earth and Moon positions. Alternate methods of determining the position would include some direct range measurement, perhaps by radar or stadiametric techniques, noting that the Moon should present the same disc diameter to the libration station as to the Earth.

5.8.2 Corrections

Although expected to be inherently stable, the libration points may present problems in oscillatory motion, especially on initial establishment. The oscillations evidenced by the Trojan asteroids indicate the basic stability of the points, but also that some troublesome oscillating effects may be encountered.

To counteract such oscillatory tendencies it may be desirable to increase the angular momentum of the stations. This may be accomplished by extending the station momentum centroid away from its center of gravity. It will quite likely also be necessary to damp out the initial oscillations by some "brute force" method such as sighting "fixed" stars, and inparting correcting impulses to keep these stars "fixed" in station coordinates.

5.9 LUNAR DESCENT

The general problems of guidance during lunar descent are those of the descent transfer, landing site selection, steering to it, soft impact control and roving vehicle control. Return from the Moon to Earth may be expected to be a direct ascent from the lunar surface, and either (1) direct transfer to Earth or Earth-orbit, or (2) rendezvous with a lunar orbiter for subsequent transfer to the Earth.

A typical lunar descent is illustrated in Fig. 5–34. In this figure, it may be seen that the direct approach with thrust reversal and braking is not postulated. Rather, it is expected, in early lunar explorations, to enter the lunar central force field with a hyperbolic trajectory, establish high-altitude orbit about the moon, and enter a transfer ellipse when opposite the landing site. When pericynthion of the descent transfer is reached, rocket braking is to be employed, and the vehicle is to hover under rocket lift, and maneuver to the desired landing site. The lack of atmosphere indicates that control rockets will also be used to match the hover with the surface rotational movement. Basic tools for accomplishing this mission appear to be altimeter, rate of descent (or climb) measurement, ground velocity measurement, and perhaps a horizon sensor. V/H (velocity to altitude ratio) sensors and inertial platforms might be employed as well.

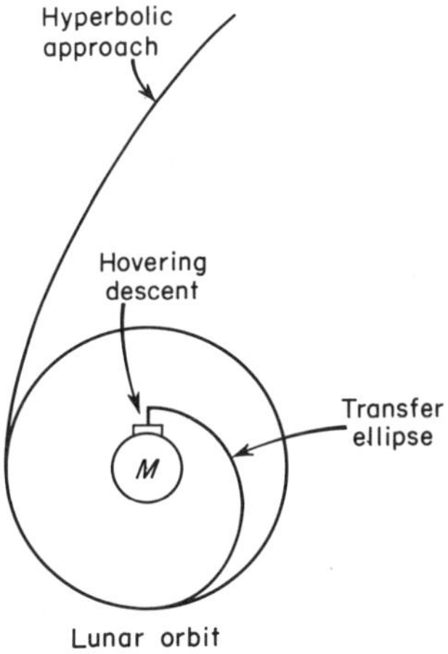

FIG. 5-34. Typical Lunar Descent

5.9.1 Altimeters

The basic way to establish an orbit ephemeris about a body is with an altimeter. This device can be used to measure distance from the surface and, knowing the body's eccentricity, an orbit may be computed. Surface irregularities may be faired out in this computation. If fully automatic operation is required, a horizon sensor might be employed to establish the local vertical or nadir for altimeter direction.

The descent transfer ellipse may be measured directly by an altimeter as well. Because the Moon rotates so slowly, the descent transfer might be biased off one or two orbits away from the desired landing site. This would permit time to establish that sufficient full reserve is available for return from the lunar surface prior to braking and descent from the actual pericynthion. Thus, the pericynthion altitude could be measured on at least one orbit and if below a certain altitude, the hovering descent could be initiated on the orbit of closest pericynthion to the desired landing site.

Altitude control during hovering descent is a necessity whether in piloted or automatic control. Three altimeters with fine directed beams may be used with offset attitudes to establish the landing orientation relative to the surface. Figure 5–35 illustrates such a situation. Doppler measurements, with such altimeters, may be made to provide much additionally significant capability.

5.9.2 Doppler Measurements

The combination of a doppler measurement with the direct altimeter illustrated in Fig. 5–35 will provide a direct reading rate of descent (or climb) indication for the astronaut. The desirability of such a device is without question since altimeters and rate of climb meters are among the most fundamental instruments used in all aircraft control. A doppler

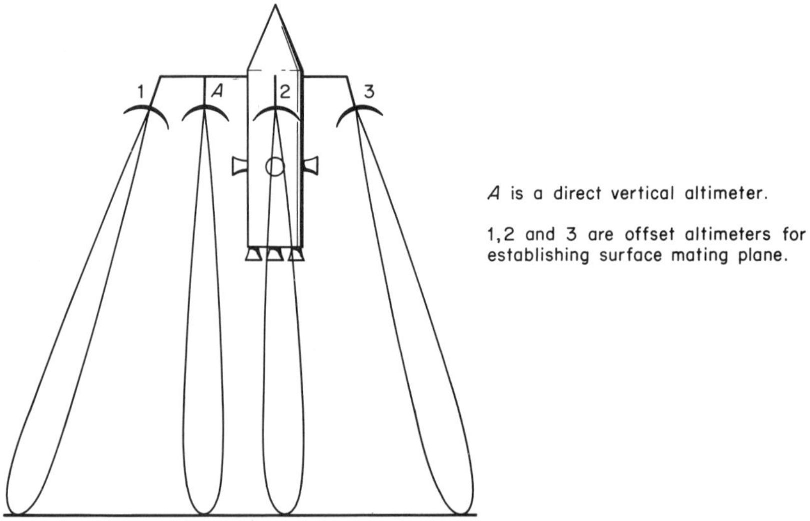

FIG. 5-35. Lunar Altimeters During Touch-Down.

measurement of altitude rate may be combined with the altitude to provide a time to touch-down indication, and to soften impact just as the rendezvous docking impact may be softened (Section 5.6.5). Such measurements and control computations will almost surely be needed for automatic lunar operations.

The offset altimeters may combine doppler measurements to good advantage as well. By suitable design, these doppler measurements can be used to provide a ground velocity and drift angle computation and indication just as aircraft-borne doppler navigators do (Section 4.3.3). Thus a single set of radar altimeters with doppler measurement capabilities may be employed to provide altitude, rate of descent, ground velocity, drift angle, and surface orientation; quantities sufficient to completely control a lunar landing automatically. In addition, the same equipment may be employed for orbit establishment and for automatic rendezvous operations.

Alternates to this scheme would appear to require some of this equipment (at least the altimeter) in addition to some other sensor. Reference (*21*) indicates the use of a television system to establish V/H, and from this to achieve attitude control of the satellite. Horizon sensors may be used at high altitudes. Some lunar Schuler-tuned inertial platform erected to a horizon sensor reference might be used as well, provided the tuning could be varied with altitude during descent so that the computer ephemeris data may be used to program the platform orientation during free flight.

5.9.3 Roving Vehicle Control

The guidance of a surface vehicle on the Moon from an Earth-bound station portends a peculiar set of problems. The situation is illustrated in Fig. 5–36. The unusual elements to this basic control system are, (1) the unknown character of the Lunar terrain, (2) the use of a variety of unusual sensors to effect control program changes, (3) the very long time lag in the 2-way transmission system. These elements are so novel as to portend some considerable development effort.

5.10 PLANETARY DESCENT

Control of space vehicle descent to some planets differs from the lunar descent problem primarily in the fact that an atmosphere will be encountered; thus operations may be performed with aerodynamic vehicles within this atmosphere. Descents to Mercury, Pluto and perhaps Uranus and Neptune will be like lunar descent because the atmosphere is insignificant. Descent to Venus, Earth, Mars, Jupiter, and Saturn will differ primarily in the atmosphere constituent properties, the pressure-altitude profile, the gravitational parameter, and the thermal effects on atmospheric entry. Entry or re-entry of the Earth's atmosphere is the most widely studied of these problems.

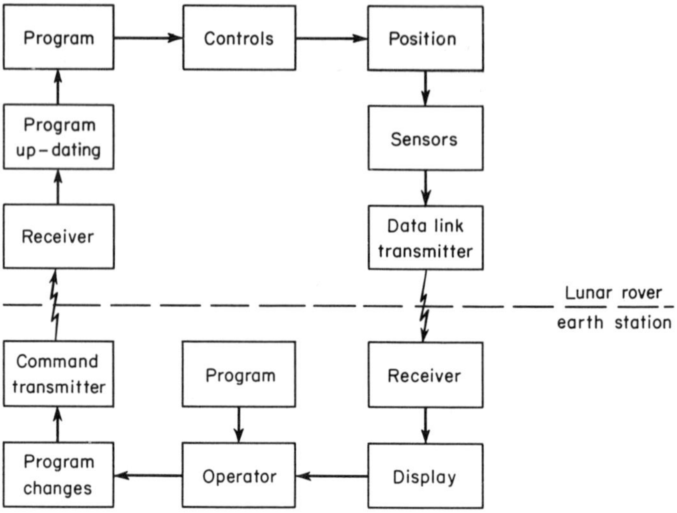

FIG. 5-36. Roving Vehicle Control.

5.10.1 Altimeters

Although initial re-entry vehicles are generally controlled by the world-wide tracking network, it is to be expected that self-contained instrumentation must be used to establish the entry trajectory into the various planetary atmospheres. As is the case for lunar descent, the altimeter may be expected to be a primary instrument. A microwave altimeter may be used to establish a planetary orbit, and to control a descent transfer to the upper fringes of the particular atmosphere involved. The same altimeter may be employed on down to the actual landing. A group of three altimeters may be used to generate ground-plane orientation for touch-down just as on the Moon. Three-component doppler data may be used to provide first-order damping for landing (zeroing of rate of descent, forward and transverse velocities). Thus, a common system may be employed to land on the various planets and the Moon.

Barometric altimeters may be employed as an aid to aerodynamic flight control in a planetary atmosphere. Both ram and static pressure devices may be used to provide data on the efficacy of the aerodynamic lifting and control surfaces. These data may also be used to direct the controls to keep the vehicle from excessive aerodynamic heating, although direct thermal measurements would seem to be a more likely control sensor for this purpose.

5.10.2 Doppler Measurements

As indicated in Section 5.10.1, "air"-speed indication may be obtained by ram and static pressure measurements in any planetary atmosphere.

The air-speed measurements will require careful calibration, and the uncertainty of local meteorological winds indicates that some direct measure of ground velocity will be required, particularly on final let-down.

The doppler navigation and rate of descent system described in Section 5.9.2 for lunar landings should be directly useful in the same manner for planetary descent. Thus, a 4-beam microwave system may be designed to provide ground speed, drift angle, rate of descent (or climb), altitude and surface orientation, and the same system may be used in both lunar and planetary landings and take-off. Such a system will be useful on Earth-entry as well as for other planets. However, on Earth-entry, this system will quite likely be employed as a primary reference system only until the local traffic control system can take over and assign a suitable flight corridor.

5.10.3 Site Selection

The use of an aerodynamic vehicle in a (foreign) planetary atmosphere should provide the opportunity for some reconnaissance and landing site selection. For a manned vehicle, the site selection will obviously be accomplished by the man. For an automatic landing, the site selection may be accomplished by establishing some pre-determined landing site criteria such as coordinates in four dimensions (i.e. time is taken as the

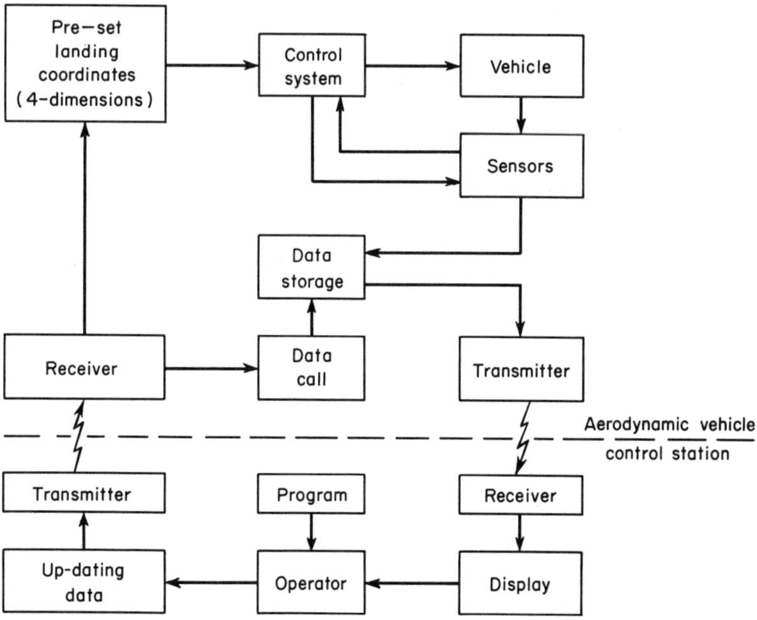

FIG. 5-37. Planetary Landing Site Selection System.

fourth dimension), landing site orientation with respect to the local vertical, landing site smoothness and landing site hardness.

As for the Moon, much of this information may be gathered on Mars, and to a limited (by weather) extent on the Earth by television systems that operate at visual wave-lengths. For Venus, Jupiter, Saturn, etc., the use of microwaves is indicated. Operation of such a system by remote control from an Earth or lunar base would appear to be extremely difficult because of the communication time lags, the narrow bandwidth necessary to maintain the contact range, and shadowing or eclipsing effects of the various bodies.

A more likely system is illustrated in Fig. 5–37 which is almost identical to the lunar roving vehicle control system. In this system, the controls are used to position the vehicle and the sensors. Vehicle movement has its effect on the sensors as well. The sensor data are stored for call down by the command link to ensure that the planet eclipsing effects to not cause data to be lost. The control station may be aboard a manned vehicle in orbit about the planet being explored, or on the Earth or lunar surface. Such a system should make it feasible to explore a potentially hostile (to man) environment prior to actual manned-vehicle descent.

Surface corrugation may be explored to some extent by APR (Airborne Profile Recorder) radar systems. Actual surface hardness and fine corrugation may be done with roving land vehicles just as indicated for lunar operation. These vehicles may be used to establish the local vertical orientation on a surface plane with simple pendelum devices. Such a remotely controlled surface explorer appears quite likely as a prelude to actual manned landings.

5.10.4 Atmospheric Flight Control

The aerodynamic characteristics of each planetary atmosphere must be established for each airfoil to be employed. Such things as variable geometry vehicles (Ref. (*22*)) have been proposed. Control of the L/D (Lift to Drag) ratio for the aerodynamic vehicle may become a primary control problem. Atmospheric entry is bounded by thermal heating on one extreme, and by insufficient atmosphere to brake the vehicle on the other. Precise measurement of both these effects may be attempted to define an entry corridor for a particular vehicle.

In order to land at a particular site on a target planet, it will be requisite to know the amount of flight maneuverability and range available subsequent to atmospheric entry. Based on a particular airframe and control surface design, envelopes of the maneuverable range may be developed for use in calculating the optimum descent transfer imitation.

The descent of non-aerodynamic shapes or capsules into foreign atmospheres will likely be similar to the current descent of Mercury capsules. Ablative heat shields and parachute systems will be used. The function of the guidance system will be to predict to some (rather poor) accuracy the dispersion pattern of the parachute system based on some atmosphere entry trajectory. Parachute deployment may be based on a barometric or radar altimeter, and some logic circuits may be used to narrow the dispersion prediction based on the altitude of parachute deployment.

5.10.5 Surveillance and Mapping Mission Guidance

The comprehensive exploration of foreign planets will require some programmed guidance of surveillance and mapping missions. Sensors will operate at visual and/or microwave frequencies depending on atmospheric characteristics. Permanent records of the terrain features will be required together with some coordinate system identification for the record. These data will be conveyed or transmitted by radio to some control center which is in charge of the operation. Ultimately, mapping data must be pieced together in some map format.

The establishment of a local planetary coordinate system will be of some concern. The poles and equator are desirable references. The prime longitude will be somewhat arbitrary, but will likely be taken with respect to some prominent surface feature in each case. It will be desirable to establish surface radio beacons as local reference monuments. Surveying these beacons will likely be accomplished by some counterpart of Earth-satellite geophysical references. Such references may be placed in synchronous equatorial orbits with the same period as the planet surface thus providing a planetary coordinate reference system.

Surveying and mapping from satellites will be coarse information by Earth standards because of the atmospheric opaqueness and refraction. The planet curvature will be a further limitation requiring some sophisticated data corrections, and perhaps detailed measurements to determine the constants to be used in the data correction formulas.

Surveying and mapping over a restricted area by an aerodynamic vehicle in a manner similar to Earth-aircraft surveying and mapping is quite feasible. The surveying of radio beacons may be accomplished by "line-crossing" techniques which measure the minimum distance between the two beacons. Knowledge of this distance plus the altitude of the measurement permits one to describe an arc on the planet "geoid" that defines a new beacon's location with respect to an old. By striking several arcs from "known" locations, a new reference point may be surveyed in. Satellite reference beacons should provide a good augmentation to this system, because a single range from a known satellite position will describe an ellipse (or at least a closed figure) on the surface. Range or range difference (satellite loran) measurements on two or more satellites will permit the surveying of any number of surface points.

Standard mapping flights with a powered or gliding vehicle may be made with some reference to surface and/or satellite references. These flights will be in the nature of several parallel legs with a common length. Camera side coverage, altitude, and side-lap requirements (usually 10 per cent) will dictate the leg separation. The navigational accuracy required in mapping missions on the Earth is one of the most difficult to obtain, and may be expected to be so for other planets as well. Generally, the aircraft navigation system will try to make the best combination of available state-of-the-art positional, velocity and inertial references.

References to Chapter 5

1. George Swanlund, "Analysis Techniques to Determine Guidance Computation Requirements for Space Vehicles," *ARS Journal*, Vol. 32, No. 5, May 1962, pp. 755–761.
2. Harry O. Ruppe, "Minimum Energy Requirements for Space Travel," Army Ballistic Missile Agency, Redstone Arsenal, Report No. DSP-TR-4-59, ASTIA No. AD230925.
3. L. Rider, "Characteristic Velocity Requirements for Impulsive Thrust Transfers Between Non Co-Planar Circular Orbits," *ARS Journal*, Vol. 31, No. 3, March 1961, pp. 345–351.
4. L. Ting, "Optimum Orbital Transfer by Several Impulses," *Astronautica Acta*, Vol. VI, Fasc. 5, pp. 256–265.
5. K. A. Ehricke, *Space Flight*, Vol. II, D. Van Nostrand Co., Inc., Princeton, N. J., 1962.
6. A. S. Boksenbom, "A Method of Graphical Trajectory Analysis," Proceedings of National Specialists Meeting on Guidance of Aerospace Vehicles, IAS, Boston, Mass., May 25–27, 1960, pp. 209–221.
7. H. J. Kelley, "Gradient Theory of Optimal Flight Paths," *ARS Journal*, Vol. 30, No. 10, October 1960, pp. 947–954.
8. C. R. Faulders, "Optimum Thrust Programming of Electrically Powered Rocket Vehicles in a Gravitational Field," *ARS Journal*, Vol. 30, No. 10, October 1960, pp. 954–960.
9. A. S. Locke, *Guidance*, D. Van Nostrand Co., Inc., Princeton, N. J., Chapter 12.
10. D. J. Povejsil, R. S. Raven, and P. Waterman, *Airborne Radar*, D. Van Nostrand Co., Inc., Princeton, N. J., 1961.
11. M. L. Nason, "A Terminal Guidance Law Which Achieves Collision Based on Coriolis Balance Techniques," Pre-print No. 61–40 of the 7th Annual Meeting of the American Astronautical Society, Dallas, Texas, Jan. 1961.
12. E. C. Lineberry, R. F. Brissenden and M. C. Kurbjun, "Analytic and Preliminary Simulation Study of a Pilot's Ability to Control the Terminal Phase of a Rendezvous with Simple Optical Devices and a Timer," NASA Technical Note D-965, October 1961.

13. M. L. Nason, "Terminal Guidance Technique for Satellite Interception Utilizing a Constant Thrust Rocket Motor," *ARS Journal*, Vol. 30, No. 9, Sept. 1960, pp. 902–904.
14. J. M. Eggelston and H. D. Beck, "A Study of the Positions and Velocities of a Space Station and a Ferry Vehicle During Rendezvous and Return," NASA Technical Report R-87 of 1961.
15. T. M. Carney, "An Automatic Guidance System for Rendezvous with a Satellite," NASA Technical Note D-923 of August 1961.
16. E. C. Lineberry and E. C. Foudriat, "Study of an Automatic System for Control of the Terminal Phase of Satellite Rendezvous," NASA Technical Report R-128 of 1962.
17. E. W. Purcell and W. B. Cowan, "Relating Geodetic Latitude and Altitude to Geocentric Latitude and Radius Vector," *ARS Journal*, Vol. 31, No. 7, July 1961, pp. 932–934.
18. N. A. Renzetti, "DSIF in the Ranger Project," *Astronautics*, September 1961, pg. 34.
19. R. B. Barrar and R. Deutsch, "Determination of Satellite Trajectories from Track-While-Scan Radar Measurements," IRE Transactions on Military Electronics, Vol. 1, MIL-5, No. 4, October 1961, pp. 306–311.
20. P. B. Richards, "Preliminary Orbit Determination of a Non-transmitting Satellite Using the DOPLOC Tracking System," *ARS Journal*, Vol. 31, No. 2, December 1961, pp. 1729–1733.
21. P. A. Button, P. E. Mallory and S. B. Boor, "V/H Satellite Altitude Control," Proceedings of the National Specialists Meeting on Guidance of Aerospace Vehicles, IAS, Boston, Mass., May 1960, pp. 114–119.
22. R. E. Ricles, "Return from the Moon: Variable Geometry for Landing after Re-entry," *Space/Aeronautics*, January 1962, pp. 60–63.

CHAPTER 6

VEHICLE CONTROL

6.1 INTRODUCTION

The control of space vehicles is an electronics problem only in the system aspect. That is, electronic system engineers should have some fundamental understanding of the control principles on which non-electronic devices operate in order to integrate them into the overall guidance and control system. Therefore, this chapter has been designed to include some fundamental treatment on control device operating principles and the orientation control of such devices.

Fundamental to vehicle guidance and control is the broad field of feedback control systems. Accordingly, up-to-date mathematical theory of feedback control systems has been included in the chapter. Although feedback control theory is used extensively in the vehicle control design it is emphasized that this theory is directly applicable as well to guidance communication and circuit design, and is included in this chapter primarily as a matter of convenience.

6.2 UNPOWERED FLIGHT

While a vehicle is in unpowered flight in free space, there is no available atmosphere against which aerodynamic control surfaces may react. With the possible exception of solar sail operations we may therefore say that an unpowered vehicle has no particular method of re-orienting itself or changing its course. The fundamental law governing the orientation of a vehicle in such a situation will be the conservation of angular momentum. That is, a vehicle in free orbit about a planet will retain its orientation at injection into orbit (last power application) unless it has an angular velocity about some axis. As illustrated in Fig. 6–1, an angular rate may be imparted to a free flying vehicle to maintain the spin axis of the vehicle in a constant orientation in inertial space. Although this figure illustrates stabilization about or of the vehicle longitudinal axis, it should be remembered that the vehicle body is essentially acting as the rotor in a frictionless gyroscope, and may be stabilized around any axis. This indicates that the equation of motion is,

$$\vec{H} = \sum_{i=1}^{3} I_i \omega_i \vec{i} \tag{6–1}$$

where $\vec{H}$ is the total angular momentum, $\vec{i}$ defines a set of three orthogonal vector coordinates, ω_i is the angular velocity around vector $\vec{i}$, and I_i is the moment of inertia around vector $\vec{i}$.

To re-orient a free flight vehicle, a forcing function in the form of an applied torque must be introduced. The momentum addition resulting from a vector torque, $\vec{T}$, is,

$$\vec{T} = \frac{d\vec{H}}{dt} \tag{6–2}$$

Thus, an initial attitude in inertial space will be continuously changed during the application of the applied torque, and the attitude at any time, t, can be determined from,

$$\vec{H}_t = \sum_{i=1}^{3} I_{i_0}\omega_{i_0}\vec{i} + \int_0^t \vec{T}\,(dt) \tag{6–3}$$

The conversion of T into orthogonal components takes the general form of,

$$\vec{T} = \sum_{i=1}^{3} T_i \tag{6–4}$$

where,

$$T_1 = I_1\dot{\omega}_1 + \omega_2\omega^3(I_3 - I_2)$$
$$T_2 = I_2\dot{\omega}_2 + \omega_1\omega_3(I_1 - I_3)$$
$$T_3 = I_3\dot{\omega}_3 + \omega_1\omega_2(I_2 - I_1) \tag{6–5}$$

The use of this equation has been illustrated further in Section 3.2.1 for the development of gyroscopic equations.

It should be noted that the conservation of angular momentum allows one to make a change in the angular rate by repositioning the inertial centroid. The term, $I\omega$, may be noted as,

$$I\omega = mr^2\,\omega \tag{6–6}$$

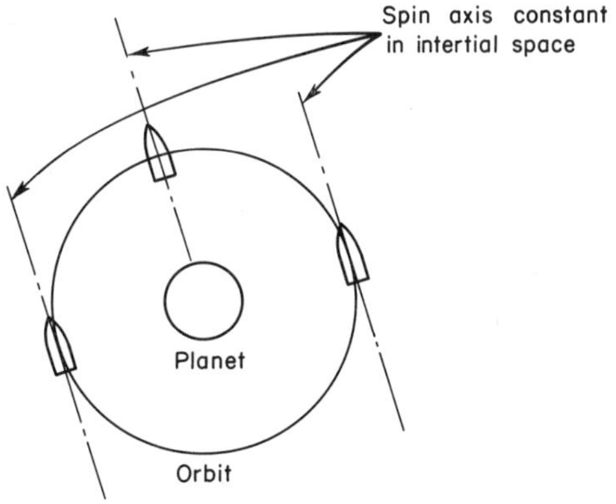

FIG. 6-1. Spin Stabilization of Vehicle Attitude in Inertial Space.

Thus, by changing the value of r, the radius at which the mass, m, is effective we may change the angular rate, ω. A good analogy of this effect is the way a spinning skater may change his spin rate by extending and retracting his arms. This technique is used to "de-spin" a space vehicle as illustrated in Fig. 6–2. This figure illustrates the release of masses which are tethered to the vehicle by long cords thus tending to reduce the angular rate of the vehicle.

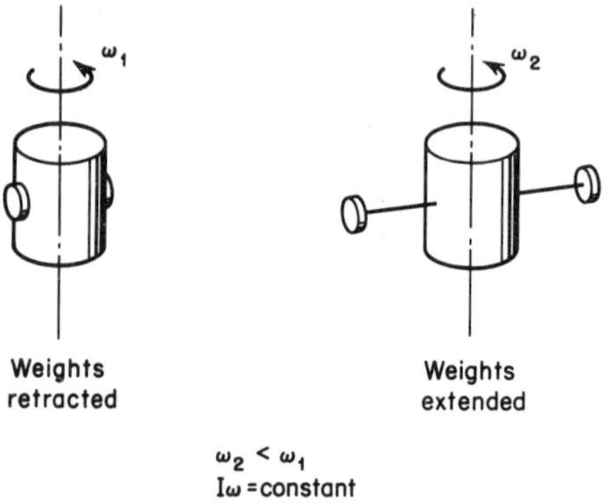

FIG. 6-2. Weight Extension to Control Spin Rate.

6.3 POWERED FLIGHT

Powered flight often calls for re-orienting the direction of flight in accordance with some guidance or control requirement. The easiest way to accomplish this, although not always the most efficient or feasible, is through some auxiliary system which re-orients the vehicle and its engine as a unit. Other approaches are to swivel the engine or its nozzle, or to introduce some exhaust redirecting media in the nozzle. A final method is to have a sophisticated flight path computer which times the engine impulses in discrete steps according to their predicted effects on the flight path.

6.3.1 Normal Jets and Rockets

A vehicle may be attitude stabilized through the use of normal jets or rockets. As illustrated in Fig. 6–3, one may impart a rotation through a small impulse through B and C (or F and G for another axis). When the desired new orientation is approached a countering impulse may be imparted through A and D (or E and H). Note that these impulses apply torques about the c.g. (center of gravity) of the vehicle, thereby setting up a new angular momentum as indicated by Eq. (6–4) and (6–5), and requiring neutralizing torques or momentum subtraction to stabilize on a new orientation. Steam jets (the expulsion of water vapor) and rockets of various kinds are suitable devices for this operation.

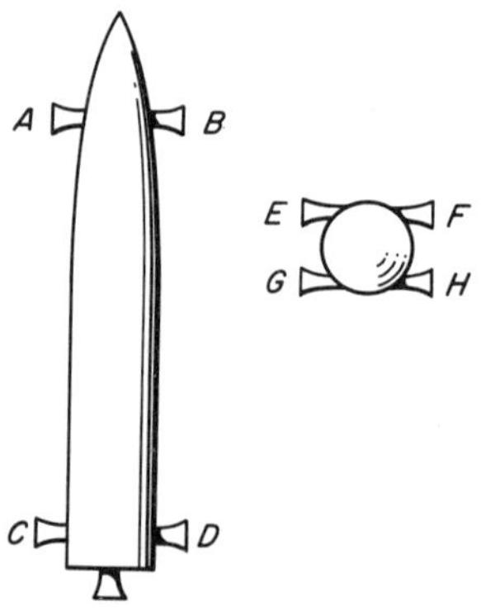

FIG. 6-3. Normal Impulse Attitude Control.

Another device for changing angular spin rate is the flywheel. A flywheel is oriented with each control axis and is speeded or slowed to change the angular spin of the vehicle about the axis. This occurs because the total angular momentum is constant, and momentum changes of the flywheel must result in compensating changes for the rest of the vehicle. Such flywheels are usually single degree of freedom devices; a set of three is used for full control. Ref. (*1*) describes a spherical device technique.

Jets and rockets may be applied to a vehicle so that their net effect is to apply a vector force through the c.g. The purpose of this is actual re-direction of the vehicle flight path through the vector sum of the applied thrusts. Figure 6–4 illustrates a vehicle concept which uses six such rockets. This type of vehicle control is particularly well suited to co-orbital rendezvous and inspection activities. The control computations are simple and direct because each control nozzle pair may be associated directly with a co-axially oriented control sensor (or pair of sensors for full hemisphere coverage).

As an alternative, one may re-orient a vehicle to direct its main rocket, and then re-orient to redirect the rocket for braking. The reaction times for the latter will be quite critical in rendezvous docking. Mating orientation requirements dictate close control systems of the type illustrated in Fig. 6–4 for at least some axes.

FIG. 6-4. Control Through Normal Rockets.

6.3.2 Exhaust Direction

Re-direction of the main rocket exhaust is an obvious way to "steer" a powered rocket. This may be accomplished in several ways. The most straightforward technique is to swivel the nozzle or even the entire engine (combustion chamber) as illustrated in Fig. 6–5. The effect of a thrust, T, in this case is a longitudinal thrust of,

$$T_L = T \cos \gamma \tag{6–7}$$

The control torque will be

$$T_C = TD \sin \gamma \tag{6–8}$$

where D is the distance between the c.g. and the engine reaction face (usually, the forward wall of the combustion chamber).

To obtain an effect similar to swivelling the nozzle or engine, control devices may be employed directly in the nozzle to re-direct the rocket thrust. One such system is jet vane or "jetavator" control of the thrust direction. As illustrated in Fig. 6–6, an aerodynamic control surface is inserted in the nozzle exhaust. This vane is hinged, and is directed to deflect the gases so as to cause a thrust moment or torque about the vehicle center of mass. The vane may be cylindrical in shape so as to provide control in any lateral direction. A variation of the jet vane concept is the insertion of small pressure nozzles within the main nozzle to introduce

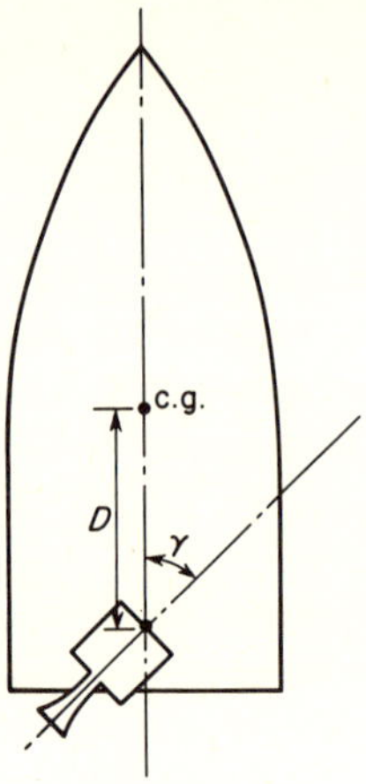

FIG. 6-5. Gimballed Engine.

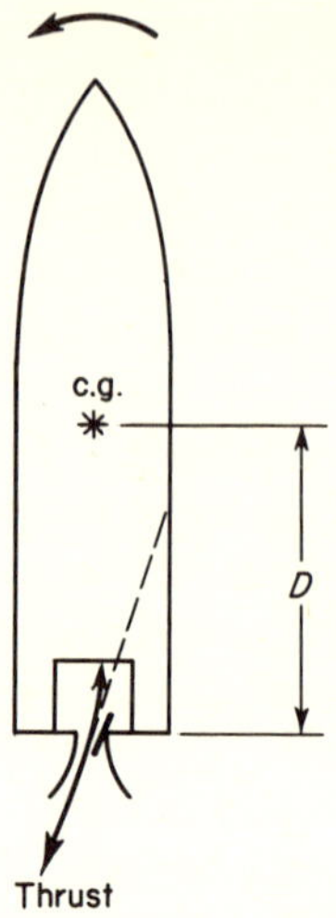

FIG. 6-6. Jet Vane Control.

a pressurized control gas so that the net effect of the main nozzle will be unevenly distributed; thus a control torque will be developed about the center of mass.

6.3.3 Impulse Control

Impulse control is an effective method of controlling both the vehicle flight path and its attitude. By impulse control we mean control not only of the impulse net vector, but of its magnitude, initiation, and duration.

If we think of an impulse as an instrument for control as well as a propulsive device we may see that the time of initiation, magnitude, and duration of an impulse are all useful control variables. Thus, if we have a vehicle in orbit, and the vehicle has an angular momentum in inertial space, a control impulse must be initiated and terminated (duration) in some relation to the orientation program. That is, a continuous control impulse may cause the vehicle merely to helix about the original orbit and orientation unless it is continually redirected in accordance with the vehicle control system attitude history. Also, a control impulse at perises will have a different net effect on the vehicle flight path than the same control impulse at apses, i.e., impulse control must include some knowledge of the vehicle future ephemeris and orientation in order to be effective. Flying in accordance with some guidance rule such as proportional navigation or steady line-of-sight rate merely dictates that the equations of these navigational techniques are the prediction of future position and motion. Reference (*2*) describes some preliminary test work in continuously controlling thrust magnitude by varying the fuel flow rate with a hydraulic system.

6.4 NUCLEAR POWERED FLIGHT CONTROL

The control of nuclear powered engines presents some unconventional problems to the control engineer. Reactor rod position may be a controlled quantity based on neutron count which is, in turn, dependent on propellant and temperature reactivity in addition to rod position. Figure 6–7 illustrates a functional block diagram for a nuclear rocket control system. Command signals are compared with a thrust (or pressure) sensitive device in the exhaust nozzle to control the mass flow to the nozzle from the reactor heat exchanger, and to control the reactor level of activity through operation of a "poison" rod. The reactor heat exchanger is seen to develop a mass flow output to the nozzle based on the reaction level. The propellant temperature and pressure are controlled quantities in going from the heat exchanger to the nozzle. The positive feedback of propellant and temperature reactivity within the reactor is depicted. A neutron counter is used in a negative feedback loop to control this "runaway" tendency. Mohler (Ref. (*3*)), has indicated that the control rod may be eliminated (except for on-off control) in high power nuclear rockets because

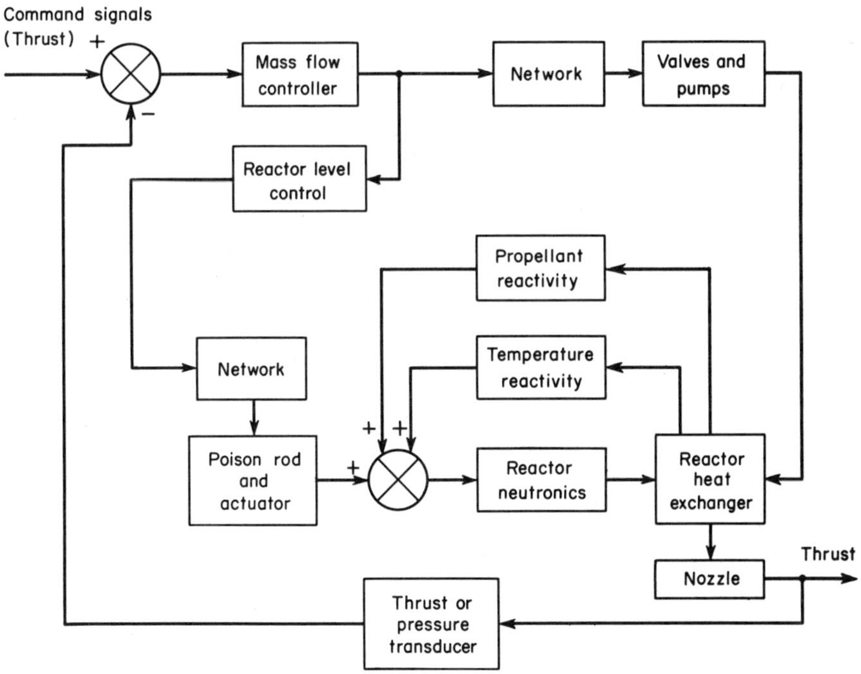

FIG. 6-7. Nuclear Rocket Control System.

the reactivity build-up is reduced as more power is drained off the reactor for propulsion, and the mass flow system may become the only control necessary. Further work in this area is clearly indicated.

6.5 ION ENGINE CONTROL

The control of ion engine mass flow rate is relatively simple. The thrust of an ion engine, Fig. 6–8, is directly proportional to the square of the accelerating potential, V, and the inverse square of the distance between the accelerating and the emitting electrodes, d. Or,

$$T = K\left(\frac{V}{d}\right)^2 \tag{6-9}$$

Thus, the thrust, T, of the ion engine may be varied just as the electron current in a thermionic vacuum tube by varying an accelerating voltage.

Ion beam shaping can be accomplished by the use of focusing electrodes. Figure 6–8 illustrates the Pierce gun which is designed to produce a rectilinear flow of an ion beam.

In order to maintain the space vehicle at a constant potential it will be necessary to emit a current of negative electrons equal in electrical current magnitude to the current of positive ions. The actual charge neutralization

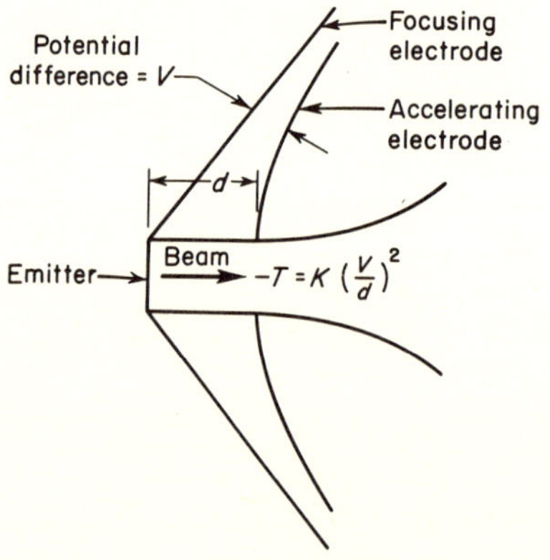

FIG. 6-8. The Pierce Gun.

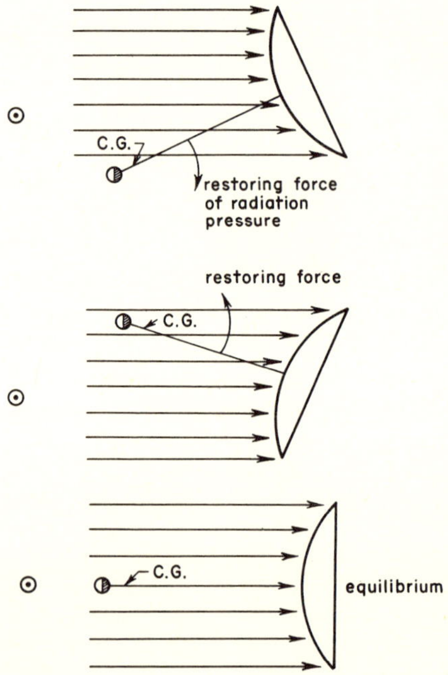

FIG. 6-9. Solar Sail Attitude Control.

system must be designed into the ion engine so as to neutralize the positive ion charge soon (much less than d-distance) after passing the accelerating electrode. Otherwise, the attraction between the accelerator and the ions which have passed it will tend to neutralize the thrust obtained originally.

6.6 SOLAR SAILS

The radiation pressure of the sun has been described as a function of the distance from the sun in Section 2.4.5. This radiation pressure may be collected in a large "sail," and used to propel a vehicle outward from the sun. The force, F, available depends on the sail area, A, its orientation, α, and the pressure density, ρ,

$$F = KA \cos \phi\rho \tag{6–10}$$

where K is an efficiency term which includes the reflection coefficient and form factors. ϕ is taken as zero when the sail is normal to the radiation (sight-line or radial to the sun). Fig. 6–9 illustrates a solar sail in three positions.

The pressure density is related to radial distance from the Sun, r, by,

$$\rho = \frac{P}{4\pi r^2} \tag{6–11}$$

where P is the total solar radiation which may vary with time. Thus,

$$F = \frac{KAP \cos \dot{\phi}}{4\pi r^2} \tag{6–12}$$

From Eq. (6–12) we may observe that a solar sail will become less effective as we travel outwardly from the sun. The question arises as to how one returns from a "sailing" mission, and, indeed, such sails may only be used with unmanned probes of the outer solar system. An interesting speculation is that solar sailing might be used as a return power system for flights to Venus or Mercury. The use of solar sails to establish the local vertical in heliocentric space is illustrated in Fig. 6–9. The sail in this illustration is oriented normally to the radiation source (Sun) by virtue of the mass to which it is tethered. Thus, solar sails may offer a potential as a weather vane device to continually sense the radial vector to the sun. Many other devices can do this, but the weather vane should require no internal power to operate once erected.

6.7 ATTITUDE MAINTENANCE

The attitude of a space vehicle may be maintained with respect to inertial space, the predominant force field center or some local body horizon, etc. An important design rule is to design an attitude maintenance system which is closely related to the governing payload and/or flight dynamics so as to minimize coordinate transformation computation and coupling errors. Various devices which may be used to accomplish this (that is, to sense the vehicle attitudes, rates and accelerations) have been described in Sections 3.2, 4.3.3, and 6.6. If the vehicle attitude can be directly sensed and controlled in each of three orthogonal references, a simple attitude control system may be developed in the general form depicted in Fig. 6–10. Within the control dynamics we may include rate and acceleration feedback loops. Also, the prime control reference sensor may be, for example, a rate sensor, and commands or programs developed in rate terms only.

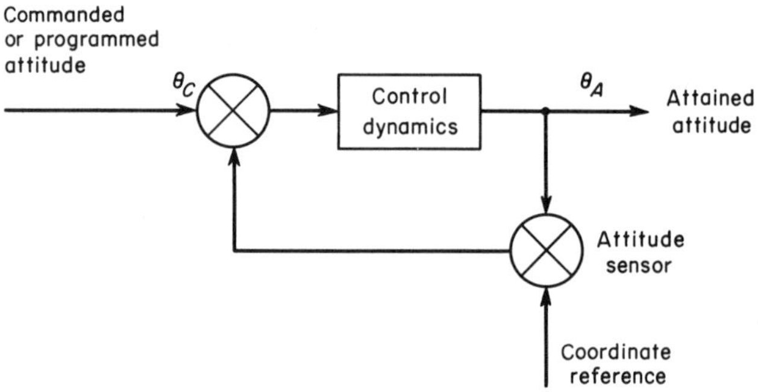

FIG. 6-10. One-Axis Attitude Control System.

Cross-coupling between the orthogonal systems should be minimized as much as possible. In some cases, such as free gyro precession, it is used as the prime indicator of attitude since it cannot be eliminated directly and is a direct and sensitive linear measure of attitude.

When the direct measurement and control of attitude independently in each of three axes is not feasible, some coordinate resolver may be added to the system as illustrated in Fig. 6–11. This system allows for the measurement of attitude externally in ρ, ϕ, ψ, coordinates as well as internally in x, y, z coordinates. If one measurement loop is known to be superior, the other loop may be eliminated unless desired for some other reason such as insurance of better control system response, operation through a "dead" zone of the superior reference, etc. In the case illustrated in Fig. 6–11, the ρ, ϕ, ψ, coordinates might refer to a radar coordinate set used in landing, while the x, y, z coordinates might refer to an internal inertial set. Thus, various coordinate sets and controls might be switched in as appropriate according to their judged validity and appropriateness to the mission phase.

If measurements are external, communication or data links may be added to the basic systems as appropriate. A computer might be added as

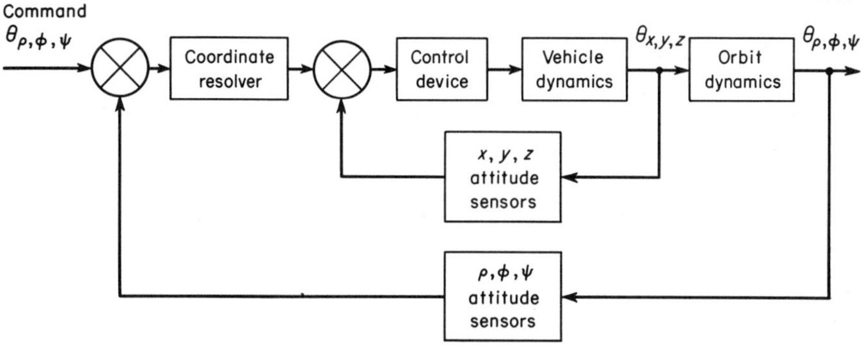

FIG. 6-11. Three Dimension Attitude Sensing with Coordinate Resolution.

well to vary the system response as a function of some controlled variable or a measurement. An example would be a non-linear control over closure time prediction in rendezvous docking as described in Section 5.6.4.

6.8 FEEDBACK CONTROL SYSTEM THEORY

Feedback control system theory is based on measurement of a system output, and comparison of this output with an input to generate an error function. The output is changed only in response to the error function. The basic block diagram for such a system is depicted in Fig. 6–12.

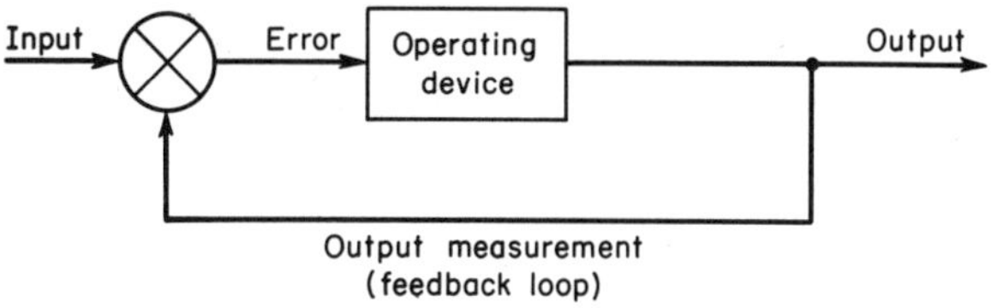

FIG. 6-12. Basic Feedback System.

The usual treatment of such systems is to consider all elements as linear, and to investigate the system stability, gain, and frequency (or phase) response in relation to the system requirements. The use of components in feedback systems which have sampling rates or are of discontinuous operation has given impetus to further advances in analytic techniques. The employment of feedback systems in multi-dimensional systems where the action of orthogonal systems introduces cross-coupling produces additional problems. Non-linear devices may be encountered in some systems, and some method is needed for handling such devices. Finally, feedback systems which adapt their response characteristics to other than the usual criteria of direct output-input comparison are coming into widespread use, and deserve some understanding.

6.8.1 Linear Systems

Locke (Ref. (*4*)), has presented the mathematical basis for analysis of feedback systems, their stability, gain, and phase margins. In very brief form we may say that a linear system is one in which the response of operating devices in the system may be described by linear integro-differential equations. That is, the input θ_i, and output, θ_0, of any single device in the system may be described by an equation of the form,

$$\sum_{i=0}^{n} a_i \frac{d^i\theta_0(t)}{dt^i} + \sum_{j=0}^{m} b_j \int^{j} \theta_0(t)(dt)^j = \theta_i(t) \qquad (6\text{–}13)$$

The response of individual devices is generally taken as the Laplace transform input-output ratio. Thus, in the complex frequency domain, the response or transfer function of a device is manipulated algebraically into the form,

$$\frac{\theta_0(s)}{\theta_i(s)} = \frac{\sum_{i=0}^{m} a_i s^i}{\sum_{j=0}^{n} b_j s^j} \qquad (6\text{–}14)$$

where $\theta(s)$ is the Laplace transform of $\theta(t)$,

$$\theta(s) = \int_0^\infty \theta(t)e^{-st}dt \qquad (6\text{–}15)$$

The transfer function, Eq. (6–14), may be multiplied directly by the input function to arrive at the output function. Thus any group of series related devices will have an overall transfer function which is the product of their individual transfer functions,

$$G(s) = \prod_{i=1}^{n} F_i(s) \qquad (6\text{–}16)$$

Tables 6–1 and 6–2 present some useful Laplace transform pairs for various time functions and operations.

If we solve the polynomials in s described by Eq. (6–14), we may express any transfer function as,

$$G(s) = \frac{\prod_{i=0}^{m} (s - c_i)}{\prod_{j=0}^{n} (s - d_j)} \qquad (6\text{–}17)$$

The manipulation of a transfer function from a series polynomial ratio (Eq. (6–14)) to a factor product ratio (Eq. (6–17)) is strictly algebraic, and

TABLE 6-1 FUNCTION—TRANSFORM PAIRS

No.	$F(s)$	$f(t) \qquad 0 \leqslant t$
1	$\frac{1}{s}$	1, or unit step function $u(t)$ at $t = 0$
2	$\frac{1}{s^2}$	t
3	$\frac{1}{s+\alpha}$	$e^{-\alpha t}$
4	$\frac{\omega_0}{s^2+\omega_0^2}$	$\sin \omega_0 t$
5	$\frac{s}{s^2+\omega_0^2}$	$\cos \omega_0 t$
6	1	unit impulse at $t = 0$
7	$\frac{s+\omega_0}{s^2+\beta^2}$	$\frac{1}{\beta}(\omega_0+\beta^2)^{\frac{1}{2}} \sin(\beta t+\psi)$ $\psi = \tan^{-1} \beta/\omega_0$
8	$\frac{1}{s(s^2+\beta^2)}$	$\frac{1}{\beta^2}(1-\cos \beta t)$
9	$\frac{1}{(s+\alpha)^2+\beta^2}$	$\frac{1}{\beta} e^{-\alpha t} \sin \beta t$
10	$\frac{s+\omega_0}{(s+\alpha)^2+\beta^2}$	$\frac{1}{\beta}[(\omega_0-\alpha)^2+\beta^2]^{\frac{1}{2}} e^{-\alpha t} \sin(\beta t+\psi)$ $\psi = \tan^{-1} \beta/(\omega_0-\alpha)$
11	$\frac{s+\alpha}{(s+\alpha)^2+\beta^2}$	$e^{-\alpha t} \cos \beta t$
12	$\frac{1}{s[(s+\alpha)^2+\beta^2]}$	$\frac{1}{\beta_0^2}+\frac{1}{\beta\beta_0} e^{-\alpha t} \sin(\beta t-\psi)$ $\psi = \tan^{-1} \beta/-\alpha$ $\beta_0^2 = \alpha^2+\beta^2$
13	$\frac{s+\omega_0}{s[(s+\alpha)^2+\beta^2]}$	$\frac{\omega_0}{\beta_0^2}+\frac{1}{\beta\beta_0}[(\omega_0-\alpha)^2+\beta^2]^{\frac{1}{2}} e^{-\alpha t} \sin(\beta t+\psi)$ $\psi = \tan^{-1} \beta/(\omega_0-\alpha) - \tan^{-1} \beta/-\alpha$ $\beta_0^2 = \alpha^2+\beta^2$
14	$\frac{1}{(s+\alpha)^n}$	$\frac{1}{(n-1)!} t^{n-1} e^{-\alpha t}$

techniques are generally presented in any basic reference book on feedback control theory (e.g., Ref. (*5*)).

When a system is of the general feedback type depicted in Fig. 6–13, the overall system transfer function becomes,

$$\frac{\theta_0}{\theta_i} = \frac{G(s)}{1 + G(s)H(s)} \tag{6–18}$$

The overall gain and frequency response of such a system are thus defined. The stability of the system is dependent on the characteristic equation which is determined from the denominator of the closed loop transfer function (Eq. 6–18).

TABLE 6-2 OPERATION—TRANSFORM PAIRS

No.	$f(t) \qquad 0 \leqslant t$	$F(s)$
1	$af_1(t) \pm bf_2(t)$	$aF_1(s) \pm bF_2(s)$
2	$f'(t) = \frac{df(t)}{dt}$	$sF(s) - f(0+)$
3	$f^{(-1)}(t) = \int f(t)\,dt$	$\frac{1}{s}[F(s) + f^{(-1)}(0+)]$
4	$f(t-b)$ for $f(t-b) = 0, \quad 0 < t < b$ b is a positive real number	$e^{-bs} F(s)$
5	$e^{-bt} f(t)$ b is complex with non-negative real part	$F(s + b)$
6	Convolution Theorem: $\int_0^t f_1(t-\tau) f_2(\tau)\,d\tau$	$F_1(s)F_2(s)$
7	$f^{(n)}(t) = \frac{d^n f(t)}{dt^n}$	$s^n F(s) - s^{n-1} f(0+) - s^{n-2} f'(0+) - \cdots - f^{(n-1)}(0+)$
8	Final Value Theorem: $\lim_{s\to 0} sF(s) = \lim_{t\to\infty} f(t)$ provided $sF(s)$ is analytic on the axis of imaginaries and in the right half plane	
9	Initial Value Theorem: $\lim_{s\to\infty} sF(s) = \lim_{t\to 0} f(t)$	

The characteristic equation is,

$$1 + G(s)\,H(s) = 0 \tag{6–19}$$

If the characteristic equation is placed in a factorial form (solved algebraically) we will have defined the "zeros" of the characteristic equation. These "zeros" are the "poles" of the closed loop transfer function. That is, such a "zero" defines a complex frequency which is characteristic of the system response to excitation by a step input containing all frequencies. The imaginary part of the complex frequency defines the period of the frequency response to unit step excitation. The real part of the complex frequency defines the damping response of the frequency. If the real part is positive, the frequency is damped naturally. If the real part is negative, the frequency will "run away" naturally, i.e. oscillations will build up naturally until the system attains operation in some saturated or limiting condition no longer defined by the transfer function at hand.

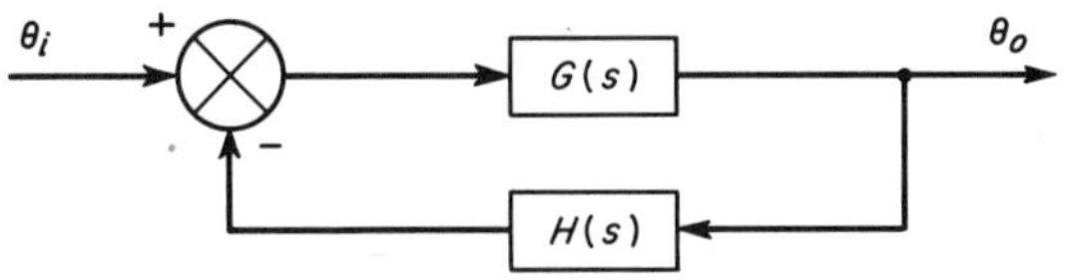

Fig. 6-13. Single Loop Feedback System with Feedback Element.

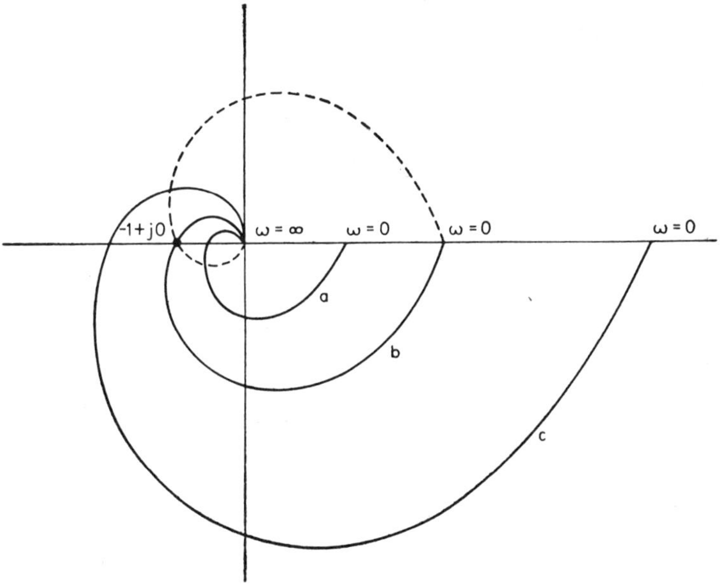

Fig. 6-14. Example of Stable and Unstable G(s) H(s) Loci.

If the real part of the frequency is zero, sustained steady-state oscillations will occur as is required in electronic oscillator design.

Various techniques are available to analyze the performance of such systems. Routh (Ref. (*4*), pp. 201–203) established a method for determining the absolute stability (but not the damping factors) of the characteristic equation without actually solving for the roots of the polynomial. Bode (Ref. (*6*)) has done extensive work in analyzing the performance of feedback circuits (systems) using frequency response plots of the transfer function steady-state gain and phase relations. The most powerful tool in such an analysis is the Nyquist plot of the open loop steady state transfer function in the complex plane. The steady-state transfer function is obtained by substituting $s = j\omega$ in the s-domain transfer function. This provides a direct relation between the measurable and calculable response of a device when subjected to any particular steady state frequency, ω_A (a substitution of $s = \omega_A$). The Nyquist criteria for stability and gain and phase margins are determinable by plotting the open-loop steady-state characteristic equation function, $G(j\omega)\ H(j\omega)$, for all values of $j\omega$ from $-\infty$ to $+\infty$, and relating the plot

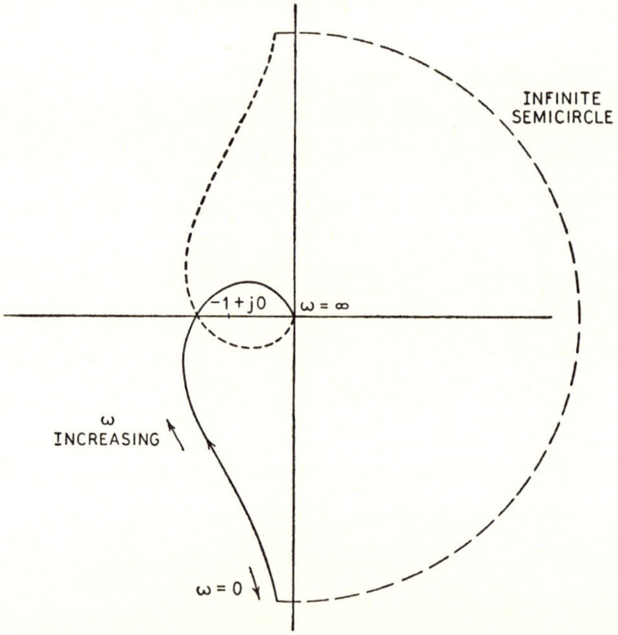

FIG. 6-15. Unstable Position Control Servo System with an Open-Loop $G(s)H(s) = \dfrac{K}{s(s + \omega_0)(s + \omega_1)}$.

to the $-1 + j0$ point on the complex plane. Figure 6–14 illustrates a complex plane plot of,

$$G(j\omega)\ H(j\omega) = \frac{K}{(j\omega + \omega_0)\ (j\omega + \omega_1)\ (j\omega + \omega_2)} \qquad (6\text{–}20)$$

for various values of K. Note that a system with this characteristic equation may be stable or not depending on the value of K. For $K = a$ the system is stable, for $K = b$ the system will be a steady oscillator, and for $K = c$ the system will build up oscillations until the characteristic equation no longer applies. In any system, a Nyquist plot may be examined for stability by comparing the rotation of a vector from the $-1 + j0$ point to the open-loop steady-state characteristic transfer function plot with the number of "poles" in the same transfer function. If we fix one end of the vector at the $-1 + j0$ point, and allow the other end to travel with increasing frequency from $-j\infty$ through 0 to $+j\infty$ we may count the number of counterclockwise revolutions. This number must be equal to or greater than the number of poles in $G(s)\ H(s)$. Figures

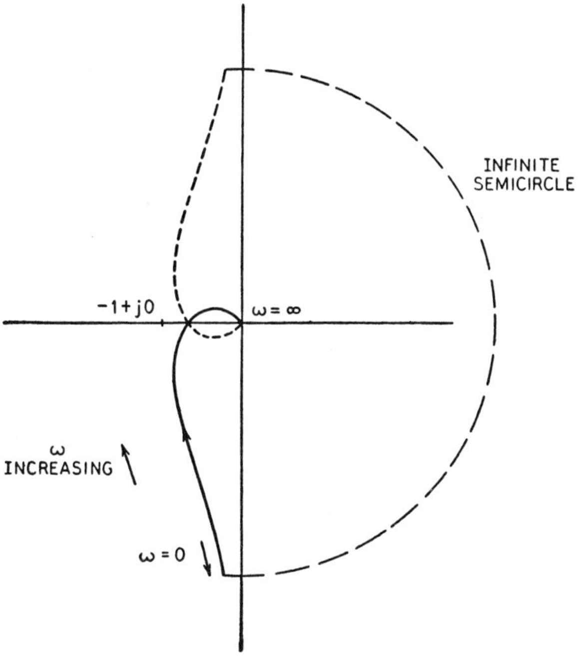

FIG. 6-16. Stable Position Control Servo System with an Open-Loop $G(s)H(s) = \dfrac{K(s + \omega_2)}{s(s + \omega_0)(s + \omega_1)}$.

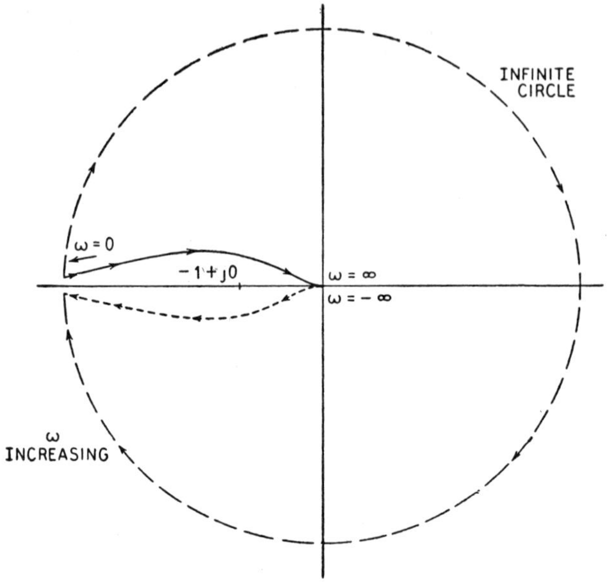

FIG. 6-17. Unstable Velocity Control Servo System with an Open-Loop $G(s)H(s) = \dfrac{K_1(s + \omega_0)}{s^2(s + \omega_1)(s + \omega_2)}$.

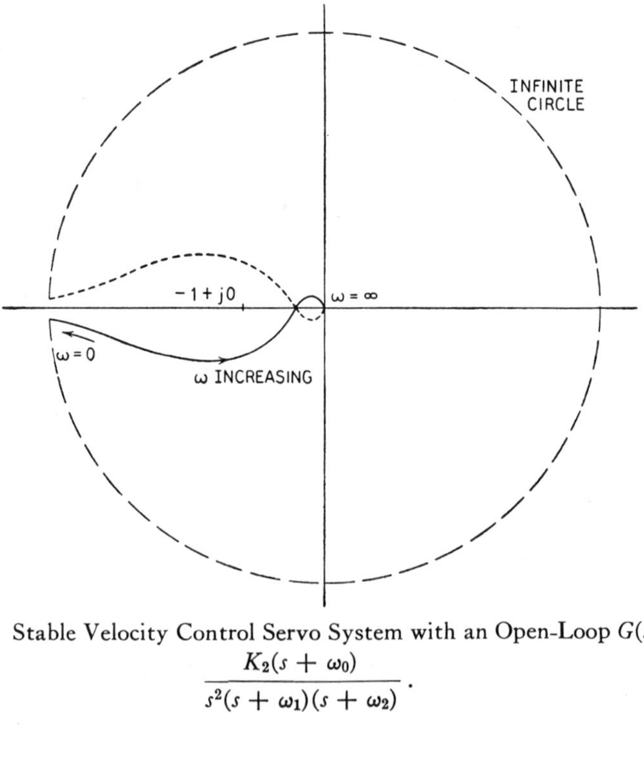

FIG. 6-18. Stable Velocity Control Servo System with an Open-Loop $G(s)H(s) = \dfrac{K_2(s + \omega_0)}{s^2(s + \omega_1)(s + \omega_2)}$.

6–15 through 6–18 illustrate various plots for some unstable and stable systems.

The gain and phase margins of a system are determinable directly from the Nyquist plot of the open-loop response. As shown for a typical $G(s)\ H(s)$ locus in Fig. 6–19, the gain margin, a, is the separation between the $-1 + j0$ point and the $G(j\omega)\ H(j\omega)$ plot crossing of the real axis. The phase margin, γ, is the phase angle between the negative real axis and the $G(j\omega)\ H(j\omega)$ crossing of the unit circle. In attenuation/phase (Bode) diagrams, these same margins are directly measurable as the number of db below unity gain when 180° phase relation exists, and the phase difference from 180° at which unity gain exists.

Based on the Nyquist analysis of system performance, a technique has been developed so that a system may be synthesized based on a definition of the transfer function (which includes the characteristic equation) required. This technique is called the "root locus" method. In brief, the definition of the response or transfer function characteristic permits one to set the gain, K, portion of $G(s)\ H(s)$ for any characteristic function in order to satisfy some criteria as to gain or phase margin and response

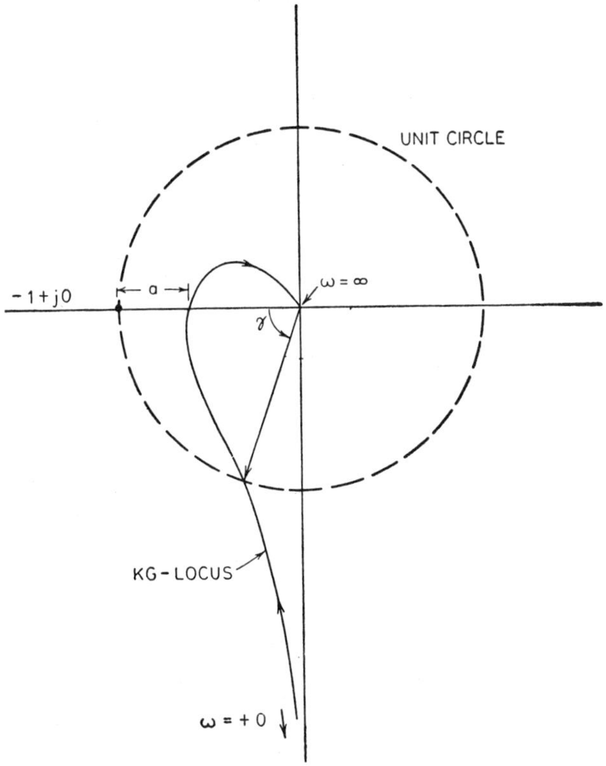

FIG. 6-19. Typical $G(s)H(s)$ Locus Illustrating Gain and Phase Margin.

characteristics. A complex-plane plot is made of the locus of values for $G(s)\ H(s)$ which have 180° phase shift using K, the gain factor, as the variable parameter. A typical plot is illustrated in Fig. 6–20. This plot is the locus of poles of the closed loop transfer function, $\dfrac{G(s)}{1 + G(s)H(s)}$ as the gain constant K takes all positive real values from 0 to $+\infty$. That is, for,

$$G(s)H(s) = KF(s) \tag{6–21}$$

$$1 + KF(s) = 0 \tag{6–22}$$

$$F(s) = -\frac{1}{K} \tag{6–23}$$

$$\arg F(s) = 180° \tag{6–24}$$

Thus, as K is increased from 0 to $+\infty$, the poles of the closed-loop transfer function will progress away from the poles of the open-loop function, and towards either an open-loop zero or infinity. Where the increasing K loci from two poles meet and split off into a complex conjugate pair as illustrated in Fig. 6–20, the indicated poles for the closed-loop system will be this pair, and the denominator of the closed-loop transfer function will have a quadratic factor $(s^2 + 2\xi\omega_n + \omega_n{}^2)$ where ξ is less than unity. In such an expression ξ is the damping ratio. A loci of constant ξ is a straight line from the origin. Thus, one may select a damping ratio, ξ, and plot it on the root locus of a system, and from the intersection with the root-locus establish the gain constant which will provide that damping ratio. The response or decay time, σ, of the closed-loop system is the abscissa of this intersection, and the natural frequency, ω_n, will be the ordinate. The predominant roots in determining system response are those closest to the imaginary (ordinate) axis.

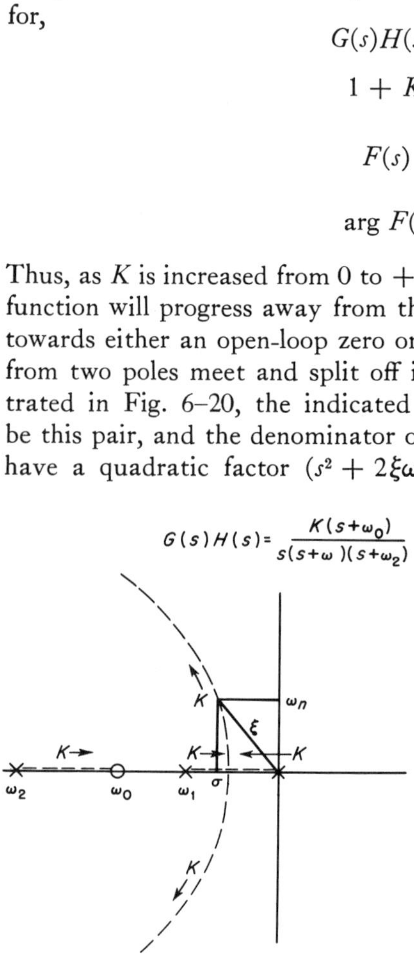

FIG. 6-20. Typical Root-Locus Plot.

6.8.2 Sampled-Data Systems

Sampled-data systems are those feedback control systems in which the signal is sampled only periodically at regular intervals, and the information is contained in the amplitude of the samples. The basic system is depicted in Fig. 6–21. The switch in the schematic is closed for short intervals and open for relatively long periods. Thus the sampled function, $e^*(t)$, is related to the continuous input function, $e(t)$, as indicated in Fig. 6–22. The relation of $e^*(t)$ to $e(t)$ may be written as,

$$e^*(t) = e(t)\, i(t) \tag{6–25}$$

where $i(t)$ represents a train of unit impulses occurring every T interval.

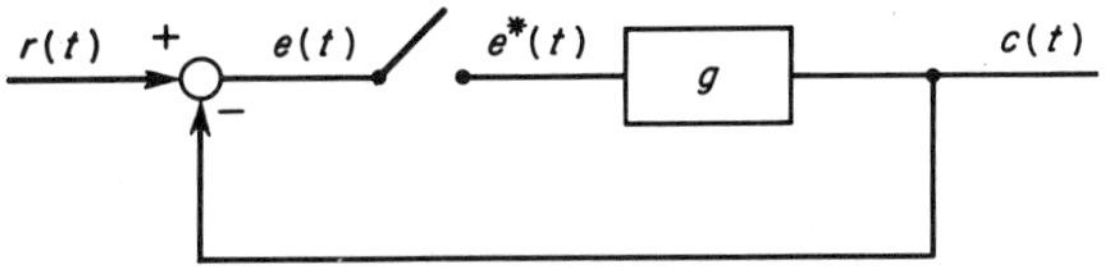

FIG. 6-21. Basic Sampled-Data System.

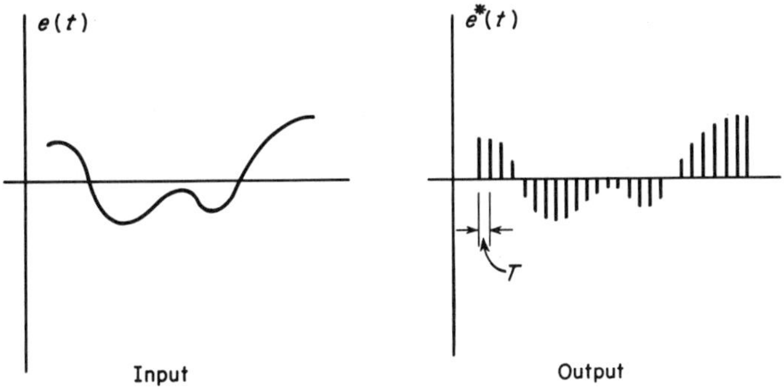

FIG. 6-22. Sampler Input-Output Characteristic.

The impulse function is,

$$i(t) = \sum_{n=0}^{\infty} u_0(t - nT) \tag{6–26}$$

and the sampled function,

$$e^*(t) = \sum_{n=0}^{\infty} e(nT)u_0(t - nT) \tag{6–27}$$

The Laplace transform of this equation is,

$$E^*(s) = \sum_{n=0}^{\infty} e(nT)e^{-nTs} \tag{6–28}$$

Another derivation of $E^*(s)$ (Ref. (7)) yields,

$$E^*(s) = \frac{1}{T} \sum_{n=-\infty}^{\infty} E(s + jn\omega_s) \tag{6–29}$$

The latter form illustrates that $E^*(s)$ is a periodic function with a period $j\omega_s$. If the input frequency content is lower than one-half the sampling

TABLE 6–3. LAPLACE AND z-TRANSFORMS

Row	Column 1	Column 2	Column 3	Column 4
	Laplace transform	Time function	z transform	Description of time function
a	1	$u_0(t)$	1	Impulse function at $t = 0$
b	e^{-nTs}	$u_0(t - nT)$	$\frac{1}{z^n}$	Impulse function at $t = nT$
c	$\frac{1}{1 - e^{-Ts}}$	$i(t)$	$\frac{z}{z - 1}$	Train of impulses at sampling instants
d	$\frac{1}{s}$	$u(t)$	$\frac{z}{z - 1}$	Step function
e	$\frac{1}{s^2}$	t	$\frac{Tz}{(z - 1)^2}$	Ramp function
f	$\frac{1}{s^3}$	$\frac{1}{2}t^2$	$\frac{1}{2}T^2\frac{z(z - 1)}{(z - 1)^3}$	Quadratic or acceleration function
g	$\frac{1}{s + a}$	e^{-at}	$\frac{z}{z - e^{-aT}}$	Exponential function
h	$\frac{a}{s^2 + a^2}$	$\sin at$	$\frac{z \sin aT}{z^2 - 2z \cos aT + 1}$	Sinusoidal function
i	$\frac{1}{a - (1/T) \ln a}$	$a^{t/T}$	$\frac{z}{z - a}$	Constant raised to power t
j	$\frac{b}{[s - (1/T) \ln a]^2 + b^2}$	$a^{t/T} \sin bt$	$\frac{za \sin bT}{z^2 - 2az \cos bT + a^2}$	Sine wave multiplied by $a^{t/T}$
k	$\frac{s - (1/T) \ln a}{[s - (1/T) \ln a]^2 + b^2}$	$a^{t/T} \cos bt$	$\frac{z(z - a \cos bT)}{z^2 - 2az \cos bT + a^2}$	Consine wave multiplied by $a^{t/T}$
l	$F(s + a)$	$e^{-at}f(t)$	$F(e^{+aT}z)$	Effect of multiplication by e^{-at}

frequency, the system will function as intended. For input frequencies equal to or greater than the sampling frequency, distortion of the signal content will be encountered. Recovery of a signal from the samples may be accomplished by a holding circuit ("boxcar" detector) which holds the last pulse amplitude until driven to a new (higher or lower) value by the next pulse.

The analysis of sampled-data systems is facilitated by the use of z-transforms. The sampled-data Laplace transform, $C^*(s)$ may be converted to the corresponding z-transform, $C(z)$, by making the substitution,

$$z = e^{sT} \tag{6–30}$$

Table 6–3 presents a group of common Laplace transforms, the related time-functions, and the corresponding z-transforms when the data is sampled.

The z-transform of a system will depend on the location of the sampler within the system. Table 6–4 illustrates several basic sampled-data

TABLE 6–4. OUTPUT TRANSFORMS FOR BASIC SAMPLED-DATA SYSTEMS

	System	Laplace transform of output $C(s)$	z transform of output $C(z)$
1		$R^*(s)$	$R(z)$
2	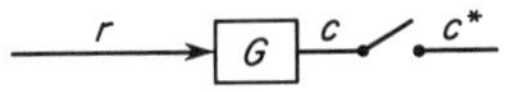	$GR^*(s)$	$GR(z)$
3	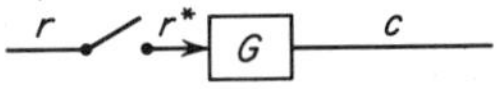	$G(s)R^*(s)$	$G(z)R(z)$
4	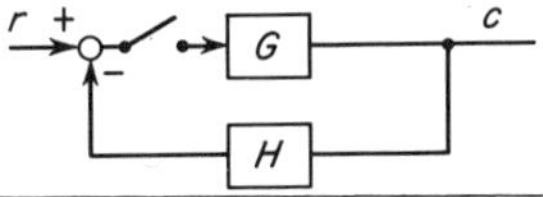	$\frac{G(s)R^*(s)}{1 + HG^*(s)}$	$\frac{G(z)R(z)}{1 + HG(z)}$
5	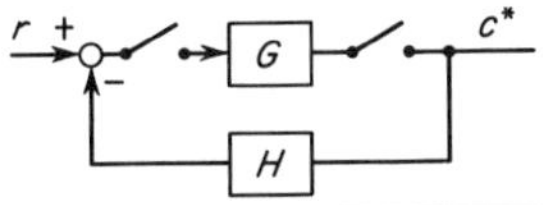	$\frac{G^*(s)R^*(s)}{1 + H^*(s)G^*(s)}$	$\frac{G(z)R(z)}{1 + H(z)G(z)}$
6	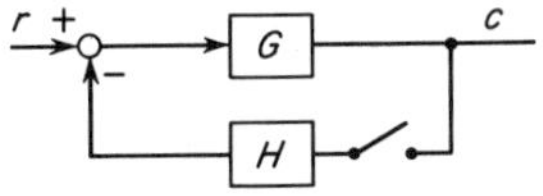	$G(s)\left[R(s) - \frac{H(s)RG^*(s)}{1 - HG^*(s)}\right]$	$\frac{RG(z)}{1 - HG(z)}$
7	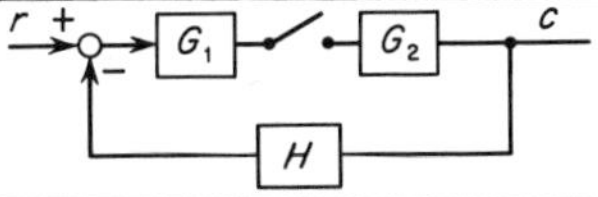	$\frac{G_2(s)RG_1^*(s)}{1 + HG_1G_2^*(s)}$	$\frac{G_2(z)RG_1(z)}{1 + HG_1G_2(z)}$

systems, their Laplace transforms, and z-transforms. The distinction between $G_1G_2(z)$ and $G_1(z)G_2(z)$ is illustrated by taking two functions,

$$G_1(s) = \frac{1}{s + \beta} \tag{6–31}$$

and

$$G_2(s) = \frac{1}{s + \gamma} \tag{6–32}$$

The z-transforms of these functions are,

$$G_1(z) = \frac{z}{z - e^{-\beta T}} \tag{6–33}$$

and

$$G_2(z) = \frac{z}{z - e^{-\gamma T}} \tag{6–34}$$

Combining these *individually* sampled functions yields,

$$G_1(z)\, G_2(z) = \frac{z^2}{(z - e^{-\beta T})\,(z - e^{-\gamma T})} \tag{6–35}$$

If the two functions are sampled as a unit, the z-transform will be,

$$G_1G_2(z) = \frac{z}{z - e^{-\beta T}} - \frac{z}{z - e^{-\gamma T}} \tag{6–36}$$

Thus, knowledge of the sampler location in the system and number of samplers is necessary to determine the proper z-transform as is illustrated further in Table 6–4.

The utility of the z-transform is that the transformation $z = e^{sT}$ maps the left half of the s-plane into a unit circle in the z-plane. Thus, stability of a system may be readily determined by the location of the zeros of the closed loop denominator z-transform relative to the unit circle. A system with a transfer function of,

$$C(s) = \frac{G(s)R^*(s)}{1 + HG^*(s)} \tag{6–37}$$

or,

$$C(z) = \frac{G(z)R(z)}{1 + HG(z)} \tag{6–38}$$

will be stable if the function $1+HG(z)$ has no zeros outside the unit circle in the z-plane. Further z-transform analysis and synthesis techniques may be found in Ref. ((*7*), Chapter 9) which illustrates Nyquist and root-locus technique application by graphical methods.

6.8.3 Discontinuous Systems

An "on-off" or "bang-bang" response presents a problem in discontinuous system operation. Basically, the system must be considered as several discrete systems. That is, when the system is "on" in one direction it has one configuration, when "off" a second, and when "on" in the other direction a third. The gain is generally constant, but the "live" and "dead" zones of the system may vary in duration. In general, one might expect such a system to decrease an initial error with long period operation with short "dead" zones between "live" operating periods, and with the operating periods becoming progressively shorter until some minimum duration operating period is attained. This time sequence is illustrated in Fig. 6–23 with a corresponding error reduction. Thus, a series of hard right and hard left (or up and down) pulses might be employed to maneuver a space vehicle to some desired orbit. The odd-numbered "live" zones should have second order variation in position error with thrust (accelerative) correction while the even-numbered "dead" zones should have linear variation in position error. The determination of the duration of the corrective pulses will generally be made by human judgement or a computer solving the problem geometry.

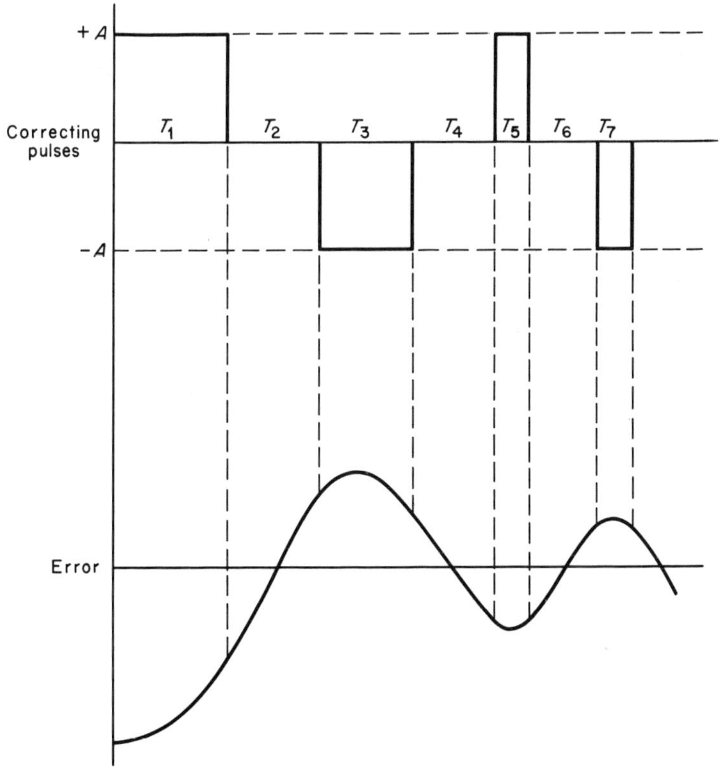

FIG. 6-23. Discontinuous System Time Sequence and Error Reduction.

6.8.4 Multi-Dimensional Systems and Cross-Coupling

In multi-dimensional systems every attempt is generally made to minimize cross-coupling between orthogonal component systems. However, the use of a common facility (e.g., communication link, power source, frequency multiplexed amplifier, etc.) will sometimes introduce significant cross-coupling between what should be orthogonal systems. The precession of gyro gimbals may be thought of as one particular form of cross-coupling although this is directly predictable from the vector cross-product relation indicated in Section 3.2.1.

The general transfer function relation for a two-dimensional system with linear cross-coupling is illustrated in Fig. 6–24. The stability of such a system is amenable to direct analysis only if the system is symmetric and the cross-coupling transfer function, $F(s)$ is expressable as a linear differential equation relationship in the time domain (Ref. (*8*)). For asymmetric systems, and systems operating in more than two dimensions with inter-dimensional cross-coupling the stability problem analysis becomes even more difficult to treat. As demonstrated in Ref. (*8*), it is even feasible to introduce cross-coupling to stabilize an otherwise unstable system, and conversely, cross-coupling may make a stable system become unstable.

6.8.5 Non-Linear Systems

The use of non-linear elements in a system complicates the analysis considerably. Firstly, the use of linear devices implies a mathematically reversible process. That is, with a linear device, knowledge of the output

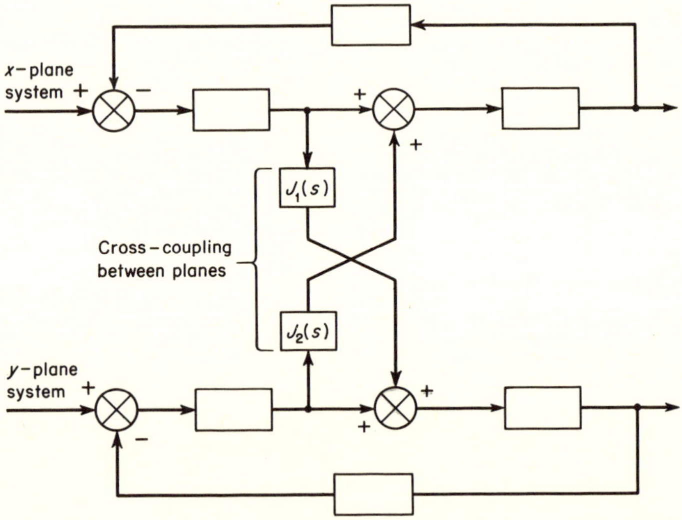

FIG. 6-24. Two-dimensional System with Linear Cross-Coupling.

defines the input and vice versa. In a non-linear device knowledge of the input allows one knowledge of the output, but knowledge of the output does not determine the input. For example, a logarithmic i.f. amplifier in which the ouput voltage is proportional to the input percentage modulation does not allow one to determine the voltage level of the input based on output voltage.

Much effort is going into specialized treatment of special types of non-linear systems (e.g. Ref. (*9*) through (*13*)). The methods are either peculiar to the particular system and its requirements, i.e., an attempt to bound the region of operation by linearizing assumptions as to a boundary system configuration, or an attempt to analyze in an assumed linear but closely held region of operation. Reference (*9*) describes a technique of defining the system stability over its linear region of operation, and then shifting the poles and zeros on a root-locus plot to explore the non-linear effects.

Reference (*10*) examines the proposition that "if a linearized system is stable for all points of the state space, the original non-linear system is asymptotically stable in the large." By examples the statement is proven not true in general for both total and incremental linearizations, thus indicating that some very careful analysis, and often simulation of the non-linear system will be required to justify its design.

Reference (*11*) explores the application of z-transform methods to non-linear, sampled-data systems which can be described by non-linear difference equations. Reference (*12*) describes an analysis of an on-off velocity servo system. Reference (*13*) studies the stability of non-linear systems as a function of input signal level.

The techniques of non-linear system analysis are thus specifically oriented to particular problems at this stage, and no general analytic procedure is available.

6.8.6 Adaptive Systems

Adaptive control implies some sort of "learning" criteria on the part of a control system. That is, the response of the control system is optimized while functioning on the basis of some measurement of the controlled variable or the error function.

One general type of adaptive system is the model reference system illustrated in Fig. 6–25. In this system, a model of the desired system performance acts on the error signal in the desired manner. The adaptive computer compares the model action with the forward loop servo system action, and adapts the forward loop gain and/or frequency response to mate with the model's.

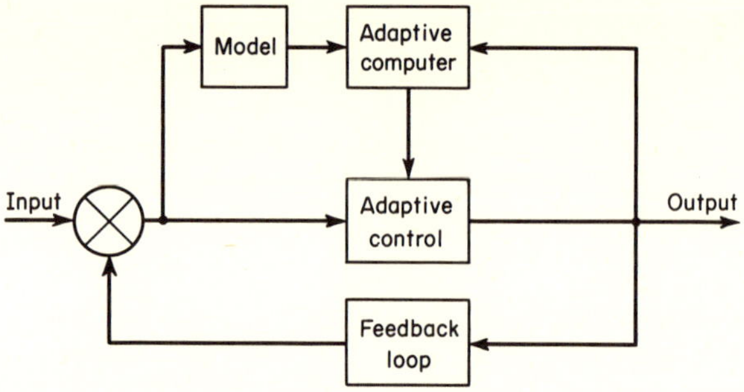

FIG. 6-25. Model Reference Adaptive Control System.

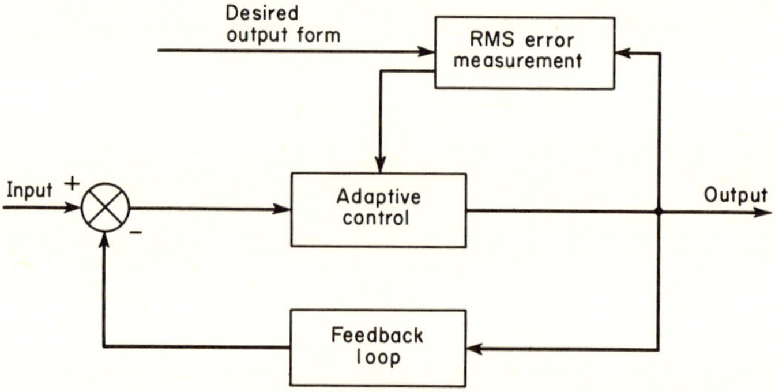

FIG. 6-26. RMS Minimizing Adaptive Control System.

A more general type of system is illustrated in Fig. 6–26. In this system, the rms error of the output function is determined in relation with the desired output form. This, for example, may indicate a minimizing of high frequency, low frequency or phase shift errors. The adaptive control adjusts the system response characteristics, gain or spectral, until the rms error by the programmed criteria is minimized. This general system permits continual adjustment of the system response to "best fit" the input spectral characteristics, and even permits a variation in the "best fit" such as a variable spectral weighted mean square criterion.

REFERENCES TO CHAPTER 6

1. R. D. Ormsby and M. C. Smith, "Capabilities and Limitations of Reaction Spheres for Attitude Control," *ARS Journal*, Vol. 31, No. 6, June 1961, pp 808–812.
2. J. J. Rodden and R. J. Pollak, "Servo Control of a Variable Thrust Rocket," *ARS Journal*, Vol. 30, No. 10, October 1960, pp 964–967.
3. R. R. Mohler, "Stability and Control of Nuclear Rocket Propulsion," IRE Transactions on Automatic Control, Vol. AC-7, No. 2, March 1962, pp 86–96.

4. A. S. Locke, *Guidance*, D. Van Nostrand Co., Inc., Princeton, N. J., Chapter 6 (Mathematical Groundwork) and Chapter 7 (Servo System Theory).
5. M. F. Gardner and J. L. Barnes, *Transients in Linear Systems*, Vol. I, John Wiley & Sons, Inc., New York, Chapter 6.
6. H. W. Bode, *Network Analysis and Feedback Amplifier Design*, D. Van Nostrand Co., Inc., Princeton, N. J.
7. J. G. Truxal, *Automatic Feedback Control System Synthesis*, McGraw-Hill Book Co., New York, 1955.
8. D. B. Newman, "The Analysis of Cross-Coupling Effects on the Stability of Two-Dimensional, Orthogonal, Feedback Control Systems," IRE Transactions on Automatic Control, September 1960, pp 314–320.
9. Z. V. Rekasius and J. E. Gibson, "Stability Analysis of Nonlinear Control Systems by the Second Method of Liapunov," IRE Transactions on Automatic Control, January 1962, pp 3–15.
10. S. Kodama, "Stability of Nonlinear Sampled-Data Control Systems," IRE Transactions on Automatic Control, January 1962, pp 15–23.
11. E. I. Jury, "Convolution Z-Transform Method Applied to Certain Nonlinear Discrete Systems," IRE Transactions on Automatic Control, January 1962, pp 57–64.
12. H. R. Farrah, "A Non-Linear Analysis Technique for an On-Off Servo with Velocity Inputs," IRE Transactions on Automatic Control, March 1962, pp 13–21.
13. C. N. Shen, "Synthesis of High Order Non-Linear Control Systems with Ramp Input," IRE Transactions on Automatic Control, March 1962, pp 22–37.

CHAPTER 7

POWER SOURCES

7.1 INTRODUCTION

Power sources in space may be self-contained or they may draw energy from solar radiation. The primary vehicle propulsive power may be by chemical, nuclear, electronic, or photon means. Auxiliary power may be derived from batteries, solar cells, nuclear energy, or high-energy fuels. The variety of power sources available and in use confirms that each of the various methods of power generation will have its advantages and disadvantages in the various applications, and that one should attempt a comparative evaluation of the then current states-of-the-art in the various power generation fields before selecting a power system for a particular application.

A direct comparison in terms of total energy available per unit weight or volume is one method of comparison. However, solar radiated power is available at an energy rate and the time of usage of a fixed energy supply must be considered in the comparison. Also, the rate of maximum energy expenditure, the duration of energy expenditure, the regulation required, the operating environment, and the power system weight budget may be determining factors.

Figure 7–1 illustrates the general areas of usefulness of the various space power systems which might be used to generate electrical power. The boundaries indicated are rather flexible, and change as the various states-of-the-art progress. However, this illustration indicates the general area in which one should expect to choose a type of electrical power system based on the power requirements and the duration of use.

7.2 BATTERIES

Batteries or electrochemical cells consist of an anode, a cathode, and an electrolyte. The anode is a reducing agent which gives up electrons, the cathode an oxydizing agent which receives electrons, and the electrolyte serves as an electron barrier which nevertheless permits current flow by ionic conductance within the cell.

The best batteries currently are generally held to be nickel-cadmium which produce about 12 watt-hours per pound (wh/lb). Silver-cadmium

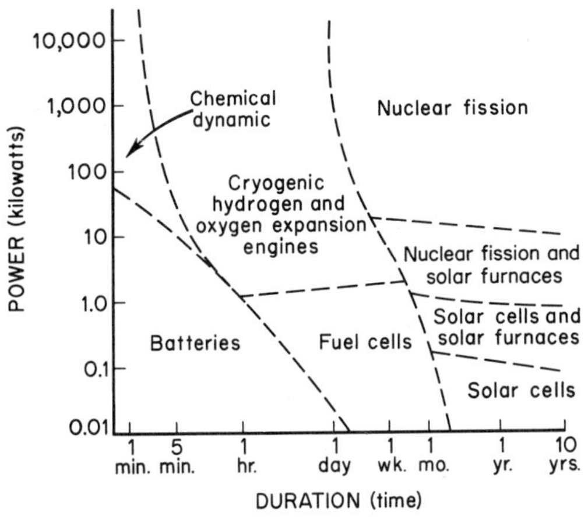

FIG. 7-1. Space Power Systems.

batteries are being developed (Ref. (*1*)) which are expected to produce 24 wh/lb. A study of new materials for battery development (Ref. (*2*)) has indicated that a magnesium/magnesium-perchlorate/mercuric-oxide cell is capable of 50 wh/lb and 3.2 wh/cu. in., and that a magnesium/magnesium-perchlorate or magnesium-bromide/*m*-dinitrobenzene (DNB) is capable of producing 90 wh/lb. Lozier (Ref. (*2*)) has estimated the current experimental and maximum electromechanical systems, as indicated in Fig. 7–2. This comparison indicates the superiority of the DNB over the others including a conventional (LeClanche) dry cell. Figure 7–3 illustrates the voltage decay of these cells through a constant load as a function of time (note: the Burgess Electro-Ore is a LeClanche cell).

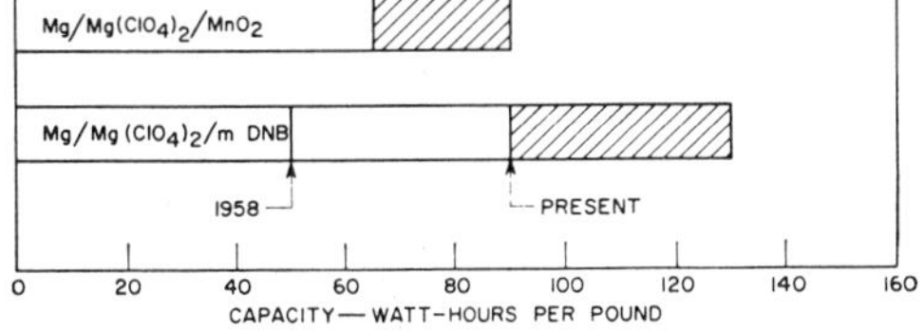

FIG. 7-2. Capacity of Various Electrochemical Systems.

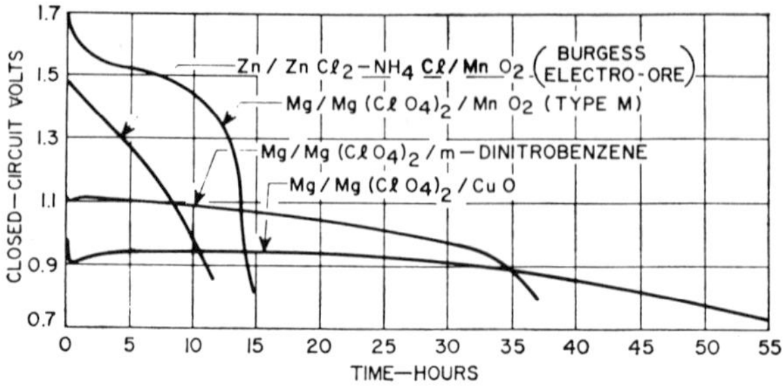

FIG. 7-3. Capacity of A-Size Dry Cells Discharged Continuously through 162/3 Ohms.

From Fig. 7–3 it is quite apparent that the magnesium-DNB and a magnesium-copper cell have much better decay characteristics than the others.

The discharge period or rate affects the capacity of a battery as well. To illustrate this, Figures 7–4 and 7–5 compare various promising magnesium batteries in terms of capacity versus discharge time. The choice of a particular battery for a given mission is thus dependent on the discharge rate in addition to capacity; weight budgets may have to be adjusted accordingly.

7.3 FUEL CELLS

A fuel cell is essentially a battery with replenishable cathode and anode materials. This permits the use of light-weight, very active cathode and anode materials. As indicated in Table 7–1, a hydrogen-oxygen fuel cell is theoretically capable of delivering better than one kilowatt-hour per pound (weight includes fuel). The energy/weight ratio improves with temperature as indicated by Table 7–1, and some provision for thermal agitation (heating) may be desirable.

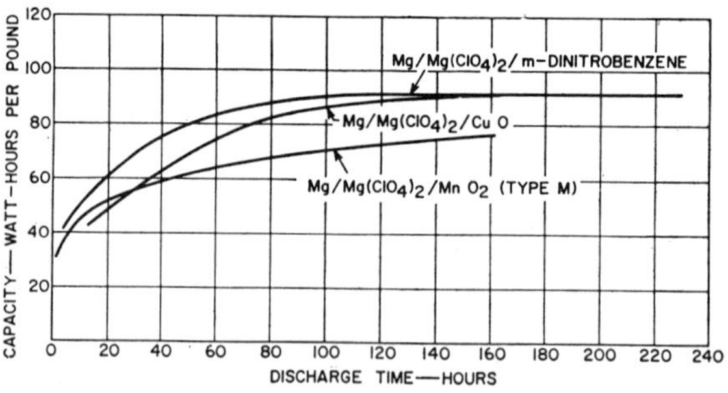

FIG. 7-4. Capacity per Unit Weight of Magnesium Primary Cells as a Function of Discharge Time.

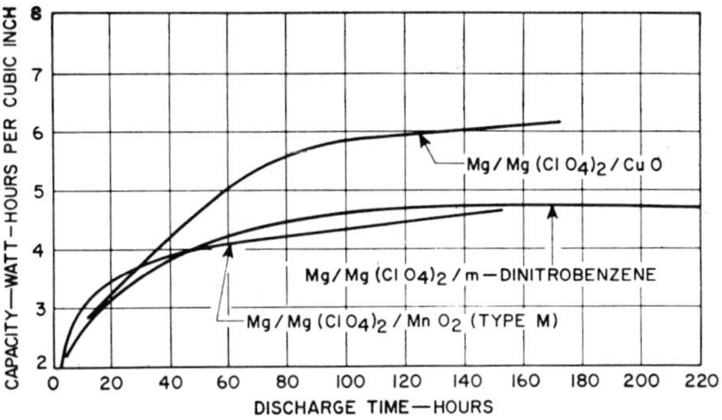

FIG. 7-5. Capacity per Unit Volume of Magnesium Primary Cells as a Function of Discharge Time.

TABLE 7–1. ENERGY/WEIGHT RATIO FOR VARIOUS EXPERIMENTAL FUEL CELLS OPERATING AT 0.75 OPEN-CIRCUIT VOLTAGE

Reactants	Cell Temperature	kw-hrs/lb
Hydrogen-oxygen	240°C	1.07
Hydrogen-oxygen	65°C	1.01
Hydrogen-oxygen	600°–800°C	1.20
Carbon-Monoxide-oxygen	600°–800°C	0.40
Methane-oxygen	600°–800°C	0.89

A simplified fuel-cell is portrayed schematically in Fig. 7–6 to illustrate the cell operation. Both anode and cathode are of porous, waterproof material (e.g., carbon or a membrane). Fuel pressurization is commonly

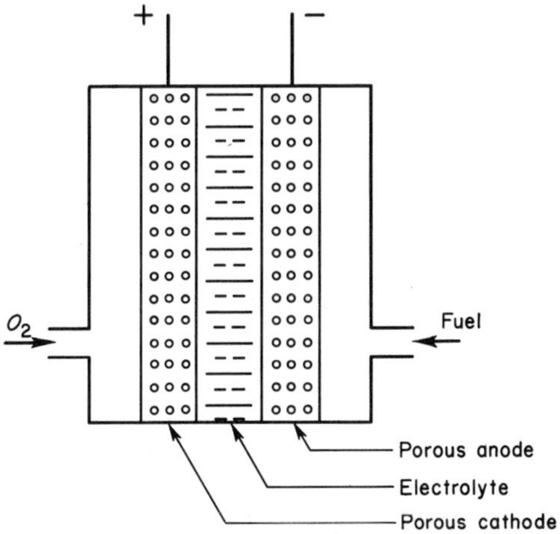

FIG. 7-6. Simplified Schematic of a Fuel Cell.

used. Sodium amalgum and metallic hydrides are being studied as possible fuels (Ref. (*3*)) for various handling and ion-exchange reaction reasons. Regenerative techniques (Ref. (*4*)) and radioisotope activation (Ref. (*5*)) are being studied as improvements as well.

Fuel cells offer a great development potential at the present time, and it is expected that they will come into greater use as their characteristics stabilize.

7.4 SOLAR CELLS

The advent of space vehicle operations has made feasible the direct tapping of solar radiation energy through the use of solar cells. Solar cells are essentially photo-electric converters of various types which can deliver electrical power in proportion to the intercepted solar radiated power. The characteristics of solar cells may be described in terms of spectral response, output power per unit radiation-collecting area, and life. Special consideration must be given to the solar eclipse during each orbit period in planning solar cell use. As a special problem, solar cell life and efficiency is impaired by particle radiation such as is encountered in the Van Allen radiation belts.

Silicon solar cells are usually made from 0.5 mm lapped slices of single crystal silicon. A typical solar cell is illustrated in Fig. 7–7. The spectral response of silicon solar cells is generally in the 0.5 to 1.0 micron region. Fifteen percent efficiency has been attained. Gallium-arsenide solar cells have a spectral response generally in the 0.6 to 0.9 micron region, and provide efficiencies on the order of 10 to 12 per cent. Life characteristics are expected to run to ten years. Gallium-arsenide cells are much more (10 to 20 times) resistant to proton bombardment than silicon cells.

A solar cell array must be oriented to face the sun or else the radiation intercepted will be reduced, as a cosine function. Eclipsing must be synchronized with power demands, or some other power system must be

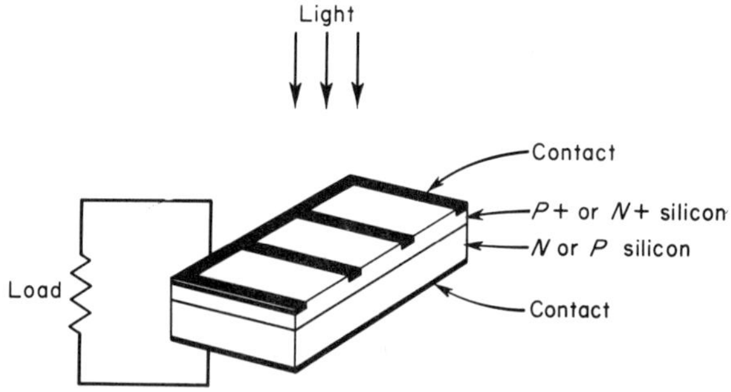

Fig. 7-7. Solar Cell Diagram.

available such as a solar cell charged battery. The storage battery size required per unit load current is related to orbit parameters by, (Ref. (*6*)),

$$\frac{C}{I_L} = \left(\frac{1 + \beta}{\alpha}\right) \frac{t_d}{t_i} \tag{7-1}$$

where C is the ampere-hour capacity of the battery, I_L is the load current, α is the maximum permissible overcharging current per ampere-hour of capacity (in hours^{-1}), β is the minimum overcharge required per ampere-hour of discharge, t_d is the maximum dark time per orbit period and t_i is the minimum illuminated time per orbit period.

The use of solar cells may be extended by employing large focusing mirrors to collect a much greater amount of solar radiated energy and to concentrate this energy on the photo electric cell. This has the advantage of working the cell at a higher efficiency because the higher density energy conversion will cause a higher operating temperature. Also, besides using the area of the large collector, even degraded by reflection and focusing factors, one may expect to obtain more power per pound than if the total collector area were composed of solar cell material for direct energy conversions, and had to support the penalty of interconnecting circuit electrical losses and weight.

Some current solar cell design figures are: at Earth radius and facing from the Sun, 6 to 9 watts per square foot at an operating temperature of 100°F and with an 85 to 90 per cent area utilization (packing factor). Spin stabilized systems generally achieve about 25 per cent utilization, so that sun-orientation will be employed generally for high power requirements (more than 100 watts). The weight of solar cell arrays is generally one to three pounds per square foot depending on the structural support used and the amount of charged-particle protection provided. Thus one may expect to obtain solar power usefully at the rate of about five watts per pound of array.

7.5 COMBUSTION DEVICES

Combustion or chemical dynamic devices are of interest generally if the energy discharge is to be of large magnitude over a short period. In such vehicles as interceptor missiles this requirement generally exists, and is met by solid propellant gas generators (if not batteries). These generators provide an exhaust jet which may be used directly for vehicle control or orientation, to pressure a hydraulic system, or to drive a turbine which operates an electric generator. Coolants are sometimes introduced in such systems to reduce the flame temperature and provide slower burning rates and better environmental stability. The use of coolants reduces the specific energy content per pound however. The best figures currently available are on the order of 980 hp-sec/lb. (Ref. (*7*)), or 203 wh/lb. The energy available from the fuel must be penalized by engine, structure and turbo-generator weight so that the combustion device is only competitive with high capacity batteries for short periods of time. The momentum imparted to a space vehicle by escaping gas from this type of power supply must be taken into account.

7.6 CRYOGENIC ENGINES

The use of cryogenic propellants opens a new domain in space propulsion. Liquid rockets offer certain advantages over solid propellant rockets in terms of high specific impulse, control of burning rate, and longer sustained burning times. By using cryogenic (super-cooled) fuels, such as liquid hydrogen which boils at −432.2°F, the greatest possible chemical specific impulse is obtainable. Of importance to the electronic engineer is the requirement for extreme thermal insulation, and the super-conductivity of materials at such low temperatures. The electrical circuits associated with cryogenic engines (e.g., starting and stopping, and perhaps electrical power generating) may have to consider the effect of these temperatures. At temperatures below 10°K for metals and 20°K for alloys the resistivity drops rather abruptly to zero. In addition, the super conductor appears to carry its current by "skin effect" in a manner similar to high-frequency conductors. This effect permits the control of current flow by magnetic fields, and special circuits are being developed which make use of these properties (Ref. (*8*)).

The derivation of electrical power from a liquid or cryogenic rocket will be generally as with the solid propellant units, that is by a turbo-electric generator. The combustion product gases may be tapped with pressure regulating valves to operate the generator for the duration of fuel oxydation (burning). It is also feasible to use a low boiling point liquid (e.g., hydrogen) to operate a turbine merely by the gas expansion. Such systems hold the potential of providing power in much greater quantities than batteries and fuel cells for the same periods.

7.7 NUCLEAR POWER

To obtain large quantities of power for extended periods, nuclear sources appear most feasible. These systems generally consist of a nuclear reactor and closed-cycle cooling system (e.g., sodium-potassium eutectic metal), a sealed working fluid gaseous system (e.g. mercury-vapor), and a turbine which drives the fluid pump and an electrical alternator. As illustrated in Fig. 7–8, the working fluid may be condensed by thermal radiation. Some typical values for the SNAP-2 Nuclear Auxiliary Power

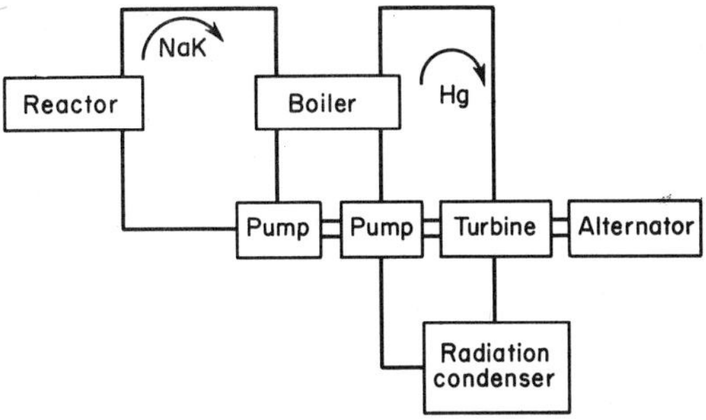

FIG. 7-8. Nuclear Auxiliary Power Unit.

Unit (APU) are (Ref. (*9*)) 3 kilowatts for one year for less than 600 lbs, or better than 43000 wh/lb. Even better figures have been quoted (Ref. (*10*)); 70 Kw for 1300 lbs. The SNAP-3 unit for the Transit IV-A satellite is only grapefruit size, is supposed to last 5 years, and weighs about one-fifth of an equivalent battery-solar cell system (Ref. (*11*)).

In using nuclear engines with manned vehicles the radiation and shielding become major design problems necessitating some great distance between the manned capsule and the engine. This will pose some design problems in distributing power and controlling the engine over long distances.

7.8 ELECTRIC PROPULSION

Electric propulsion can provide power for indefinite periods. Although of lower thrust level, the mass expended is so little that the specific impulse of electric propulsion "fuels" may be 10,000 sec (lb-sec/lb) as compared with 1,000 sec for nuclear engines (Ref. (*12*)). Three types of electric propulsion engines are in development:

(1) Electro-thermal or arc-jet engines,
(2) Electro static or ion engines, and
(3) Electromagnetic or plasma engines.

Arc jet engines achieve thrust by heating the propellant with an electric arc, and then ejecting the heated propellant through a conventionally-designed rocket nozzle. Ion engines obtain thrust by accelerating charged particles with electrostatic fields much as electrons are accelerated by grids in electronic vacuum tubes. Plasma engines achieve thrust by accelerating an ionized plasma propellant with an electromagnetic field.

Figure 7–9 illustrates an arc-jet engine. The cathode and anode are of tungsten and the insulation of boron nitride. The working fluid is injected into the engine, heated by the arc, and expelled through the nozzle. Using hydrogen as a working fluid, specific impulses up to 1000 sec have been obtained at a 30 Kw power level (Ref. (*13*)).

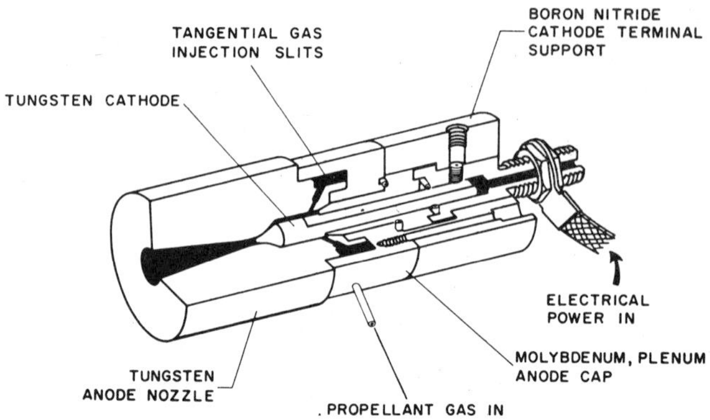

FIG. 7-9. Arc-Jet Engine.

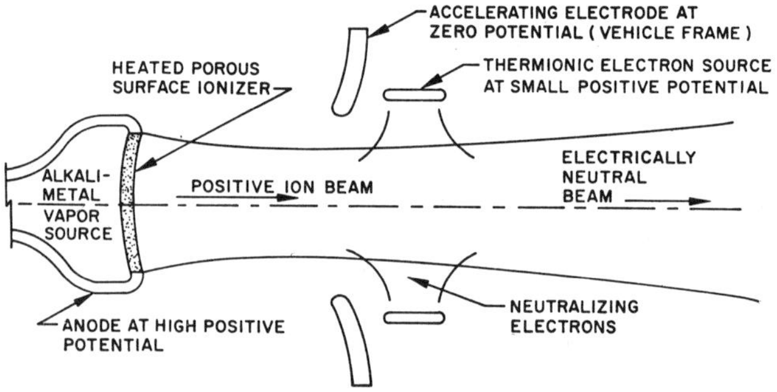

FIG. 7-10. Electrostatic Propulsion Schematic.

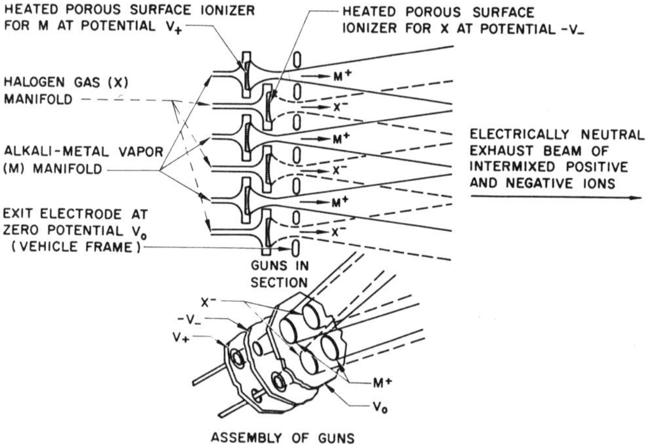

FIG. 7-11. A Recessed Checkerboard of Positive and Negative Ion Guns.

An ion engine is illustrated in Fig. 7–10. An alkali metal vapor source at high positive potential is heated to produce free positive ions (typically cesium) which are accelerated by a zero potential electrode through a hole to the outside space. Positive focusing electrodes are generally necessary although not shown in the schematic. To prevent the ions from slowing down and returning to the spacecraft the positive ion beam is neutralized by a thermionic electron source immediately after accelerating, ideally at peak velocity. As an alternative, a negative ion engine may be used in combination with a positive ion engine (Ref. (*14*)). Halogen ions have been selected to attempt mass matching between the positive and negative ions. Figure 7–11 illustrates a combination of positive and negative ion guns for this purpose. Table 7–2 indicates the positive-negative ion mass ratios feasible. For maximum efficiency it will be necessary to equalize the velocities of the ions. This can be accomplished by adjusting the accelerating potentials in proportion to the masses,

$$\frac{V^+}{V^-} = \frac{m_+}{m_-} \tag{7–2}$$

Specific impulses of 800 sec. appear feasible for ion engines (Ref. (*15*)), and measurements have been quoted as high as 11000 sec. (Ref. (*16*)).

TABLE 7–2. MASS RATIOS OF ALKALI-METAL AND HALOGEN-ION PAIRS

Positive ion mass	Negative ion mass	Ion mass ratio
m_+ (amu)	m_- (amu)	m_+/m_-
K^+ 39.1	Cl^- 35.5	1.10
Rb^+ 85.5	Br^- 79.9	1.07
Cs^+ 132.9	I^- 126.9	1.05

The plasma engine is essentially an electromagnetically controlled jet of hot gas. Heavy ions such as charged colloids are employed as the jet plasma. Colloid engines may be designed for specific impulses of 2000 to 10,000 sec. (Ref. (*17*)). Colloids may be produced by spray injection or condensation techniques. A considerable effort is underway to evaluate colloid sources and methods of injecting charged colloids into a gas.

7.9 PHOTON ROCKETS

To achieve a velocity approaching a significant portion of the speed of light will require techniques beyond those previously described. The photon rocket is sometimes suggested for this application. The principle is to eject photons from a ship and to equate the momentum of the photons in one direction to the momentum of the ship in the other.

If we convert a fuel completely into photons (no heat or other losses) and eject it in one direction we may achieve 99.5 per cent of the speed of light by expanding 95 per cent of our mass (Ref. (*18*)). To stop at some destination, attain the same velocity for return and brake to a stop at the starting point will leave only $(.05)^4$ of our original mass!

The use of photon rockets for manned spacecraft propulsion is a long way off, at a minimum. However, the conversion of a mass to energy for propulsion as in the photon rocket is a significant goal in that this would provide the highest specific impulse of all power plant systems. The photonic rocket jet would have a tremendous piercing effect, and as such would be a "death ray." That is, the conversion of mass directly to photons, and focusing it with a high energy concentration would provide a ground-based weapon capable of atomizing objects at significant distances (hundreds of miles) with a speed of light capability (Ref. (*19*)).

References to Chapter 7

1. R. C. Shair, G. Rampel and E. Kautner, "Hermetically-Sealed Nickel-Cadmium and Silver-Cadmium Storage Batteries," IRE Transactions on Military Electronics, January 1962, pp. 67–71.
2. G. S. Lozier, "New Materials for Primary Batteries," IRE Transactions on Military Electronics, January 1962, pp. 72–77.
3. H. W. Fox and R. Roberts, "Fuel Cells," IRE Transactions on Military Electronics, January 1962, pp. 46–57.
4. S. E. Voltz and M. D. Read, "Regenerative Hydrogen-Oxygen Fuel Cell," IRE Transactions on Military Electronics, January 1962, pp. 63–67.
5. M. Lang, "Radioisotope-Activated Fuel-Cell Electrodes," IRE Transactions on Military Electronics, January 1962, pp. 58–62.
6. N. W. Snyder, "Solar-Cell Power Systems for Space Vehicles," IRE Transactions on Military Electronics, January 1962, pp. 84–91.
7. B. M. Corley, "Solid-propellant Gas Generators," *Space/Aeronautics*, October 1961, pp. 72–76.
8. J. B. McFerran, "Applied Cryogenics Spurs Electromechanical Design," *Space/Aeronautics*, January 1961, pp. 67–71.
9. J. R. Welch, H. M. Dieckamp and D. J. Cockerman, "Snap 2-Nuclear Space Power System," *Astronautics*, December 1960, pg. 24.
10. B. Miller, "Nuclear-Thermionic Power Unit Proposed," *Aviation Week*, July 10, 1961, pg. 61.
11. H. Gettings, "Era of Nuclear Space APU's Arrives," *Missiles and Rockets*, July 10, 1961, pp. 12–13.
12. J. W. Stearns, "Electro-propulsion System Applications," *Astronautics*, March 1962, pg. 22.
13. M. I. Yarymovych, F. A. deWeiss, and R. R. John, "Feasibility of Arc-Jet Propelled Spacecraft," *Astronautics*, June 1962, pg. 36.
14. M. A. Gilleo and S. W. Kash, "Propulsion by Composite Beans of Negative and Positive Ions," *ARS Journal*, May 1961, pp. 621–627.
15. R. Hayes, R. N. Seitz and E. Stuhlinger, "The U.S. Ion Propulsion Program," *Astronautics*, January 1961, pg. 30.
16. A. T. Forrester and J. M. Teem, "Advances in Electrostatic Propulsion," *Astronautics*, January 1961, pg. 34.
17. A. L. Cox, "Toward Colloid Propulsion," *Astronautics*, 1962, pp. 56–59.
18. J. R. Pierce, "Relativity and Space Travel," *Proceedings of the IRE*, June 1959, pp. 1053–1061.
19. E. Saenger Interview, "Fission, Fusion, and Photonic Rockets," *Missiles and Space*, May 1961, pp. 18–21.

CHAPTER 8

PAYLOAD CONSIDERATIONS

8.1 INTRODUCTION

The variety of electronic payloads employed by space vehicles is enormous. The great variety of missions discussed in Chapter 1 and the need for superior data on the space and extra-terrestrial environment dictate these payloads. The purpose of this chapter is to discuss some of the considerations bearing on electronic payloads or systems employed in the three most common operational missions of space vehicles: weaponry, reconnaissance and surveillance, and communications. The related electronic sensors for these three systems are fuzes, spotting and measuring sensors, and data relay systems. Subsequent sections discuss some of the peculiarities of these electronic systems and the unusual demands placed on the host spacecraft for their most efficient employment.

8.2 WARHEADS AND FUZING*

Electronic fuzes are used in many missiles and projectiles. In space, we may expect to find interdictory missiles employing such fuzes of three broad, discernible categories: proximity, timed, and command. Proximity fuzes will generally be employed with missilcs bearing relatively small warheads such that a small miss distance is expected at intercept. Such missiles will generally be of the homing type in their terminal phase although beam-rider and command guided missiles may employ small warheads as well. The key point is that the warhead lethal radius or kill pattern must be of the same order of magnitude as the missile guidance accuracy. The type of proximity fuze used will depend on the warhead kill pattern; the design range of the fuze sensor being directly related to the maximum lethal range of the warhead, and the sensor directivity related to the warhead blast pattern directivity. Figure 8–1 illustrates this concept. If we postulate an interceptor missile attacking a satellite in a "hard" intercept (with a large relative velocity), we see that the optimum time for the warhead to detonate is immediately prior to intercept; thus the fuze must sense the target so as to detonate the warhead at this optimum point. The electronic fuze will be essentially a small radar set which uses a

*See also Grayson Merrill, *et. al.*, Operations Research, Armament, Launching, Princeton, N.J.: D. Van Nostrand Co., Inc., 1956.

forward-canted, saucer shaped beam of very fine cross-section. Looking at Fig. 8–1, we may say that the warhead fragments will have a blast-generated velocity, V_γ, in the direction normal to the missile axis, and also a velocity V_m due to the relative motion of the missile, V_m. Thus, the warhead fragments will be impelled with a velocity of,

$$V_w = V_m + iV_\gamma, \text{ relative to the target} \tag{8–1}$$

(Note the assumption that the vehicle longitudinal axis has been stabilized in the direction of the velocity vector. This is a requirement necessarily imposed on the vehicle stabilization and control system so that the warhead "aiming" can be predictable.)

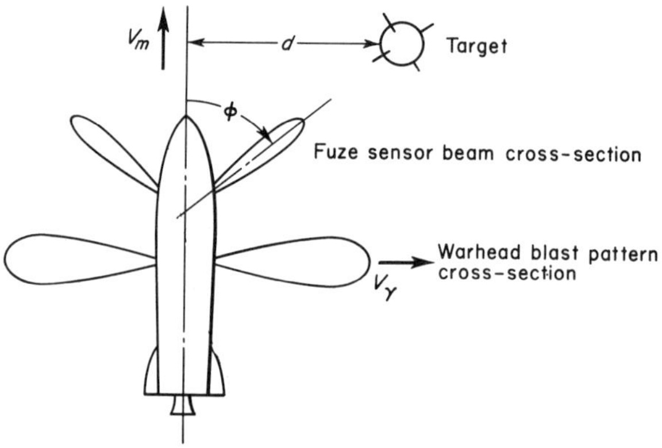

FIG. 8-1. Proximity Fuze.

The fuze antenna pattern is canted forward to provide a sensing of the target in advance of the warhead detonation. Thus, the fuze senses the target when it is illuminated by the thin, saucer-shaped beam. The optimum design relation is, assuming instantaneous warhead detonation on sighting of the target by the fuze,

$$(V_m + iV_\gamma)T = d \cot \varphi + id \tag{8–2}$$

The solution for φ is thus,

$$\varphi = \text{arc cot}\left[\frac{V_m T}{d}\right] \tag{8–3}$$

where,

$$d = V_\gamma T \tag{8–4}$$

or,

$$\varphi = \text{arc cot}\left[\frac{V_m}{V_\gamma}\right] \tag{8–5}$$

Note that the assumptions were made that the interceptor had a significant velocity relative to the target, and that the interceptor was stabilized along the velocity vector. If the relative velocity of the warhead fragments is not as predicted or if the missile velocity varies through quite a wide range the operation of this fuze/warhead system will not be optimal but will be degraded according to the random effectiveness of the warhead blast pattern in the actual intercept geometry.

The omnidirectional warhead is probably most typical of the atomic type in which kill probability is generally taken as a direct radial function. In this case, if the lethal radius can be made several times the nominal intercept miss distance, either a time fuze or command fuze may be employed. A time fuze is one which acts as either a free running clock or after time integration of an acceleration measurement to ascertain that a certain pre-computed velocity or distance has been attained.

A command fuze is one in which the warhead is detonated upon receipt of a command from an external station which has determined that the intercept location has been achieved. Such a system is illustrated in Fig. 8–2. A target tracking system and an interceptor tracking system each provide data to an intercept computer which determines when a satis-

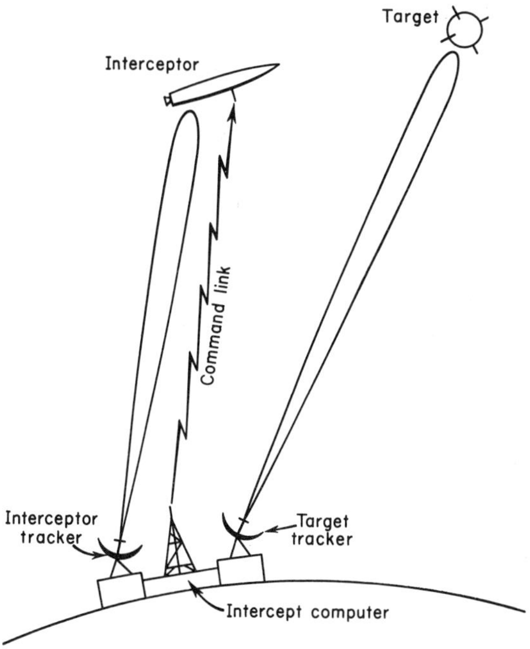

FIG. 8-2. Command Fuze.

factory intercept has been achieved and issues a detonation command over a command link. Many variations are available. A sensor such as a TV camera might be included in the interceptor, and a remote operator might determine intercept based on the TV display. The command link might be included as modulation or lack of modulation on the interceptor tracking link. The detonation command might be in the form of turning the transmitter off so that the missile would always detonate at some time, and would be capable of detonating early if the command transmitter failed. The latter system insures detonation at some time. The important concept is that a fuze system must be related to the warhead requirements and capabilities, and that these are necessarily related to the guidance accuracy during the terminal phase of flight.

8.3 RECONNAISSANCE, SURVEILLANCE, AND MAPPING SENSORS

The primary performance parameters of reconnaissance, surveillance, and mapping sensors are their resolution, coverage, and orientation. These parameters must be examined in terms of satisfying a particular mission in a useful period of time and providing the desired data in a useful and timely form. The wavelength employed by the sensor may be in any portion of the electro-magnetic spectrum, but will generally be chosen on the basis of propagation factors, the size of equipment needed to achieve the requisite performance, and the need for high or low broadcast power or quiescent (passive) operation. These sensors may be employed as guidance devices if sufficient information is obtainable in good detail and in sufficient time to effect course corrections. The analysis of sensor data is particularly important from the point of view of determining artifacts of interest, their location and deployment. Also, the screening of sensor-gathered data to arrive at only those particular data of primary

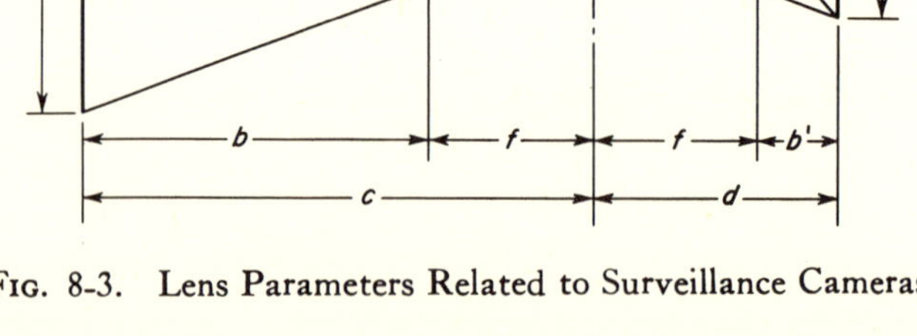

FIG. 8-3. Lens Parameters Related to Surveillance Cameras.

(or priority) interest is of extreme importance for military urgency, for communication bandwidth reduction (hence range extension) and, for selection of particular data for reexamination in greater detail.

8.3.1 Resolution, Coverage, and Orientation

Reconnaissance, surveillance, and mapping sensors may be of several types: photographic, scanning or fixed beam. In general, we may say that the vehicle motion provides scanning along its axis, and that data are taken in frames, lines, or spots in sequence.

The lens is the primary factor in determining camera picture quality. Figure 8–3 defines the basic geometric relations necessary to understand and use a lens. To focus a lens the following relation must be obtained:

$$\frac{1}{c} + \frac{1}{d} = \frac{1}{f} \tag{8–6}$$

or,

$$b \cdot b' = f^2 \tag{8–7}$$

This provides a magnification of,

$$m = \frac{-a'}{a} \tag{8–8}$$

The "f" number, F, of a lens is not to be confused with the focal length, f. The "f" number is the focal length divided by the effective diameter of the lens opening,

$$F = \frac{f}{d_{\text{eff}}}, \tag{8–9}$$

and is a measure of the lens' light-gathering power. The "f" number varies with light intensity according to the effective area of the lens opening. Thus, "f" numbers of 1, 1.4, 2, 2.8, 4, 5.6, 8, 11.2, 16, 22.4, 32, 45, etc. indicate successive decreases of 1/2 in light intensity.

The hyperfocal distance, H, is the nearest distance at which a lens can be focused so that all images out to infinity are in sharp focus.

$$H = \frac{f^2}{F \times c \times 12}, \tag{8–10}$$

where c is the diameter of the circle of confusion in inches; f is in inches, and H is in feet.

The limits of depth of field are,

$$D_n = \frac{H \times u}{H + \left(u - \frac{f}{12}\right)} \tag{8–11}$$

$$D_f = \frac{H \times u}{H - \left(u - \frac{f}{12}\right)}, \tag{8–12}$$

where H is in feet, u is the distance to which the camera is focused in feet, f is in inches; D_n is the near distance in feet, and D_f is the far distance in feet. Camera resolution is limited by film grain size or the number of lines per millimeter one can distinguish, motion blur, and image perspective angle. The ground resolution is determinable from,

$$G = \frac{h}{f \times R} \tag{8–13}$$

where G is in feet, h is the altitude in feet, f is in inches, and R is the film resolution in lines per inch.

Image blur is caused by camera movement while the shutter is open. Thus, a movement approaching G/h radians will degrade the ground resolution available, and indicates a requirement for camera mounting which is insensitive to vehicle motions of high frequency. Gyroscopically slaved camera mounting platforms are the most sophisticated devices available, although simple shock mounting may suffice for many applications.

Photo resolution is quoted in lines per millimeter generally, but is really line pairs per millimeter. Thus the comparable television resolution is,

$$\frac{N_{TV}}{L} = 2R \tag{8–14}$$

where N_{TV}/L is the number of television lines per millimeter.

Image motion compensation (IMC) is employed to improve the resolution and perspective of aerial photography, and may be necessary for low altitude aerial photo mapping of other planets and the major satellites.

$$\text{IMC} = \frac{1.6878 \times V \times f \times \sin \delta}{h} \tag{8–15}$$

where V is the true ground velocity in nautical miles per hour, h is the altitude in feet, f is the focal length in inches, and δ is the depression angle in degrees from horizontal. IMC is the true rate of image movement in inches per second, and may be applied to a camera film during the shutter-open period as a function of a V/h measurement to provide sharper images hence better resolution.

Automatic exposure control (AEC) is commonly employed in unmanned photography; it is merely the control of exposure period as a function of a light intensity measurement. Thus with AEC and ICM in combination the resolution of a camera is a direct function of film resolution, mount stabilization, lens focal length, altitude and depression angle (or object perspective).

The coverage of a camera is determinable from Fig. 8–4, which shows that,

$$W = \frac{h}{f} \times w, \qquad (8\text{–}16)$$

i.e., the ground coverage width is to the altitude as the camera frame width is to the focal length.

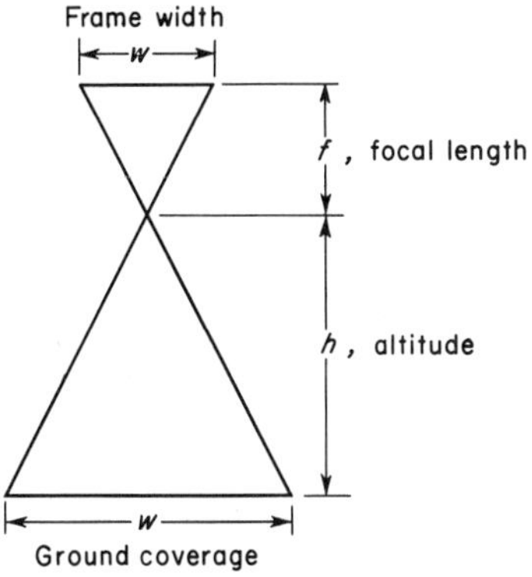

FIG. 8-4. Camera Coverage.

Television and scanning infra-red sensors scan areas as a sequence of lines. The resolution element is the spot diameter in the scan direction, and the line width or reciprocal of lines per inch in the forward direction. A continuous strip film is generally exposed, the forward motion of the strip, V_s, being related to the focal length, f, altitude, h, and ground velocity, V_G by,

$$V_s = \frac{f}{h} V_G \qquad (8\text{–}17)$$

The width of ground coverage is again evident from Fig. 8–4. The ground area is scanned at a rate of $V_G W$, so that the total memory capacity of a film may be taken in ground coverage as

$$\text{Area Capacity} = V_G W T = lw \qquad (8\text{–}18)$$

where T is the total flight time during which the strip camera is used, l is the length of film available, and w is the film width. A frame-exposed camera generally is controlled to overlap adjacent frames by 10 per cent to ensure continuous coverage or by 60 per cent to obtain continuous 50 per cent stereo overlap coverage, and the area capacity is reduced accordingly,

$$\text{Area Capacity} = W^2(1 - P) = V_G W T(1 - P) \qquad (8\text{–}19)$$

where a square format has been assumed, and P is the per cent overlap between adjacent frames.

Side-looking radars are employed as mapping sensors at low altitudes, and may prove particularly useful for mapping such planets as Venus, Jupiter, Saturn, etc. A side-looking radar generally employs a long antenna along the fuselage of the carrying vehicle, thus providing a very narrow fan-shaped beam normal to the fuselage axis. Short pulses may be transmitted, and the echoes scanned as video signals on a film to provide a continuous radar strip-map where resolution in the normal direction will be related to pulse length, τ, and resolution in the forward direction proportional to the product of angular resolution, γ, and echo time, t_e. Thus the incremental area resolution of a radar strip map may be

$$R_R = k \times \tau \times \gamma \times t_e \tag{8-20}$$

where k has been introduced to provide for resolution improvement (or degradation) dependent on the data processing techniques employed. The ability to track the pulse leading edge will improve the resolution (small k).

A synthetic antenna aperture larger than the actual may be obtained by summing the returns from several pulses in proper phase. Thus, an antenna of length L may be used to obtain an effective antenna of length nL if we may sum the individual returns, $R(t)$, as,

$$R_n(t) = \sum_{i=1}^{n} R_i(t)\ \varphi_i(t) \tag{8-21}$$

where,

$$\varphi_i = \varphi_{i+1} + T_p \tag{8-22}$$

T_p is the period between pulse return summations, and is related to the antenna length, L, and the forward velocity, V, by

$$T_p = L/V \tag{8-23}$$

$$\varphi_{n-1} = T_p \tag{8-24}$$

$$\varphi_i = (n-1)\ T_p \tag{8-25}$$

The effective time of the aperture sighting is

$$\varphi_{\text{eff}} = \varphi_{n/2} \tag{8-26}$$

Lest the above indicate that a panacea is attainable, practical applications dictate than an n of a few integers be the largest practical before the antenna segments deviate from their assumed geometric relation because of vehicle path deviations. If large values for n should become practical, another limit will be approached at which the data effective time, $\varphi_{n/2}$, will be so late that the data becomes stale. If we assume a value of $n = 5$, we may see that we can obtain an effective aperture of 5 times the normal length, and that the data will be stale by only two summing periods. By suitable data storage, retrieval, and summing we may obtain an effective aperture of this type for every pulse transmitted. Note that the summing period, T_p has been related to the ratio of antenna size to forward velocity, L/V, rather than the period between pulses, $1/PRF$. This indicates that the pulse period, $1/PRF$, may be used as T_p to determine this ratio and, thus, the minimum antenna length, L, that may be employed to obtain an effective continuous antenna for an unambiguous radar range, $1/PRF$. The flight path stability will generally be the determining factor for the maximum synthetic aperture length obtainable. The use of segmented effective apertures rather than continuous is a possibility where the pulse period, $1/PRF$, exceeds the aperture to velocity, L/V ratio. Such synthetic apertures may be considered as large segmented antennas with apparent side lobes.

Antenna segment summing periods may be varied electronically to "steer" or "squint" the beam to other than the normal to vehicle direction. Usually, the beams are steered in the forward direction to provide forward looking, "high-resolution" radar. The steering may be likened to summing pairs of beams on a common difference time, or to summing the set of beams on a common doppler frequency, the normal side-looking case being summed at zero difference or zero doppler. Lines of constant difference or constant doppler or "isodops" are cones. The intersection of these cones with a flat earth provides a set of hyperbolas as illustrated in Fig. 8–5. The targets of interest are generally well beyond the distance from

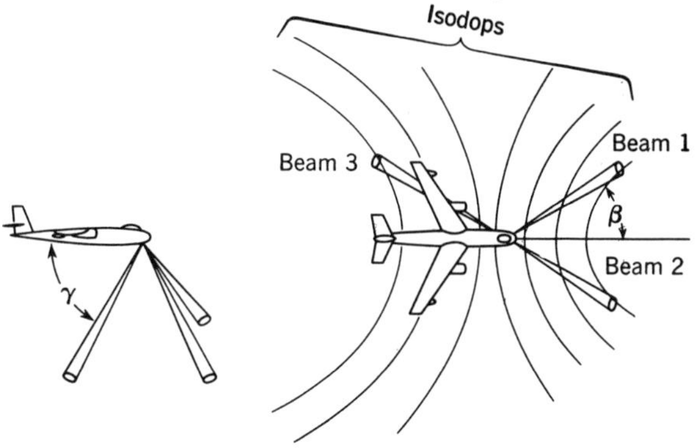

FIG. 8-5. Isodops.

the vehicle to the hyperbolic focal point so that the hyperbolic asymptotes may be assumed to simplify the data processing. Hyperbolic resolution dilution may be expected as one aims his beam farther forward. The hyperbolic dilution will be similar to that encountered in Loran computations. Thus, one may see that no amount of phase summation of side-looking antenna segments may be used to obtain a directly forward looking effective beam, or even good resolution in the forward-looking direction.

Passive radar devices may be employed to "listen" in certain directions relative to the vehicle. A pair of such devices may be oriented as illustrated in Fig. 8–6. If we consider that the stationary target, T, moves relative to the vehicle with the vehicle velocity, V, we may see that an identifiable target will be sighted by the pair of receiving beams with a period, φ, between sightings. Thus, for low altitude observations ($h \to 0$),

$$2d = \varphi V \qquad (8\text{–}26)$$

We may say that the distance of the transmitter is displaced from the ground path by,

$$a = d \cot \gamma = \frac{\varphi V \cot \gamma}{2} \tag{8–27}$$

and that the location of the target is at distance "a" normal to the flight path at a point mid-way between sightings.

For higher altitude work (as with satellites) we must include an assumption that the target is stationary on the surface at a distance equal to $\sqrt{h^2 + a^2}$ away from the vehicle. Thus, we must now say,

$$b = \sqrt{h^2 + a^2} = \frac{\varphi V \cot \gamma}{2} \tag{8–28}$$

or,

$$a = \sqrt{\left(\frac{\varphi V \cot \gamma}{2}\right)^2 - h^2} \tag{8-29}$$

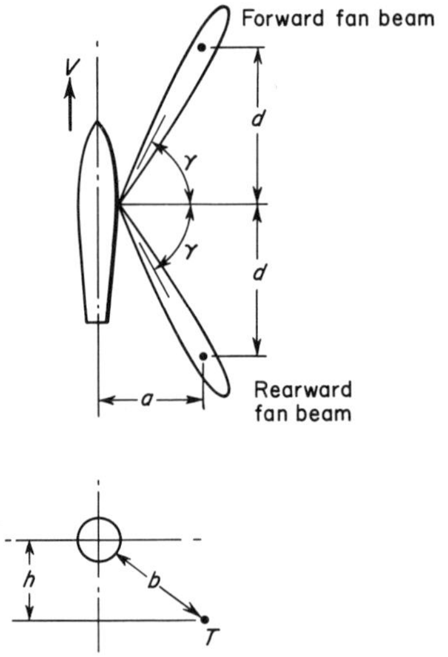

FIG. 8-6. Radar Detection System.

Now we have defined the target location as normal to the sub-orbital path of the satellite, at a distance a off the point midway between sightings. The resolution accuracy is thus limited by knowledge of φ, V, γ, and h. The direction of the target (which side of the path) may be determined by employing separate beams (antennas) on each side of the vehicle. The beam width of such systems is generally limited in resolution by knowledge of γ; in effect a "blur factor." Sightings of the same target by a single wide beam on successive satellite orbits may be used to narrow the target coordinates, provided one is sure of the target identity and the orientation of the beam at the times of sighting the target. Thus, one may think of each sighting as defining a circle of possible location on the surface, and on two or more sightings only that area common to the individual circles for each sighting can contain the target.

The orientation of payload sensors is a problem common to all types of sensors but unique to space vehicle applications. Sensors, in general, are not omnidirectional, but must be aimed at the area of interest. Satellites, and space vehicles in general, do not naturally maintain a fixed attitude relation with respect to planet surfaces or areas of interest. Thus, it will generally be necessary to have some device (e.g., horizon scanner) which may be used to orient the sensor with respect to the vertical, or else merely use the sensor when it just happens to be aimed at the area of interest (requiring a decision function to determine whether it is worthwhile to expend film or not). The orientation and direction of sensors for various purposes is a many-faceted subject which includes the use of wide area sensors to locate smaller areas for close inspection by precision sensors. The direction of such precision sensors off the sub-orbital path can involve some complex calculations; particularly if necessary to do such close inspection during another orbit or from another satellite in the inspection network.

8.3.2 Payload Sensors as Guidance Devices

Payload sensors may be employed as guidance devices provided they satisfy the following criteria:

1. Resolution must be equal to or better than the guidance precision required.
2. Coverage must be equal to or larger than the operating range of the guidance system as determined from initial or search and acquisition requirements, and maneuvers anticipated.
3. Data must be available in useful form to the guidance system in time to permit performance of decision, navigation, and control functions.
4. The reliability of the guidance system is not compromised.

In general, one might expect that payload sensors would be useful guidance sensors, although this has been the case in only a few systems in the past. Payload sensors are generally developed around the goal of obtaining "hard" copy with high resolution, and with little regard to the time requirements and data format requirements for guidance. Guidance sensors are generally developed around the goal of obtaining the required accuracy, coverage and response time with no consideration for permanent memory or hard copy. Possible exceptions are attempts to use television cameras for guidance, and to measure and record miss-distance as determined by the guidance sensor.

Another major drawback in using payload sensors for guidance has been that guidance requirements dictate that the sensor "look ahead" to determine where one is heading with relation to desired heading; payload sensors generally "look down" or "to the side" to maximize their resolution capabilities, i.e., to minimize angular perspective dilution of their resolution. For orbital altitudes, these two requirements may come together somewhat, and we may find that the combined weight and power allowances for payload and guidance sensors may most usefully be employed by a common sensor.

8.3.3 Data Analyses

The analyses of data obtained by payload sensors can run from the extreme of obtaining all the information possible ("milking" it for everything) to the other extreme of discarding everything but a very few pertinent bits of information. To obtain everything possible from a data sensor generally requires the highest resolution film, lens, aperture, processing circuits, et cetera possible. A minimum of film reading and radio communication is employed because the transfer of data generally degrades the resolution content. The obvious solution is, of course, to convey the original film from satellites and probes to Earth bases for detailed, leisurely analyses.

Where data are obtained electronically as with radar and scanning optical devices, these data may be stored in digital format in the space vehicle with minor loss in content, and subsequently transferred to Earth by radio link with reduced bandwidth if necessary (constant time-bandwidth product). Parity checks on the transmission will permit verification of its accuracy prior to erasure of the spaceborne memory device for reuse.

Another means of reducing communication link bandwidth is the in-flight analysis of the data to determine which portion is valuable and which is not. Thus, we may postulate a system which automatically rejects the bulk of information collected. Such a system might work on the basis of rejecting redundant information, and transferring only new information. In one sense, MTI (Moving Target Indication) is just such a system. The scanning and screening of data may be done in-flight to select that data of interest. An example would be a weather satellite that photographed cloud formations, and analyzed these formations for storm patterns (signatures). If by this process a storm is identified, its nature, size, location, and vector velocity might be determined and communicated to the ground.

The recognition of target signatures of particular interest is a matter of prime importance. If one can recognize a predeterminable signature as a hurricane, a missile launching, a nuclear explosion, etc.; and make this determination in a very short time, a pre-formatted message may be sent by a simple code word designation (e.g., message ACE). Such a system would thus permit a pre-determined message to be selected from a group of pre-determined messages on the ground. The bandwidth reduction thus feasible is phenomenal; the bandwidth required being a function only of the *number* of message alternates which might be transmitted by the communication link during a message period.

8.4 ORBIT RELATIONS

Figure 8–7 shows the typical reconnaissance circular orbit and three sensor surface Instantaneous Coverage Area (ICA) for an electromagnetic collector payload. A circular orbit has been chosen for simplicity and because synchronization of an elliptic orbit perigee with a particular surface area of observational interest would impose a requirement on the orbit establishment of having a 24-hour related period with perigee synchronized over the target geocentric coordinates which would disclose one's interest in the particular target area. Further, a circular orbit is the more general usage, and permits one to employ field-of-view and scanning or frame-rate figures as direct functions of satellite velocity to altitude ratio and target motions, with minor compensations or overlap tolerances (as between adjacent camera frames) to allow for orbit perturbations. The three ICA's shown in Fig. 8–7 may be taken as defining ICA's for a sensor frame sequence, or as ICA's for adjacent, common-orbit satellites in a system which attempts to obtain simultaneous or continuous-duty world-wide reconnaissance. The Total Coverage Lines (TCL) define the limit of assured coverage in one orbit, including the coverage limits for a scanning system as in radar, infra-red, and scanning camera mapping systems.

If one wishes to reconnoiter the entire Earth with one such satellite, the obvious orbit is one with $\pi/2$ radians inclination (that is, a polar orbit),

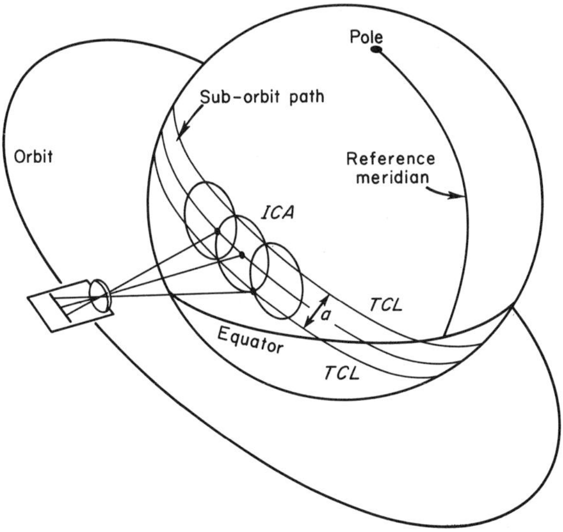

FIG. 8-7. Instantaneous Coverage Areas (ICA) and Total Coverage Lines. (TCL) for Satellite Viewing Sensors.

or at least with an inclination angle which permits the useful (distortion and resolution limited) sensor field of view to include the poles. If the polar regions are not of interest, the orbit inclination plane can obviously be brought closer to the equatorial plane. If a single satellite does not view the area of interest frequently enough, more satellites in some constellation arrangement are required.

To determine the minimum time required for global reconnaissance with a single satellite we must assume some satellite orbit period synchronization with the Earth's period. The synchronization requirement may be defined in terms of the longitudinal equatorial crossings by two equations.

$$L_n = L_{n-1} + T_s\dot{\theta} \times \frac{360}{2\pi} \text{ (degrees)} \tag{8–30}$$

and

$$L_n = L_{n-1} + \frac{a}{R \sin \delta} \times \frac{360}{2\pi} \text{ (degrees)} \tag{8–31}$$

where R is the Earth's radius, δ the orbit inclination plane, T_s the satellite period and $\dot{\theta}$ the Earth rotational velocity. Total equatorial coverage and, hence, global coverage is obtained when,

$$\frac{na}{R \sin \delta} \times \frac{360}{2\pi} = 360 \tag{8–32}$$

The upper latitude limit of coverage will be:

$$L = \delta + \frac{a/2}{R} \times \frac{360}{2\pi} \text{ degrees} \tag{8–33}$$

The minimum time for global reconnaissance with one satellite will be:

$$T_R = NT_s \tag{8–34}$$

where N is the number of orbits required. Determining the number of orbits required, N, may be done by taking the half-equator, dividing by the TCL separation, a, and multiplying by the sine of the orbit plane inclination, δ,

$$N = \pi R \frac{\sin \delta}{a} \tag{8–35}$$

The orbit period, T_s, is determined by the satellite orbit velocity, V_h, and altitude, h, which are related with the orbit angular velocity, $\dot{\phi}$, by,

$$(R + h)\dot{\phi} = V_h \tag{8–36}$$

For a circular orbit the acceleration inward may be expressed as,

$$(R + h)\ (\dot{\phi})^2 = g_h \tag{8–37}$$

From this expression we may obtain the orbit period, T_s, as

$$T_s = \frac{2\pi}{\dot{\phi}} = 2\pi \sqrt{\frac{R+h}{g_h}} \tag{8-38}$$

Recognizing that g_h will vary as a function of h, we may balance gravitational and centrifugal forces to obtain

$$\frac{km_E m_s}{(R+h)^2} = m_s g_h \tag{8-39}$$

where k is a unit-defining constant, m_E is the Earth's mass and m_s is the satellite mass. Setting $h = 0$ yields,

$$G = \frac{km_E}{R^2} \tag{8-40}$$

and

$$g_h = G\left(\frac{R}{R+h}\right)^2 \tag{8-41}$$

where G is the value of gravitational acceleration at the Earth's surface (32.2 ft/sec²). Combining Eq. (8–38) and (8–41) we obtain an orbit period expression,

$$T_s = \frac{2\pi}{R}\frac{(R+H)^{3/2}}{G^{1/2}} \tag{8-42}$$

This expression may be combined with Equations 8–34 and 8–35 to yield,

$$T_R = 2\pi^2 \frac{\sin\delta}{a} \cdot \frac{(R+h)^{3/2}}{G^{1/2}} \tag{8-43}$$

Equation (8–43) shows quantitatively how the global reconnaissance frame time may be decreased by decreasing the orbit inclination angle or the orbit altitude, or by increasing the surface coverage of the sensor.

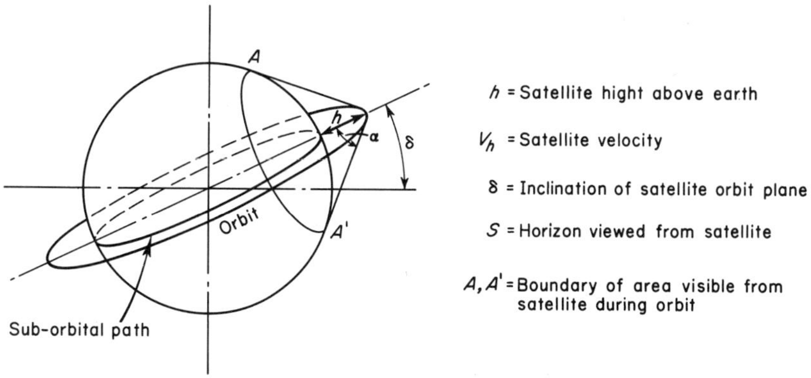

FIG. 8-8. Horizon Coverage Diagram.

For communication satellites we are interested in the horizon visible from a satellite. This will depend on its altitude as shown in Fig. 8–8. Distance to the horizon from the satellite may be determined as,

$$S = \sqrt{h\,(2R + h)} \tag{8–44}$$

The angle, α, between the horizon and true vertical will be,

$$\alpha = \sin^{-1}\frac{R}{R+h} \tag{8–45}$$

The ICA for horizon limited coverage will be,

$$ICA_H = \int_0^{2\pi}\int_0^{\pi/2-\alpha} (Rd\phi)\,(R\sin\phi d\theta) = 2\pi R^2\,(1 - \sin\alpha) \tag{8–46}$$

or,

$$ICA_H = 2\pi R^2\left(1 - \frac{R}{R+h}\right) \tag{8–47}$$

This equation shows that for an extremely high satellite, when $h >> R$, $\alpha \rightarrow 0$, and $ICA_H \rightarrow 2\pi R^2$ or the area of a hemisphere as depicted in Fig. 8–9. The usefulness of a satellite sighting at the horizon is questionable, and some slightly smaller area coverage will be available because some finite time will be required to perform the measurements required. This area curtailment will be only a chopping off of the A – A′ area by planes parallel to the orbit surface trace, and thus a useful area of the type depicted in Fig. 8–6 by the sensor total coverage lines (TCL) will be available. Various studies are available, such as Ref. (*1*) and (*2*), which relate number of satellites required for continuous coverage with maximum latitude covered, satellite altitude, minimum viewing angle for camera (or other sensor) coverage with horizon tangent equal to zero, number of satellite orbit

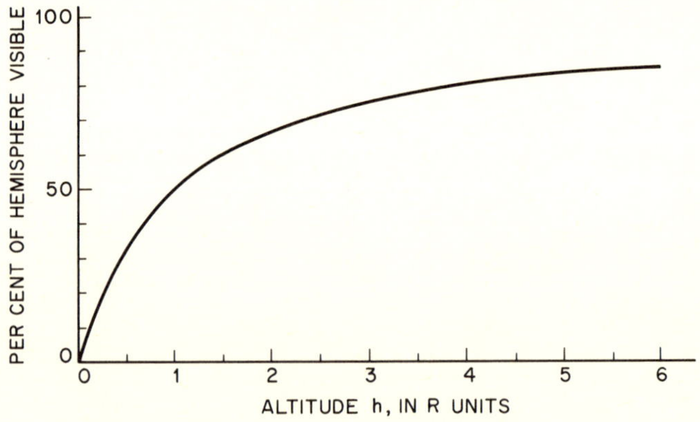

FIG. 8-9. Satellite Instantaneous Coverage Area.

planes, and number of satellites per orbit plane. Figures 8–10 through 8–13 show the result of Luders study (Ref. (*2*)).

Figure 8–10 shows the number and distribution of satellites in a polar network providing complete global coverage as a function of altitude and number of orbit planes. This figure is for a minimum sighting angle, σ, of zero degrees (tangent to horizon). The orbit planes will be oriented with $\frac{\pi}{n_1}$ radians separation between adjacent planes. One may only select networks for which the total number of satellites (ordinate) is devisible integrally by the number of orbit planes, n_1. From this set of curves, it is

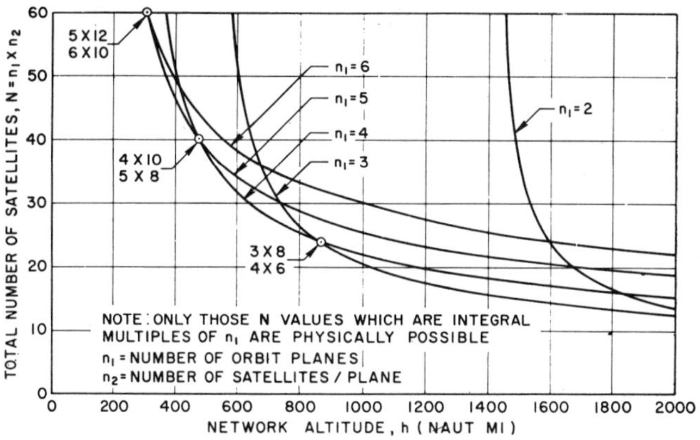

FIG. 8-10. Number and Distribution of Satellites in a Polar Network Providing Complete Global Coverage (λ = 0 deg, σ = 0 deg).

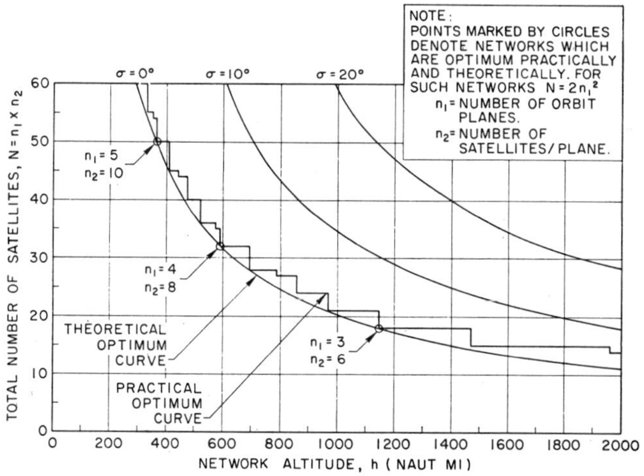

FIG. 8-11. Optimum Satellite Utilization in a Polar Network Providing Complete Global Coverage.

obvious that a choice of optimum satellite altitude will dictate the most efficient network in terms of least number of satellites required, and that the number of orbit planes required in the most efficient network is the function of altitude illustrated.

Figure 8–11 presents optimum satellite utilization curves in terms of number of satellites required versus altitude in a polar network for continuous global coverage with minimum sighting angles of 0, 10, and 20 degrees. From these curves it is quite evident that increasing the minimum sighting angle by 10 degrees requires many more satellites at a given altitude, or a very much higher altitude for a given number of satellites.

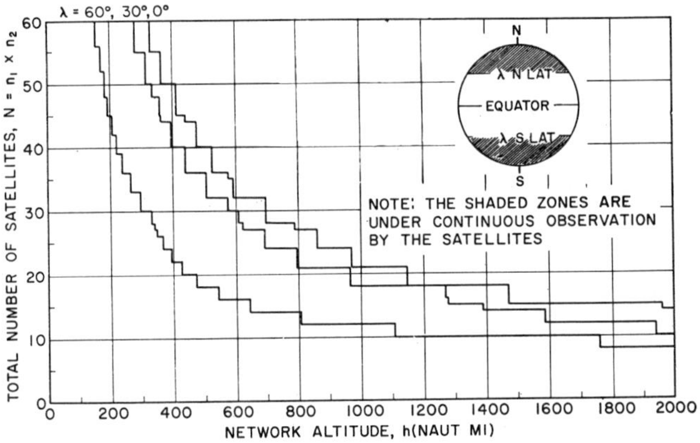

Fig. 8-12. Polar Satellite Networks Providing Continuous Coverage from Poles to Minimum Latitudes of 0, 30, and 60 Degrees.

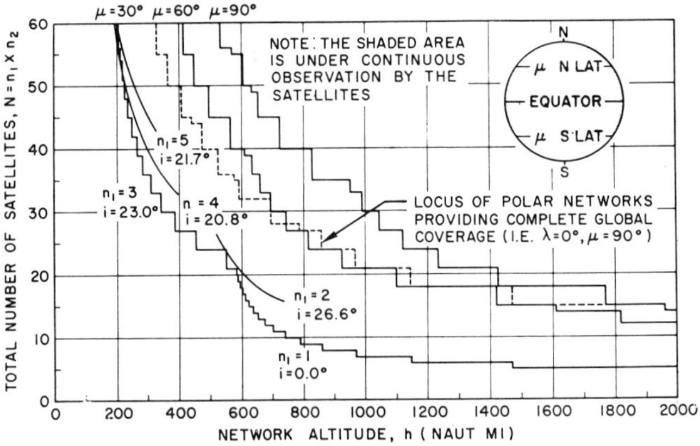

Fig. 8-13. Inclined and Polar Satellite Networks Providing Coverage from the Equator to Maximum Latitudes of 30, 60, and 90 Degrees.

Figure 8–12 indicates what can be done to reduce the number of satellites by requiring continuous coverage of the polar network only above some minimum latitude. Curves are presented for minimum sighting angle, σ, of zero degrees, and for minimum latitude, λ, of zero (equator), 30, and 60 degrees.

Figure 8–13 indicates how the number of satellites required may be reduced by using orbit planes of less than polar inclination if one needs coverage from the equator to some maximum latitude. Curves are presented for minimum sighting angle, σ, of zero degrees and for maximum latitudes, μ, of 30, 60, and 90 (polar coverage) degrees.

The concepts illustrated in Figures 8–12 and 8–13 may be further refined to provide a satellite network capable of continuous coverage within a latitude band between λ and μ. That is, an inclination of less than polar will determine a maximum latitude, and orbit plane separations may be specified for continuous coverage at the minimum latitude.

8.5 DATA TRANSFER SYSTEMS

Data transfer systems present some unique requirements and restrictions in their own right. The basic data transfer system is the communication link which has been treated in Chapter 4. Of interest is the data rate or bandwidth requirements which are necessary for data transfer, the geometrical situations which impose the requirements for relay stations and their peculiar antenna direction and guidance problems.

8.5.1 Relay Systems

Relay stations will be required because of eclipsing action or the need for signal enhancement en route. The long ranges in space indicate that very low bandwidth communication links must be employed whenever feasible, and that very directive antennas should similarly be employed. When one has done all he can with code-selection of pre-formated messages, parity checks, data screening, and selection, there may remain more information content than the available time-bandwidth product can handle with a relatively non-directional antenna. The need for signal recognition, correlation with established patterns or vocabularies en route, and the need to eliminate some large noise source from the signal path indicate the requirements for a data relay station in space. The prime problems will thus be to ensure establishment and maintenance of relay station contact with each terminal location when required.

8.5.2 Antenna Direction

The use of high-gain antennas with relay stations has been indicated to improve the noise rejection in the link, and hence the range and/or message content possible in the overall data transfer system. To operate effectively, a directive antenna must be pointed towards the (apparent) other end of the communication link. If we look at Fig. 8–14, we may see the problem illustrated. The apparent radio source is in the direction of the receiving antenna axis, but is really moving relative to the receiving antenna at an angular rate such that the transit time of the message takes the source out of the receiver beam. Once locked-on to the source, the receiver can be designed to track the apparent maximum energy direction, and so maintain contact and receive the message. The two problems are

1. How to aim the transmitting antenna.
2. How to search and acquire the transmitter with the receiver.

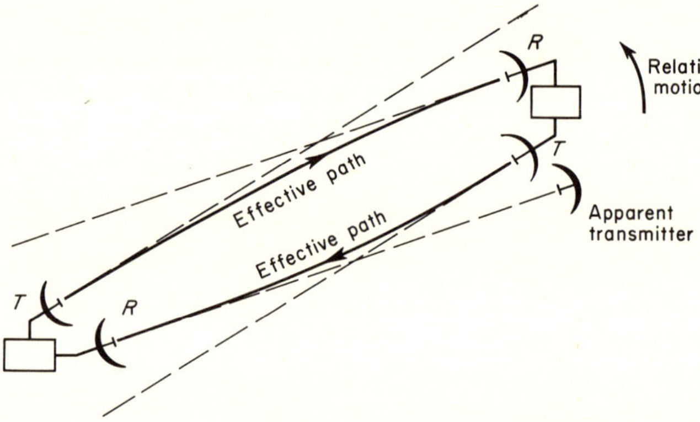

FIG. 8-14. Antenna Direction Problem.

The solution to the first problem lies in how accurately one can estimate or calculate the proper direction for antenna aiming at the transmitter location. An Earth-disc tracker might be used in cislunar space, and the transmitter beamwidth set to illuminate the Earth disc. If trying to illuminate a space relay station to bypass the sun's interference when communicating between Earth and Venus during Inferior Conjunction, one could only search and detect a similar transmitter with a narrow band receiver, and then transmit in the same general direction using a broad enough beam to allow for all computational inaccuracies. Thus, we may postulate a rather sophisticated antenna direction system. A narrow-band search, acquisition and tracking receiver may be used to provide data on the transmitter ephemeris so as to update the accuracy of a precalculated transmitter ephemeris. The updated ephemeris data may be used to calculate the reverse link transmitter direction and perhaps bandwidth or beam width as a function of a data accuracy estimate. One end of such a system is indicated functionally in Fig. 8–15. Whether a system of this sophistication will ever be instrumented is moot. However, the

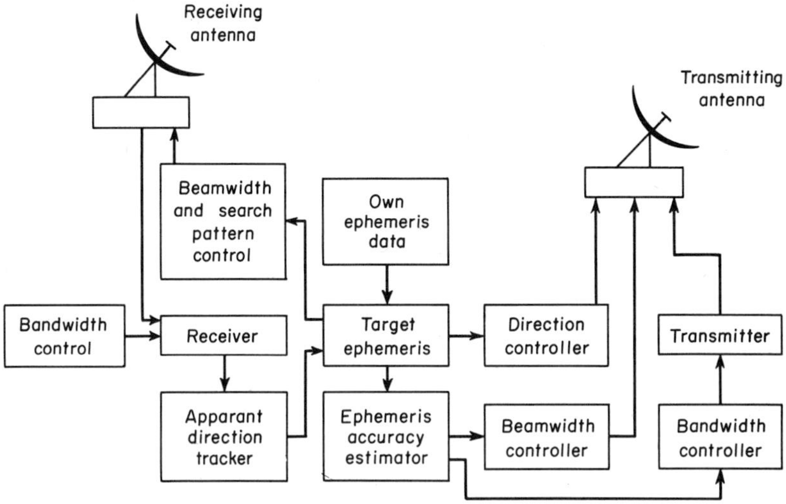

FIG. 8-15. One End of Long-Range Space Communication Link.

concepts have been presented to indicate that such solutions can be accomplished if and when needed.

8.5.3 Guidance Requirements

Figures 8–14 and 8–15 illustrate a problem in antenna direction for long-range data transfer, and a sophisticated solution. The successful solution to the problem will depend to a large extent on how well one can predict the ephemeris of the other terminus of the communication link. This solution will depend on accurate knowledge of one's own ephemeris and that of the target in order to arrive at useful estimates of the apparent direction from which to expect his transmissions and the best direction to aim transmissions so as to provide him with maximum signal power to work with in his receiver. This means that antenna pointing requirements for such data transfer systems will be stringent in terms of maintaining *pre-planned* orbits at each end of the link so that search, acquisition and lock-on functions can be performed when required, and so that a maximum amount of data (bandwidth) may be transferred when locked-on at both ends of the link.

REFERENCES TO CHAPTER 8

1. L. G. Vargo, "Orbital Patterns for Satellite Systems," AAS pre-print 60–48, Sixth National Annual Meeting, New York, Jan. 18–21, 1960.
2. R. D. Luders, "Satellite Networks for Continuous Zonal Coverage," *ARS Journal*, Vol. 31, No. 2, Feb. 1961, pp. 179–184.

CHAPTER 9

VEHICLE DESIGN CONSIDERATIONS

9.1 INTRODUCTION

Previous chapters have concentrated on the nature of space missions, environment, systems problems and general characteristics. It is the purpose of this chapter to provide an introduction to some of the more practical problems to be considered in vehicle installations of various electronic system components.

Because guided missiles and space vehicles are very similar as to the design requirements they place upon the associated electronics it is possible to rely heavily on the other volumes of this series as references; see Ref. (*1a*) through (*1f*).

9.2 ANTENNA AND LENS INSTALLATIONS

The primary electronic system devices of concern in space vehicle installations are antennas and lenses. Other electronic equipment can be protected from the external environment, but antennas and lenses are the sensory connection with the outside world, and certain properties of these devices must be given particular consideration in considering their installations. These considerations will generally center around the effects of the external environment on the materials employed, and the effects of the installation location and vehicle configuration on the performance characteristics of antenna or the lens. Reference (*1a*) provides excellent coverage of the design tradeoffs between airborne radar, parent vehicle and mission.

9.2.1 Beam-Widths

One of the prime characteristics of an antenna is its beam-width and its direction. In selecting and designing an antenna installation, care should be exercised to ensure that the vehicle body and frame do not act as inadvertent beam-forming devices. Probably the best criterion to employ for normal installations is to ensure that no major structural member intercepts the beam within its 3 db (1/2-power) beam-width. When accurate data are required, as with microwave tracking, some further consideration may be needed such as ensuring that minor lobes are not reflected by the structure. In general, an attempt should be made to extend or mount the antenna in a location where the structural frame of the vehicle is wholely to the back side of the antenna forming elements (reflectors, radiators, etc.). Actual test models are generally required to verify empirically whether the actual installation design will perform with the required accuracy.

Antennas which are required to operate only in a space vacuum can be folded compactly and extended after the vehicle is free of appreciable aerodynamic drag; this explains the spidery appearance of many space vehicles.

Figure 9–1 illustrates the distortion of a beam by vehicle obstruction in a case where part of the original beam is intercepted by the vehicle. The intercepted portion of the beam is reflected, and adds vectorially with the other portion of the beam. This vectorial addition includes phase in time which will result in corrugations in the amplitude and phase patterns of the beam as well as re-alignment of the beam axis and some change in the beam width and polar plot.

9.2.2 Near Fields

Antennas have near-field zones in which the beam is not fully formed; in effect it is out-of-focus. A restriction on antennas which are used to generate angular data as in tracking is that they not be employed in the near field. Some criterion such as $2D^2/\lambda$ or $4D^2/\lambda$ (D = aperture diameter; λ = wavelength) can be used to define the near field limit at which the beam is well enough formed to provide tracking data to the required accuracy. In order to overcome near field limitations (and the attendant dynamic range required of the receiver) one may either offset the antenna location to prevent the near field from ever being intercepted, or switch to a low gain, broad-beam "near field" antenna. Figure 9–2 illustrates

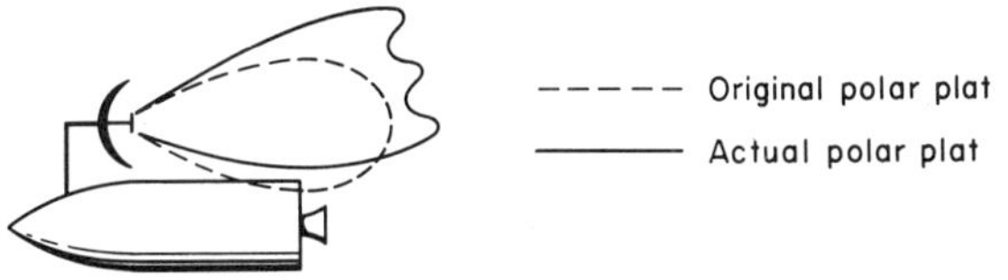

FIG. 9-1. Antenna Beam-Width Distortion.

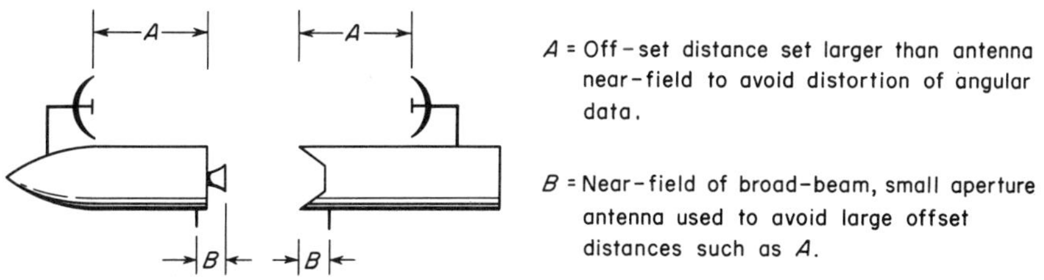

FIG. 9-2. Near Field Antenna Solutions.

these two approaches. Note that when the large offset distance necessary for a high-gain, narrow beam antenna cannot be provided, a low-gain, wide-beam antenna, which has a much smaller aperture, may be employed to reduce the near field off-set requirement. Also, one may be able to operate within the near field of a transmitter for some special but limited applications.

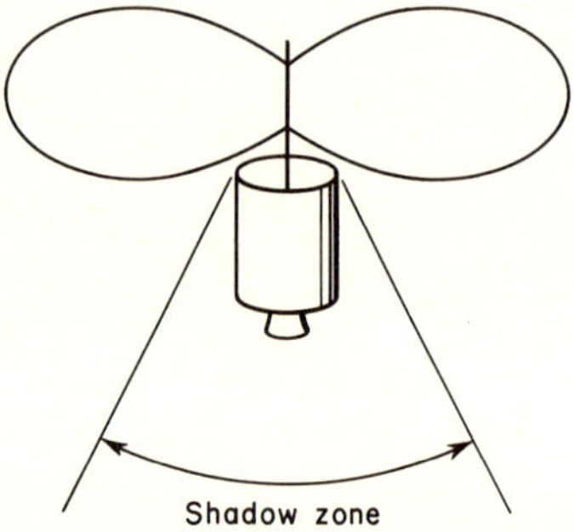

FIG. 9-3. Vehicle Shadowing of Antenna Coverage.

9.2.3 Shadowing

Beam shadowing or eclipsing is a further problem to consider in selecting an antenna site on a space vehicle. One will generally want to use as few antennas as possible to cover the required volume of space. In general, the most one can cover with a single antenna will be 2π (or a little more) sterradians or an extended hemisphere; the supporting vehicle will eclipse a significant portion of the sphere. Figure 9–3 illustrates such a coverage "shadow zone." One may expect two or more antennas to be employed when full spherical coverage is required.

9.2.4 Scanning, Tracking, and Stabilization

Antennas which are required to detect a signal which may originate anywhere in a large volume of space must either have a beam-width and receiver sensitivity to match this volume and the signal-to-noise expected, or obtain the same effect by scanning the volume with a high-gain, narrow beam-width antenna and a receiver which has the required sensitivity when coupled with the antenna. Reference (*1a*) covers the analytical approach to this problem.

Tracking a signal may be accomplished by a signal maximum seeking receiver (direction finder technique) or by more accurate multi-lobe antennas with receivers which compare the signals available from the lobes.

In scanning and tracking systems of these types the need generally arises for stabilization of the antenna on a platform or mount to create a reference for the scanning and tracking data. An alternative is to align the whole vehicle as a mount. The use of "strap-down" scanning and tracking systems which operate relative to the vehicle frame is feasible; however, the accuracy of such systems has generally been inferior to stabilized systems because of non-linearities in the measurement characteristics as one departs very far from the tracking null-point (or boresight axis) about which linearity may usually be assumed.

The weight penalties of stabilizing antennas have to be compared with the weight penalties of the more sophisticated data measuring, handling and processing techniques necessary for unstabilized systems to approach the same order of accuracy. With increasing functional density in electronic data processing systems, the day may come when sophisticated data processing of non-linear data will be preferable and more accurate than the direct stabilization of antenna systems to ensure operation in linear error regions.

9.2.5 Windows and Radomes

Windows and radomes are merely electromagnetically transparent devices which protect a basic sensor element (e.g. antenna, collecting mirror, etc.) from some external environment. The use of windows and radomes in space is not necessary for aerodynamic shielding and is probably ineffective for high speed meteorite shielding. Windows and radomes generally introduce some refraction of the electromagnetic rays passing through, and through lack of complete symmetry some distortion of predicted characteristics of the basic sensor. Windows and radomes will only be useful when it is required to protect the basic sensor from the outside environment. Examples might include sensors such as humans or mirrors with high vaporization rate materials. Distortion of windows and radomes by one atmosphere pressure differential should not be as severe as distortion encountered in high speed aircraft operating with significant ram pressure forces. Attention is invited to References (*1b*), (*1d*), (*2*), and (*3*) which cover, respectively, the environment of space, analytical design of radomes and the properties of materials in space.

9.2.6 Star Tracker Axes

Star tracker axes are important in establishing a stellar reference system for space navigation. Stars should be selected for some peculiar property either directly related to the navigation problem coordinates (e.g. coplanar or colinear with the destination during terminal guidance phase), or to establish a general coordinate system which in effect permits the navigation problem to be accomplished with reference to this coordinate system (e.g. celestial latitude-longitude system with central force field radius as the third coordinate).

A set of three stars might be employed such as Polaris, Aries and one other in the ecliptic plane (instead of the sun) at right angles to the Polaris-Aries plane. The sensitivity of star trackers to solar radiation will generally dictate that one or more of the star tracker devices be periodically (annually) shielded from direct solar view to prevent its destruction, as is the case in the present Orbiting Astronomical Observatory Satellite. In this case the system must provide some memory (e.g. gyroscope erected to the star tracker coordinates) until the tracker can re-acquire its reference, otherwise a companion star tracker may be used to track a star in the opposite direction from the eclipsed star. Reference (*1c*) provides a treatise on space vehicle dynamics in the celestial environment.

9.2.7 Horizon Scanners

Horizon scanners generally generate self or vehicle-stabilizing data continuously so as to maintain a local geometric vertical reference. Horizon scanners must operate at high altitudes to be effective for this purpose so that surface corrugations do not introduce short term changes in the vertical reference data.

The determination of the horizon by horizon scanning must not be susceptible to interference or blinding by the sun. Thus, an IR scanner will be required to be sensitive in the wavelength spectrum of internally generated heat, and not responsive to reflected solar radiation. If direct viewing of the sun may be anticipated, some shielding of the horizon scanner must be provided for this period, as well as a memory of the local vertical precession rate and acceleration (for high eccentricity orbits). Radar scanners which determine the local vertical from artificial, range-gated, horizons promise a solution to these problems.

9.2.8 Altimeters and Ground Trackers

Altimeters and ground trackers must operate outside the "shadowing" region as discussed in Section 9.2.3. The interposition of landing struts may pose an aperture blocking problem if these intercept a good portion of the beam. However, range and doppler measurements will generally not be seriously affected.

Multiple use of antenna apertures by multi-feed microwave lenses is very desirable as this will conserve antenna weight requirements. A typical installation of this type is illustrated in Fig. 9–4. Multiple beam lenses of this type have been developed and produced in large quantities and used operationally in aerodynamic vehicles. A current 10-inch lens is illustrated in Fig. 9–5. It is to be expected that weight restrictions for space vehicle installations will dictate more multiple use of lens antennas of this type to avoid the aperture blocking of multi-feed reflectors, and the use of several individual antennas.

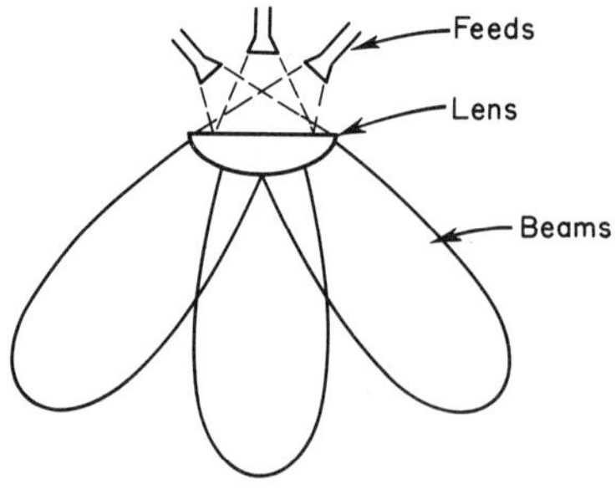

FIG. 9-4. Multi-Feed Lens.

FIG. 9-5. Ten-inch Multi-Beam Lens Antenna.

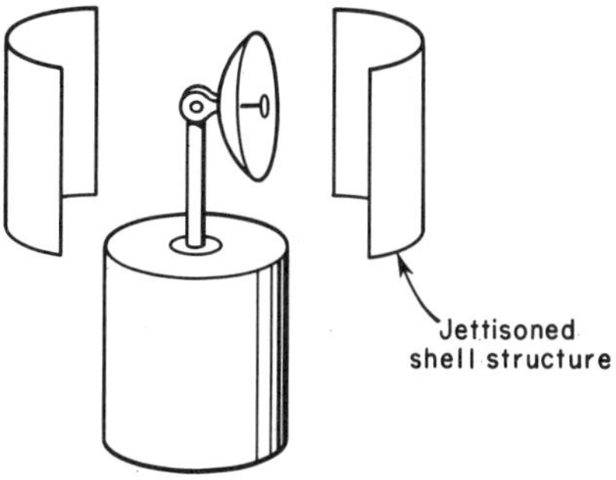

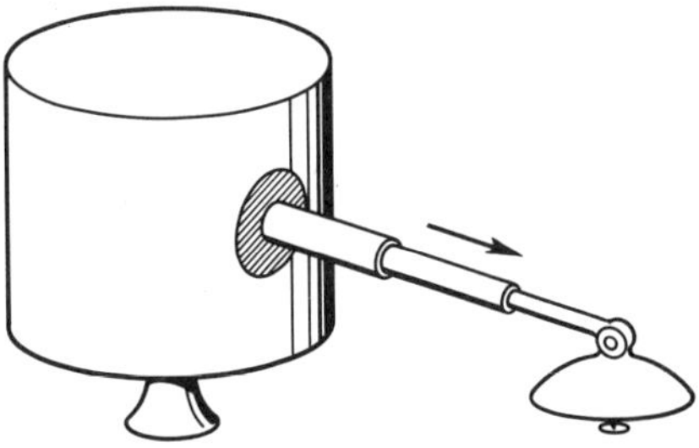

FIG. 9-6. Antenna Erection by Jettisoning Surrounding Frame.

FIG. 9-7. Antenna Erection by Extension Arm.

9.2.9 Erection Devices

When it is necessary to obtain an offset from the space vehicle body by the antenna beam, some sort of antenna erection device may be required.

The simplest and lightest may be simply to have the antenna system installed in its operating position and to jettison the surrounding vehicle shield which may have been required aerodynamically and structurally during the booster rocket phase of the flight. Figure 9–6 illustrates such an antenna.

With large, manned vehicles, it may be necessary to erect antennas by extending them in various directions as illustrated in Fig. 9–7. Such techniques are feasible, and may be accomplished by telescoping arms, locking elbow joints, parallelograms, etc. Structural weight will depend on the amount of rigidity required to make the antenna tracking data meaningful at the vehicle end in the presence of a vehicle thrust forcing function. Erection beam stiffness is a structural problem best left to structural engineers. However, such things as large-diameter, thin-walled pipes are fairly rigid, and new structural materials such as beryllium may eventually be employed as light-weight substitutes for aluminum. A preliminary design of this type of erection device has been illustrated in Ref. (*4*). Reference (*1e*) gives valuable data bearing on structural design.

9.3 INERTIAL REFERENCE INSTALLATIONS

Inertial references are peculiar in that they are sensitive to angular momentum, gravity, and force changes rather than impinging radiation. The proper installation of such devices demands that their mechanical environment be carefully controlled or predictable in a form for which compensation may be introduced as with Schuler tuning of low-altitude, inertial platforms for Earth satellite (and other vehicle) use. Reference (*1f*) provides complete coverage of inertial devices as employed in space vehicles.

9.3.1 Environmental and Structural Consideration

Inertial devices are mechanical in nature and are basically responsive to applied torques. Environmental isolation is generally necessary in the form of thermal regulation and virbration isolation to ensure good (statistically) predictable operation. Unless the inertial sensor is to be used to measure structural vibrations, it is generally desirable to install the sensor at some structural vibration node so that random vibration effects will be minimized.

Shock mounting is used generally to isolate these sensors further; in effect, as a low-pass mechanical filter on the motion at the sensor installation site.

9.3.2 Vehicle Center of Gravity Considerations

Inertial sensors are generally responsive to angular momentum changes. Thus, it is always desirable to install inertial sensors at or near the vehicle c.g. in order to minimize the effects of random angular momentum changes.

An exception to this would be inertial sensors which are used deliberately to measure these irregular changes in which case the sensor would be installed with some separation between it and the vehicle c.g. The distinction made is between an inertial sensor which is being precessed by a computer to provide a mechanical reference as to the vehicle's position in space, as opposed to a sensor which is being used to determine a vehicle's controlled attitude in space.

In trying to locate an inertial sensor near the c.g. one must consider the c.g. movement during rocket staging and fuel usage. Either multiple sensors must be used, or one must be sure that the master sensor location in the payload section will be satisfactory for all stages of the flight. Generally the payload c.g. location will also be the c.g. location for all but a small portion of the flight; however, if the inertial device is being used to control motor cut-off, nozzle attitude, etc., during early flight, some consideration must be given to the need for additional sensors.

9.4 MAGNETIC FIELD SENSOR INSTALLATIONS

Magnetic fields do not vary with steep gradients over a vehicle's dimensions so that vibration and even thermal effects are not generally serious in determining the proper installation site for magnetic field sensors. The more important problem is to be sure that these sensors are installed where the measured field will not be distorted by the vehicle or any equipment aboard. This generally means, away from ferrous materials, structural members, magnets, electric conductors, antennas, etc. Empirical tests using mock-ups and the prototype vehicles will generally be required to ensure the proper operation of magnetic field sensors.

9.5 VEHICLE ERROR, SPACE, RELIABILITY, WEIGHT, AND POWER CONSIDERATIONS

Space vehicles must have budgets for errors, internal space, reliability, weight and power. The interrelations between these are direct but not usually linear. The result is a variety of possible designs for any particular mission with "trade-off" studies to back up the design choices or orders of merit.

In general we may say that for some particular mission we must get there with an acceptable error, have high reliability, and use a minimum of space, weight, and power. Upon vehicle arrival, we must obtain the required data to some satisfactory resolution and return these data by conveyance or communication link again within some tolerable error limits, with high reliability, and with minimum space, weight, and power. As an alternate, upon arrival at our destination we may merely be required to deliver a payload with some accuracy, kill probability, or logistic survival probability.

The assignment of budgets to these various parameters is a complicated procedure involving the requirements and the current state-of-the-arts. If we may acquire power en route or at our terminus (e.g. solar radiation or lunar minerals) we may save somewhat in our energy storage requirement; this means a savings in the power, weight, and space budgets. If we are using a minimum size thermonuclear reactor we may have a fixed space and weight allocation with surplus power available. Thus the problem has many facets.

Manned space flight will generally require high survival probability hence high reliability and redundancy or back-up capability among sub-systems.

Weight is a major factor currently and will be for exploratory and logistic missions for some time to come. Probably the primary budget to work with for typical minimum energy trajectory (MET) systems is the weight budget. Examples of weight minimization state-of-the-art are Figs. 9–5 (page 375), 9–8, and 9–9.

Figure 9–5 illustrates the efficient use of a common aperture by three antenna feeds. The single aperture is used to form the three beams simultaneously, and to direct them in the various directions required. In such a system transmit-receive isolation is generally achieved by time separation as with "long-pulse" or ICW techniques.

Figure 9–8 illustrates the current state-of-the-art in solid state transmitters. The picture is of a coherent frequency multiplier which multiplies the reference signal (input) frequency by 144 to produce a coherent K-band transmission. This device weighs 3.6 lbs., requires no external power supply or bias voltages, is all solid state, and extremely reliable (MTBF = 34,000 hours). Such a device is generally most useful in CW or ICW systems because of the peak power limitations of varactor diodes. The elimination of high voltage power supply and bias voltage requirements is a major assist to weight reduction and reliability.

FIG. 9-8. Solid-State Transmitter.

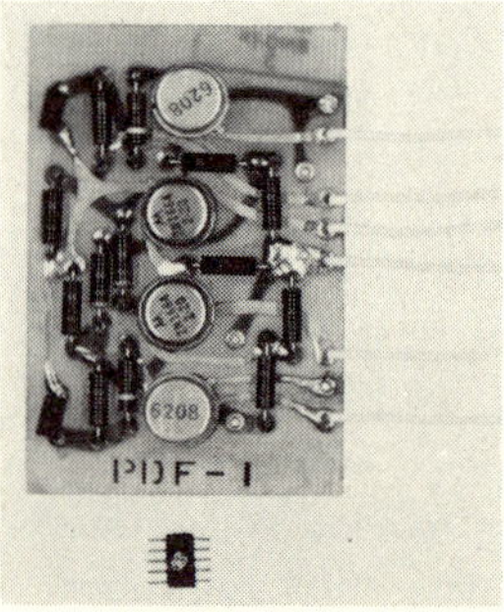

FIG. 9-9. Conversion of Standard Circuits to Integrated Circuitry.

Figure 9–9 illustrates current state-of-the-art in computer circuitry packaging, and the conversion to the next generation of integrated circuit packaging. The card and the integrated circuit module illustrated perform identical arithmetic operations. A typical navigation computer might show a 4½ to one weight improvement, a 12 to one volume improvement, and a 6 to one reliability (MTBF) improvement by converting to integrated circuitry.

If we assign a weight budget as,

$$W = \sum_{i=1}^{n} w_i \tag{9–1}$$

we may indicate the various functions performed as i functions. If these functions are serially related, or must all perform independently, we may say that the success probability or reliability budget is,

$$R = \prod_{i=1}^{n} r_i \tag{9–2}$$

The space and power budgets are sum functions of the type described in Eq. 9–1.

The error budget is the most sophisticated of all. Unrelated errors may often be summed geometrically as the root-mean-square of the contributing error,

$$E = \sqrt{\sum_{i=1}^{n} e_i^2} \tag{9–3}$$

However, this type of summation does not apply generally. If there are harmonically- (with frequency or time) related errors, their phase relation will generally add these errors directly and vectorially. The geometric relations and coordinates of the errors are important as well. The spherical error distribution of a vehicle's position at a flight phase terminus may be less important than the circular error distribution, the vehicle attitude for initiation of the succeeding flight phase, or perhaps the ratio of these two quantities. Thus, the error allocation or budget for a particular system is generally the most complex to analyze. An example of such an analysis for an antiaircraft missile is to be found in Ref. (*1d*).

Error analysis and budget or sub-system specification establishment is one of the primary functions of the electronic systems engineer. The quality of this analysis and the accompanying synthesis are the distinguishing marks of good versus bad system development. Required in this analysis are (1) a mathematical model of the proposed system, (2) identification and mathematical definition of noise sources, and (3) error-sensitivity calculations.

The mathematical model of the system represents the degree of understanding of the proposed system design. This model must reduce component systems to cause and effect functional relations; transfer functions in either the time or frequency domain.

The noise-source identification and definition is the test of the physical environment knowledge of the systems engineer, and his feeling for the sub-system dynamic ranges and minimum operating levels. Without knowledge and definition of the noise sources, and subsequent introduction of these sources into the mathematical model, the systems engineer will

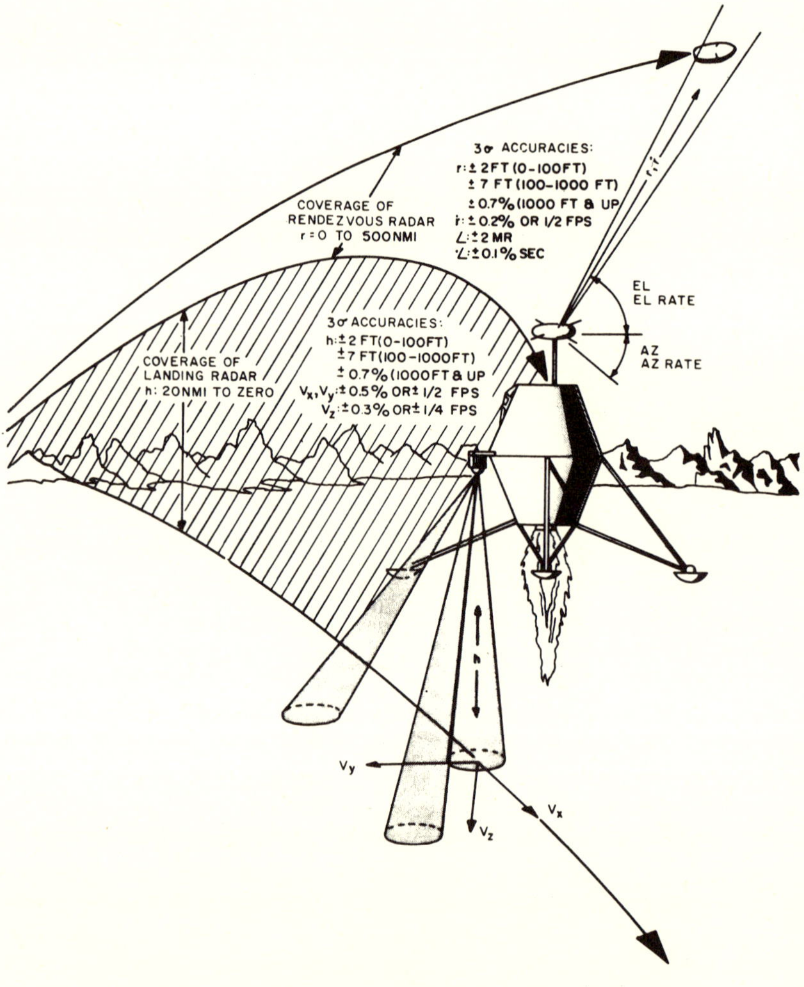

FIG. 9-10. Lunar Landing Radar Performance Summary.

not be able to predict the performance of the system with any assurance. The error sensitivity calculations involve a determination of the input effects on the output of the various noise sources. Also included in error sensitivity calculations are the effect of incremental error signals at input locations to determine their effect on output signals, displays or controllers. Closed loop system transfer functions may be used for this purpose. Partial differentials of such a function with respect to particular inputs establish the sensitivity. If the system is complex it may have to be treated as serially related sub-systems, and then over-all relations established. Computer simulation of the system is a widely used technique which allows one to explore many error sensitivities of the simulated system in a relatively short time. System simulation with available components coupled in is a further development technique which allows one to establish even better the effect of the over-all system, and possible design changes. Thus, the over-all system may be analyzed from an error stand-point; space, reliability, weight and power budgets assigned to the various sub-systems; and trade-offs made from this starting point to ascertain whether the design may be further optimized or is critical with respect to some component system or noise source.

A preliminary design of a lunar landing radar has been carried out in some detail in Ref. (*5*). This report carries the lunar landing mission through fuel optimization criteria, and pilot sensory cue requirements to the ultimate preliminary design of the radar. Figure 9–10 is a performance summary for such a landing radar, and includes rendezvous radar performance as well. Note that one-sigma ranging accuracies of better than one foot are available, and range rate to better than one-fifth foot-per second. Figure 9–11 is a block diagram of such a system. Note the use of frequency multipliers for transmitter and local oscillator to enable coherent Doppler measurements on the carrier. Also, note the common lens antenna for this three-beam, ICW system. Such a system can be assembled at less than 30 pounds using present hardware techniques.

The integration of several radar functions into a single multi-purpose set offers considerable advantages and savings in weight, space and power requirements. Reference (*6*) illustrates a design approach in which a central computer may perform all computations, a single lens antenna system may perform the radar sightings in all required directions, a single coherent transmitter may be employed for all frequency channels, and a single display unit may be used to combine all displays in a contact analog for a pilot. Such concepts are really in the embryo stage at present, but the optimization of multi-purpose radar design offers a considerable challenge for the future and promises considerable improvement in space, weight and power requirements.

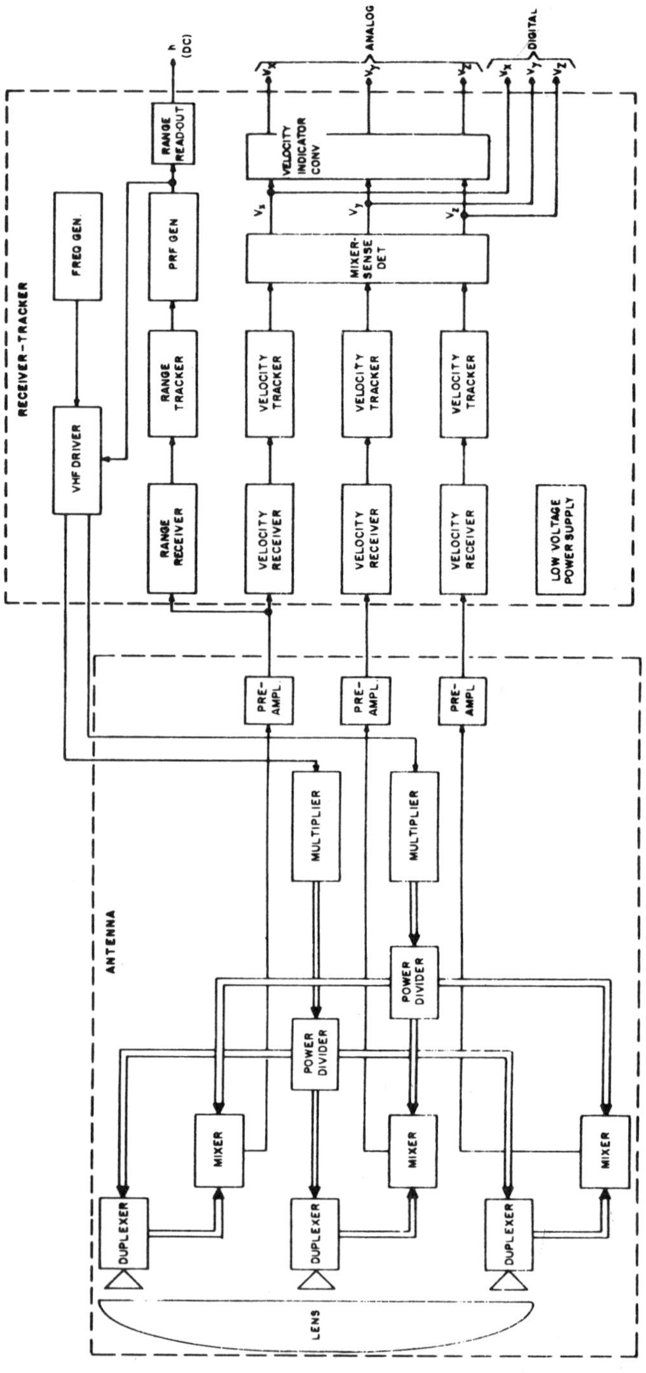

FIG. 9-11. Lunar Landing Radar Block Diagram.

References to Chapter 9

1. *Principles of Guided Missile Design*, D. Van Nostrand Co., Inc., Princeton, N. J.
   (a) Povejsil, *Airborne Radar*
   (b) Ehricke, *Space Flight*, Vol. I
   (c) Ehricke, *Space Flight*, Vol. II
   (d) Jerger, *Systems Preliminary Design*
   (e) Besserer, *Missile Engineering Handbook*
   (f) Parvin, *Inertial Navigation*
2. L. D. Jaffe and J. B. Rittenhouse, "Behavior of Materials in Space Environments," Jet Propulsion Laboratory Technical Report No. 32-150 of November 1, 1961.
3. L. D. Jaffe, "Effects of Space Environment upon Plastics and Elastomers," Jet Propulsion Laboratory Technical Report No. 32-176 of November 16, 1961.
4. D. B. Newman and H. E. Prew, "Spaceborne Radar for Rendezvous Guidance and Lunar Landing," IRE Ninth East Coast Conference on Aerospace and Navigational Electronics, October 1962, Baltimore, Maryland.
5. "Applications of Doppler Radar in LEM Lunar Landing and Rendezvous," LFE-Electronics Technical Report No. TP-316 of December 19, 1962.
6. D. Newman and J. Burrows, "Integrated Avionics Design," LFE-Electronics Technical Memorandum No. TP-314 of November 1962.

INDEX